STATISTICAL
MODELING
TECHNIQUES

STATISTICS: Textbooks and Monographs

A SERIES EDITED BY

D. B. OWEN, Coordinating Editor
Department of Statistics
Southern Methodist University
Dallas, Texas

OTHER VOLUMES IN PREPARATION

STATISTICAL MODELING TECHNIQUES

SAMUEL S. SHAPIRO
Florida International University
Miami, Florida

ALAN J. GROSS
Medical University of South Carolina
Charleston, South Carolina

MARCEL DEKKER, INC.   New York and Basel

Library of Congress Cataloging in Publication Data

Shapiro, Samuel S. [date]
    Statistical modeling techniques.

    (Statistics, textbooks and monographs ; v. 38)
    Includes index.
    1. Mathematical statistics. I. Gross, Alan J.,
[date]. II. Title.
QA276.S46   001.4'34   81-12526
ISBN 0-8247-1387-7   AACR2

MARCEL DEKKER, INC.
270 Madison Avenue, New York, New York 10016

Current printing (last digit):
10 9 8 7 6 5 4 3 2 1

PRINTED IN THE UNITED STATES OF AMERICA

83 006014

This book is dedicated to the memories of William Shapiro, Gus Rose Shapiro, Sadie Gross, and Jack Gross; parents who have given so much and demanded so little.

Statistical models have become an important factor in evaluating
the complex systems that are ubiquitous in today's world. Systems
to which statistical models have been applied include electronic
systems such as large computers, weapons systems, and biological
systems. It is apparent that the aforementioned systems are in
themselves so complex that very often relevant information concerning them can be obtained only by means of a statistical model.
For example, a clinical diagnosis of a patient with a given set of
symptoms may be made in terms of a probability statement--a patient
with an elevated blood pressure, high cholesterol count, and who
is short of breath may with high probability have a heart attack
within a year's time. The purpose of this book is to acquaint the
reader with current statistical modeling techniques that are useful in evaluating complex systems.

The book has been written for students in the mathematical,
biological, physical, engineering, management, and social sciences
who desire to learn the basic concepts involved in the use of
statistical models. The primary tool for understanding the material
in this book is the undergraduate calculus sequence. The book is
self-contained, statistically, but the material in Chapters 2, 3,
and 4 may be supplemented with an undergraduate level statistics
book.

Chapter 1 discusses the basic ideas underlying statistical
models and their applications. Contained in Chapter 1 is some of
the basic terminology such as the definitions of both a model and

a statistical model. Other basic concepts such as parameters,
unbiasedness, populations, and samples are also defined.

Chapter 2 contains a brief overview of the basic concepts of
statistical theory. The topics that are discussed include
probability; probability distributions in a single variable;
probability distributions in more than one variable; marginal and
conditional probability distributions, and concomitant to these,
the concept of statistical independence; moments and moment-
generating functions; and fundamental ideas in statistical estimation, confidence intervals, and tests of hypotheses. The reader
who would like a more detailed treatment of these concepts should
refer to the bibliography at the end of the chapter.

Chapter 3 is concerned with models for measurement in the
continuous case. Emphasis is placed on demonstrating how each
model arises in nature. The models that are treated include the
exponential, Weibull, gamma, extreme value, lognormal, beta, and
logistic models.

Chapter 4 concerns itself with models for measurement in the
discrete case. Again, the emphasis is on showing how these models
arise in nature. Those models that are discussed include the
binomial, Poisson, hypergeometric, geometric, negative binomial,
multinomial, and compound models.

Often in an experimental situation the experimental setup
may be so complex that the only way in which a model can be selected is to find one that fits the data collected. Such models are
called empirical models and are examined in Chapter 5. Among the
empirical procedures that are included in Chapter 5 are systems
for classifying distributions, i.e., classifying distributions by
their moments or percentiles; the generalized lambda family; the
Johnson system; and the Pearson family.

Once a model has been chosen to fit the data from a particular
experimental setup one should try to determine, within the limits
of probability, how well that particular model fits the data.
Chapter 6 is concerned with this issue as it devotes itself to
testing the assumptions of applying a particular model to a given

set of data. The particular tests that are presented include
regression tests, the distance tests, and the chi-square goodness-
of-fit test.

Chapter 7 applies all the techniques of the previous chapters
and two new ones, Monte Carlo simulation and propagation of moments,
to the analysis of complex systems. A case study is considered
involving the simulation of the maximum clad temperature of a
nuclear reactor.

The instructor who wishes to use this book for a one-semester
course may omit all of Chapter 2 provided that the students have
had a course in beginning statistical theory. Other sections that
may be omitted at the instructor's discretion include Sections 4.4
and 4.5 on the hypergeometric, geometric, and negative binomial
models, Section 5.5 on the Pearson family, and Section 6.4 on the
chi-square goodness-of-fit test if the students have had this
material in a basic statistical theory course.

Finally, problems for solution are included at the end
of each section of each chapter. The problems are designed to
test the reader's understanding of the concepts presented in each
section. Many of these problems are computational in nature as it
is felt that "learning by doing" is an effective method of instruc-
tion.

S. S. Shapiro

A. J. Gross

ACKNOWLEDGMENTS

We wish to thank the *Biometrika* Trustees for their permission to
reproduce the following tables:

(1) Fit of Johnson S_U distribution [Table A.4] from *Biometrika*,
     Volume 52 (1965)

(2) Coefficients for the W test of normality [Table A.6] and
     percentiles for the W test of normality [Table A.7] from
     Biometrika, Volume 52 (1965)

(3) Percentiles for the standardized Pearson distribution [Table
     A.5] from *Biometrika Tables for Statisticians*, Volume 1 (1966)

   We are grateful to John Wiley and Sons (New York) for permission to publish the normal [Table A.1] and chi-square [Table A.2]
percentiles from *Statistical Tables and Formulas* by Hald and to reproduce Example 3.3.3 which appeared as Examples 4.2.2 and 4.4.2 in *Survival Distributions: Reliability Applications in the Biomedical
Sciences* by Gross and Clark (1975).

   We are also indebted to the Board of Directors of the
American Statistical Association for their permission to reproduce
the following tables:

(1) Percentiles of the Kolmogorov-Smirnov test [Table A.9] from
     the *Journal of the American Statistical Association*, Volume
     46 (1951)

(2) Coefficients for fitting the generalized lambda distribution
     [Table A.3] from *Technometrics*, Volume 21 (1979)

(3) Percentiles of the Shapiro-Wilk test for exponentiality [Table
     A.8] from *Technometrics*, Volume 14 (1972)

     We also thank Dr. Michael Stephens for supplying the percen-
tiles for the modified version of the Anderson-Darling and
Cramér-Von Mises statistics [Table A.10], and Dr. Neil Cox for
the equations involving the propagation of system moments. We
also thank all those whose research work contributed to the material in this book and who are referenced in the text.

     Finally, we are most appreciative of the efforts of Ms. Jeanne
Disney and Ms. Marie Seabrook in preparing the manuscript. Ms.
Seabrook typed the final copy.

Chapter 1

BASIC CONCEPTS OF STATISTICAL MODELS

1.1 INTRODUCTION

The modern day world is an enormous system of complexities. Every-day questions that at one time appeared to have routine and simple answers have grown in a concomitant fashion to the world so that almost without exception these questions have complex answers that are no longer absolute. Often, answers to important contemporary questions may only be answered within the framework of probability. For example, the internal core temperature of a nuclear reactor must not exceed $2000^{\circ}F$. If the reactor meets all specifications when it has been tested, one may only be able to conclude the probability is 0.005 that the core temperatures will exceed $2000^{\circ}F$. In view of recent reactor problems how can one know whether the reactor meets the required safety specifications? Also, is 0.005 a safety margin of error, considering only a single reactor is under consideration, whereas there are roughly 70 nuclear power plants in the United States alone?

To amplify further on this problem and to illustrate some of the techniques that are subsequently discussed in greater detail, consider the following realistic problem concerning nuclear power plants. (A part of this problem is analyzed in greater detail in Example 7.4.3.) Engineers who are members of a nuclear agency are asked to determine the effect on the safety of a nuclear reactor if certain design and operating restrictions that are reflected in the minimum design specifications are related. The nuclear power

companies have petitioned for these relaxations in order to reduce
the costs of construction and operation of the associated nuclear
power plant. If these relaxations are agreed upon, it is possible
that the safety margins of the power plant will be lowered so that
the probability that a temperature or pressure exceeds a maximum
allowable value will be too large; e.g., the probability the temperature exceeds 2000°F will exceed 0.005. This, in turn, increases
the chances of a nuclear accident beyond some reasonably low value.

   In order to reduce the complexity of the problem, consider only
one particular temperature whose maximum value may not exceed 2000°F
for safe operations. The problem, simply stated, is to determine
the probability that this temperature exceeds 2000°F under the relaxed specifications. Engineers have available to them a highly
complex computer program that can trace the temperature through a
range of operating conditions. This temperature is a function of
the design parameters and the laws of nature. One might believe
that the above problem is simple and all that is needed is to run
the computer program under the relaxed specifications and observe
the maximum value of the temperature. If it is less than 2000°F,
the design is satisfactory. Unfortunately the problem is far more
complex. Actually, it is not possible to determine the exact temperature at a given condition. First, the computer program cannot
exactly model the operation of a nuclear reactor; scientists do not
know the exact equations that describe the physical laws governing
the behavior of the reactor. Second, the temperature depends on
various coefficients that are properties of the materials, areas,
flows, etc., that cannot be measured exactly. Third, the temperature depends on a large number of interactions, some of which cannot
be described under the present state of knowledge. Finally, a computer run can only approximate an actual temperature. This should
not be surprising. Given the same start-up conditions the temperature behavior of a reactor is different for each run, whereas a computer program yields precisely the same answer with the same input
values.

Without going into detail at this time, it is possible to analyze the above problem.   The technique is to express the uncertainty concerning each of the parameters in terms of a statistical
model.   The next step is to combine all these uncertainties to
determine a statistical model for the output temperature.   The last
step is to use the final model to predict the probability that the
maximum allowable temperature is exceeded.   The estimate of probability can be used by the agency to decide whether to relax the
specifications.

The above example is just one of the many problems that can be
analyzed using statistical models.   Such problems arise in business
where the items of principal concern are the corporation and its
profit margin rather than a nuclear reactor and it temperature.
Other similar examples include the output of a chemical process,
the blood pressure at a particular point in a circulatory system,
the behavior of an animal under a set of stimuli, and the gross
national product for a national economy.

The objective of this book is to present some techniques
that can be used in the analyses of the above and similar problems.
The remainder of this chapter defines some of the commonly used
terms in the book.   Chapter 2 provides a brief review of probability and probability functions for those readers not previously
familiar with this material.   Chapters 3 and 4 describe some of the
statistical models that are in common use.   Chapter 5 discusses
methods of fitting models to those cases where either data or the
system moments are available.   Chapter 6 describes a number of
statistical tests that can be used to determine, in those cases
where data are available, whether a chosen model fits the data.
Chapter 7 discusses the techniques of propagation of moments and
Monte Carlo simulation in the analysis of systems.

1.2 BASIC CONCEPTS

Certain generic terms that define objects or phenomena under study
are used throughout this volume.   These terms are now defined.

Definition 1.2.1 A *system* is the object or phenomena under study.
It represents the largest unit being considered.

A system can refer to a nuclear reactor as above, a corporation, chemical process, mechanical device, biological mechanism,
a society, an economy, or any other conceivable object that is
under study.

Definition 1.2.2 The *output* of a system is a response that can be
measured quantitatively or enumerated.

The temperature of a nuclear reactor, the profit of a corporation, yield of a chemical process, torque of a motor, life span of
an organism, length of the forearm of a member of a tribal society,
scores on a psychological test, and the inflation rate of an
economy are all examples of system outputs.

Definition 1.2.3 A *model* is a mathematical idealization that is
used as an approximation to represent the output of a system.

Models can be quite simple or highly complex. They can be
expressed in terms of a single variable, many variables, or even a
set of nonlinear differential equations where the solution is the
model equation.

Definition 1.2.4 A *statistical model* is a mathematical formulation
that expresses in terms of probabilities the various outputs of a
system.

Thus, a statistical model is used in those cases where the
output cannot be expressed as a fixed function of the input variables. The following example illustrates this point.

Example 1.2.1 The system under study is the measurement of the length
of a metal bar. It is desired to determine the true length of the
bar. The output is the resultant measurement in inches. In order
to determine a model for this system, it is necessary to consider
closely the actual operation. In fact, it is useful, where possible,
to observe the performance of the system. Five measurements are
made: 2.103, 2.105, 2.102, 2.105, and 2.109. First, note that the

output of the system varies with each operation (measurement).
Therefore, a model is selected that can represent this variability.
It is assumed that the actual length of the bar is fixed and that
the variations in the readings are due to errors in measurement.
Thus, the system output combines the true size of the bar and the
measurement error. The model

$$y_i = \mu + e_i \tag{1.1}$$

expresses this combination in a simple way, where

y_i = the output, i.e., the ith measurement

μ = true size of the bar

e_i = measurement error for the ith trial

In order that (1.1) be a statistical model, it is necessary to
select a probability distribution to represent e_i; this is discussed
in subsequent chapters.

Definition 1.2.5 A parameter of a model is a constant that expresses
a characteristic of the system.

In equation (1.1) the term μ is a parameter that describes the
length of the bar and is the characteristic of interest.

The model developed in Example 1.2.1 is quite simple, but it
has many of the features of more complex examples. First, the
model is an idealization of the system; it is not exact. Usually,
the more complex the system, the less exact the model. This is
true because of the difficulty in reducing the output of complex
systems to mathematical expressions and the inability to manipu-
late highly complex expressions mathematically. A common dilemna
is: The more the model is simplified, the easier it is to analyze
but the less precise the results. Even in Example 1.2.1 simplifying
assumptions have been made. For example, the model assumes that μ
is a fixed constant; however, the bar size may actually change from
measurement to measurement as a result of the measuring process. A
second point is that the model, expressed in (1.1), is not an equation in the usual mathematical sense. One does not solve for y_i
knowing μ and e_i; in fact, one never knows μ or e_i; one only

observes y_i.

A simple model has been described. Useful information can be abstracted from it. For example, in Example 1.2.1 the true length of the bar was the parameter of interest. Since every output of the system is different (this is true if the bar is measured precisely enough), it should be clear that merely a single measurement by itself does not provide enough information concerning length. In order to discuss this point in more detail, the following definition concerning a characteristic of the measuring device is necessary.

Definition 1.2.6 A measuring device is said to be unbiased if the average measurement error is zero.

This definition implies that sometimes the e_i are positive and sometimes negative, but taken over a large number of observations e_i will average zero. Definition 1.2.6 may be used to obtain an estimate of the parameter μ in (1.1). If all the y_i's are averaged then applying the model, the $\mu + e_i$'s are averaged and since the average of the e_i's is zero, this average is simply μ. Of course, one has to do this over a very large number of observations to get the errors to average to exactly zero. There is a limited number of data points in any real problem, and hence only an approximation to μ is obtained. Thus, an estimate of the variability of the measuring error is required.

Definition 1.2.7 The *variance* of the measuring error e_i is given by

$$\text{Variance } e_i = \sum_{i=1}^{N} \frac{e_i^2}{N} = \sigma_e^2$$

where the summation is taken over all possible measurements.

The measure of variance is discussed in a more precise manner in Chapter 2. This quantity is unknown since the e_i's are unknown. However, for a sample of size n, it can be estimated by

$$s_e^2 = \sum_{i=1}^{n} \frac{(y_i - \bar{y})^2}{n-1} \tag{1.2}$$

The use of n - 1 instead of n is due to a theoretical consideration and is explained in Chapter 2.

Having estimates of μ and σ^2 is still not sufficient. It is important to know how the values of e_i are spread out over all possible values. Are these values symmetric about the average, zero? Or is there greater probability of getting a negative error than a positive one? Such questions are of interest.

Definition 1.2.8 The *statistical model* for the measurement error describes how the values of the error are distributed over all its possible values.

The concept of a statistical model is one of the important subjects of this book. It is discussed in great detail in subsequent chapters.

Estimates of μ and σ^2 and the statistical model for e_i provide the desired information concerning the size of the bar and the probability that it is smaller than or exceeds a given limit.

1.3 USES AND SELECTION OF STATISTICAL MODELS

Some of the uses of statistical models have been indicated in the previous sections. Other uses and methods of selection are illustrated below.

Statistical models are used to predict the performance of "conceptual" systems, i.e., systems that have not been built. Often this area is called *tolerance analysis*, or uncertainty analysis. The tolerance or uncertainties are expressed in terms of statistical models, and the distribution of the output is obtained.

Example 1.3.1 Consider the simple circuit shown in Figure 1.1.

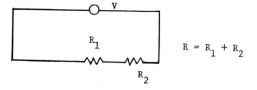

$$R = R_1 + R_2$$

Figure 1.1 Simple resistor circuit.

The specification for the output of the battery V is 12 ± 0.5 volts; it is 10 ± 0.5 ohms for the resistor R_1; and it is 5 ± 0.3 ohms for the resistor R_2. The equation of circuit is $I = V/R$. Knowledge of the variation in the current, given the tolerances for the components of the system, is required. A statistical model can be constructued to represent the circuit and answer the question about the variability of the system output.

Statistical models are also used to represent the behavior of a system based on a limited number of measurements and a knowledge of the failure mechanism. In this application a model is selected based on the knowledge of the phenomena, and the data are used to determine the values of the parameters of the model.

Example 1.3.2 Suppose an organism dies once it is attacked k times by a certain virus. The time to death of 25 organisms when placed in an environment containing the viruses is observed. It is known that a gamma model can be used to represent the phenomona. The data confirm that the model

$$f(x) = \frac{1}{18} x \, e^{-x/9} \qquad x > 0 \qquad\qquad (1.3)$$

may be used to represent the time to death in hours. One may then estimate the probability of survival of an organism for at least 5 years.

Statistical models can be used to summarize a set of data. In many cases at the end of an experiment the researcher has a large number of observations. One convenient method for summarizing these numbers is to find one member of a general family of distributions that can represent the data. There are several such families whose members cover a wide range of shapes. The resulting model can then be used for purposes of extrapolation or interpolation in determining the probability of occurrence in some specific interval of values.

Statistical models are selected in a variety of ways. The best method is to have a good understanding of the physical

properties or principles of operation of the system and to choose a
model that has been derived from these same principles.   For example,
it is shown later that if the probability of the occurrence of an
event in a given time interval is proportional to the length of the
time interval and does not depend on what occurred previously, then
an exponential model of the form

$$f(x) = \lambda e^{-\lambda x} \qquad x > 0, \ \lambda > 0 \qquad\qquad (1.4)$$

can be used to represent the system output.   Thus, this model is
chosen when such a phenomenon is the underlying principle for
system behavior.

In many cases, however, one does not really understand the
basic principles of the system.   Perhaps the second best method is
to use a model that has been used successfully in the past when
dealing with related systems.   In such cases one usually checks on
the validity of such an assumption when data are available or determines the sensitivity of the output to the selection of a specific
model.   A third procedure is to make the selection on a trial-and-
error basis, where a number of models is selected and the one that
appears best by some criteria is selected.   This procedure can be
used only when there are measurements available to evaluate the
selection.   Another procedure when data are available is to use a
generalized, flexible family of models and fit one of these to the
observations.   Finally, when no data are available, it is possible
to study the conceptual model on a computer and generate synthetic
observations.   Then one can use one of the previous methods to select a model.

1.4   DEFINITION OF TERMS

Statistics has a language of its own.   Several of the special terms
have been previously defined.   Addtional terms follow.

Definition 1.4.1   A *population* is the group under study with the
given characteristic (to be measured or counted).   It is the conceptual entity for which predictions are to be made.

Thus, in the context of this study it is the totality of
similar system responses. If a tolerance analysis is being performed,
e.g., in manufacturing, the population is all possible outputs of
systems that are fabricated under these specifications. It is the
totality of responses of organisms in a biological experiment sub-
jected to the given environment. It is the output of all similarly
designed reactors in the nuclear reactor problem.

Definition 1.4.2 A *sample* is a subset of the population that has
been selected for study.

In most cases one considers random samples that have been
drawn so that each possible combination of elements from the population has an equal probability of being selected. If 25 television
sets are randomly selected and their failure times observed, then
these 25 measurements are a random sample from the population of all
possible failure times for this model TV.

The concepts presented in this chapter are important and should
be in the focus of the reader as the subsequent chapters develop.
The primary purpose of this book is to investigate the properties
of models developed from basic principles such as the model (1.4).
On the other hand, when it is not possible to develop a statistical
model from basic principles, the other methods, e.g., using a model
that has worked with previously related systems or by trial-and-error
selection from a group of candidate models, should be used. Once
a model has been selected, it is important to reiterate at this point
the necessity of comparing the output of the model to the actual
data that are collected. That is, the experimenter must check to
determine how well the chosen model fits his or her real-world data.

Finally, the experimenter should not be reluctant to change a
model. For example, in developing a model to forecast inflation,
such a model may have worked well between 1960 and 1973. However,
after the 1973 oil crisis, the need to readjust such a model has
become clearly and painfully evident.

Chapter 2

CONCEPTS OF STATISTICAL THEORY

2.1 INTRODUCTION

The purpose of this chapter is to present, briefly, an overview of
the statistical concepts necessary to understand the models that
constitute the remainder of the book. All this material is available in more detail in other elementary books on statistical and
probability theory. The complete list of such books is lengthy,
but some of the more pertinent books include Bennett and Franklin
(1954), Feller (1957), Miller and Freund (1965), Hahn and Shapiro
(1967), Hoel (1962), Hogg and Craig (1970), and Johnson and Leone
(1964). The reader who is interested in gaining a thorough founda-
tion in basic statistical and probability theory would be well ad-
vised to pursue this objective in one of these books.

2.2 PROBABILITY

Probability can be considered as the measure of likelihood. For
example, when the weather forecaster says the probability of rain
is 0.8, one can interpret this as meaning that given five such
identical weather patterns the likelihood is that four of these
will produce rain and one will not produce rain. More abstractly,
the probability of the occurrence of a physical phenomenon is the
number that describes in accordance with fixed conventions the likelihood of the occurrence of that phenomenon. In order to give a
more precise definition of probability, the following definitions
are necessary.

Definition 2.2.1   An *experiment* is a sequence of (possibly) an un-limited number of trials.

Definition 2.2.2   An *event* is an outcome of an experiment.

Definition 2.2.3   The set of all points (sample points) that re-presents the outcomes of an experiment is called the *sample space*, or event space of the experiment.

Definition 2.2.4   (Finite Probability)   Suppose an experiment can produce n distinct outcomes, all of which are equally likely to occur.   Suppose, further, r of these events are termed favorable and n − r, unfavorable.   The probability of a favorable event (un-favorable event) is r/n [(n − r)/n].

Example 2.2.1   Consider the simple experiment of tossing an unbiased coin three times.   If the value 1 is assigned whenever "heads" occurs and the value zero is assigned when "tails" is observed, the *sample space* describing this *experiment* is $(1,0,0)$, $(1,1,0)$, $(1,1,1)$, $(1,0,1)$, $(0,1,1)$, $(0,0,1)$, $(0,0,0)$, $(0,1,0)$.   The outcome $(1,0,0)$ is one of the eight equally likely events.   Finally, the *probability* that exactly two heads occurs is 3/8 since there are three favorable events $(1,1,0)$, $(1,0,1)$, and $(0,1,1)$ among the eight events that constitute the sample space.

In applying probability, one is usually concerned with a re-lated number of events rather than merely a single event.   For instance, in the previous example there were three distinct events that satisfied the criterion of being favorable.   Thus, the follow-ing ideas from set theory are needed to consider multiple events.

Definition 2.2.5   Let E_1, E_2, ..., E_k be any k events (sets) in an event space S.   Then:

(i)   The event consisting of all sample points belonging to either E_1 or E_2, ..., or E_k is called the *union* of the events E_1, E_2, ..., E_k and is denoted as $\bigcup_{i=1}^{k} E_i$.

(ii)   The event consisting of all sample points that belongs to E_1 and E_2, ..., and E_k is called the *intersection* of the events

E_1, E_2, ..., E_k and is denoted as $\bigcap_{i=1}^{k} E_i$.

(iii) The *null event* is the event $\emptyset$ which contains no sample points. $\emptyset$ is also termed the *vacuous event*, the *empty event*--the event that never occurs, e.g., all Sundays that fall on Monday.

(iv) The *universal event* is the same as S, the *sample* space. It is the set of all sample points in the sample space.

(v) If E is any event, then the *complement* of E, denoted as $\bar{E}$, is the event consisting of all sample points in the sample space S that do not belong to E.

(vi) The event E_1 is said to be a *subevent* of E_2, denoted as $E_1 \subset E_2$, if every sample point of E_1 is also a sample point of E_2. If at the same time $E_2 \subset E_1$ and $E_1 \subset E_2$, then E_2 and E_1 contain the same sample points, i.e., $E_1 = E_2$.

(vii) The event $E_1 - E_2$ is the set of all sample points that belong to E_1 but not to E_2. Equivalently, $E_1 - E_2 = E_1 \cap \bar{E}_2$.

(viii) Two events E_1 and E_2 are said to be *mutually exclusive* if their intersection is the null event, i.e., if $E_1 \cap E_2 = \emptyset$.

Definition 2.2.6 Let E be an event in an event space S. P(E) is the probability the event E occurs. If S consists of n points (or equivalently n equally likely events) and if E consists of $r \leq n$ of these points, then $P(E) = r/n$. Clearly, $P(\emptyset) = 0$, $P(S) = 1$, and since $\emptyset \subset E \subset S$ for any event E in sample space S, $0 \leq P(E) \leq 1$.

Theorem 2.2.1 (Addition Theorem for Finite Probability) Let E_1 and E_2 be any two events in a sample space S. Then

$$P(E_1 \cup E_2) = P(E_1) + P(E_2) - P(E_1 \cap E_2) \qquad (2.1)$$

Proof. Suppose E_1 consists of r_1 points and E_2 consists of r_2 points. Further, if r_{12} is the number of points common to both E_1 and E_2, then, clearly,

$$P(E_1 \cup E_2) = \frac{r_1 + r_2 - r_{12}}{n}$$

However, $P(E_1) = r_1/n$, $P(E_2) = r_2/n$, and $P(E_1 \cap E_2) = r_{12}/n$, and the proof is complete.

The addition theorem can be extended to any finite sequence of events E_1, E_2, ..., E_k. This extension is stated but not proved as Theorem 2.2.2.

Theorem 2.2.2 Let E_1, E_2, ..., E_k be any finite sequence of events in a sample space S.   Then

$$P\left(\bigcup_{i=1}^{n} E_i\right) = \sum_{i=1}^{n} P(E_i) - \sum_{i \ne j} P(E_i \cap E_j) + \sum_{i \ne j \ne k} P(E_i \cap E_j \cap E_k)$$

$$+ \cdots + (-1)^{n+1} P\left(\bigcap_{i=1}^{n} E_i\right) \qquad (2.2)$$

Note: It is convenient to write the intersection $E_i \cap E_j$ as $E_i E_j$, and $E_i \cap E_j \cap E_k$ as $E_i E_j E_k$. This convention is adopted for future use.

If the events E_1, E_2, ..., E_n are mutually exclusive, then Theorem 2.2.2 yields the following simple, but very important corollary.

Corollary 2.2.2 Let E_1, E_2, ..., E_n be any finite sequence of mutually exclusive events.   Then

$$P\left(\bigcup_{i=1}^{n} E_i\right) = \sum_{i=1}^{n} P(E_i) \qquad (2.3)$$

Equation (2.3), when extended to the infinite case, is one of the defining properties of a probability measure.   In fact, putting together the extension of Corollary 2.2.2 and Definition 2.2.6, a more general definition of a probability function is now given.

Definition 2.2.7 Suppose S is a sample space and P is a function defined on S with the following properties.

 (i) $P(\emptyset) = 0$, where $\emptyset$ is the null set.

 (ii) $P(S) = 1$.

 (iii) $0 \le P(E) \le 1$, where E is any event in S.

 (iv) $P(\bigcup_{i=1}^{\infty} E_i)$, where E_1, E_2, ... is any

sequence of mutually exclusive events in S; then P is said to be a *probability function* defined on the event space S.

Example 2.2.2 Let S be the sample space of Example 2.2.1. Let
E_1, E_2, ..., E_8 be the eight mutually exclusive events $(1,0,0)$,
$(1,1,0)$, ..., $(0,1,0)$ which comprise S. Let H_1 be the event
"exactly one head in three tosses of a coin." Then, clearly,

$$P(H_1) = P(E_1 \cup E_6 \cup E_8) = P(E_1) + P(E_6) + P(E_8) = \frac{3}{8}$$

by (iv) of Definition 2.2.7 or Corollary 2.2.2. Furthermore, it
follows that by defining P(E) to be the probability that any event
$E \subset S$ occurs (including $\emptyset$, the null event), properties (i) to (iv)
of Definition 2.2.7 hold and hence P is a probability function on S.

Definition 2.2.8 Let E_1 and E_2 be any two events in an event space
S, and suppose $P(E_1) > 0$. The *conditional probability* that event
E_2 occurs given the occurrence of E_1, written $P(E_2|E_1)$, is defined
as

$$P(E_2|E_1) = \frac{P(E_1 E_2)}{P(E_1)} \tag{2.4}$$

Note, that from (2.4) it follows that

$$P(E_1 E_2) = P(E_2|E_1)P(E_1) \tag{2.5}$$

To illustrate the application of conditional probabilities, consider
the following examples.

Example 2.2.3 Consider a deck of 52 ordinary playing cards. Suppose two cards are drawn from the deck without replacement. To
compute the probability that both cards are spades, consider the
following: Let E_1 be the event the first card is a spade and $E_2|E_1$
be the event the second card is a spade given the first card is a
spade. Clearly, $P(E_1) = 13/52 = 1/4$. If a spade is the first card
drawn, then 51 cards remain in the deck (prior to the second draw),
12 of which are spades. Thus, $P(E_2|E_1) = 12/51 = 4/17$. Hence,
$P(E_1 E_2) = 1/4 \times 4/17 = 1/17$.

As a slightly more complicated example, the following problem is presented. Two urns each contain 10 balls. In the first urn there are 7 red balls and 3 white balls and the second contains 6 red and 4 white balls. Suppose an urn is chosen at random by tossing an unbiased coin and from it a white ball is drawn. What is the probability that the white ball came from the first urn? To solve this problem, let W be the event the ball is white, and U_i the event the ball was selected from the ith urn, $i = 1, 2$. The following probabilities are readily computed: $P(W|U_1) = 3/10$, $P(W|U_2) = 2/5$, and $P(U_i) = 1/2$, $i = 1, 2$, since either urn is equally likely to be selected. Further, $P(WU_1) = P(W|U_1) \times P(U_1) = 3/20$, and $P(WU_2) = P(W|U_2)P(U_2) = 1/5$. Now the probability that the ball drawn is white regardless of the urn is then $P(W) = P(WU_1) + P(WU_2) = 7/20$ since $W = WU_1 \cup WU_2$, where WU_1 and WU_2 are mutually exclusive. Thus, since $P(U_1|W) = P(WU_1)/P(W)$, $P(U_1|W) = 3/7$.

The second part of Example 2.2.3 introduces an important concept in both theoretical and applied probability. This is known as Bayes' theorem after its discoverer Sir Thomas Bayes (1702-1761), a British minister who was also an amateur mathematician.

Theorem 2.2.3 Suppose the sample space S can be partitioned into k mutually exclusive regions (events) R_1, R_2, ..., R_k and suppose, further, that $E \subset S$, $P(E) > 0$, is an arbitrary event in S. Let $P(E|R_i)$ be the probability that event E occurs, given the occurrence of R_i. Then $P(R_i|E)$, the probability that event R_i occurs given the occurrence of E is

$$P(R_i|E) = \frac{P(E|R_i)P(R_i)}{\Sigma_{i=1}^{k}P(E|R_i)\ P(R_i)} \qquad i = 1, 2, ..., k \qquad (2.6)$$

Proof. Since R_1, R_2, ..., R_k are mutually exclusive, $\cup_{i=1}^{k}R_i = S$, it is not difficult to see that $E = \cup_{i=1}^{k}ER_i$ and that ER_1, ER_2, ..., ER_k are mutually exclusive. Thus, $P(E) = \Sigma_{i=1}^{k}P(ER_i) = \Sigma_{i=1}^{k}\ P(E|R_i) \times P(R_i)$. Furthermore, $P(R_i|E) = P(R_iE)/P(E) = P(E|R_i)P(R_i)/P(E)$, and hence (2.6) follows.

Definition 2.2.9 The events E_1 and E_2 are said to be statistically *independent* if $P(E_2|E_1) = P(E_2)$. That is, the probability that the event E_2 occurs is unaffected by the occurrence (or nonoccurrence) of E_1.

It is clear that Definition 2.2.9 implies that if E_1 and E_2 are independent, then

(i) $P(E_1E_2) = P(E_1)P(E_2)$

and

(ii) $P(E_1|E_2) = P(E_1)$

and conversely. In fact, the following statements are equivalent: (a) events E_1 and E_2 are independent; (b) $P(E_1E_2) = P(E_1)P(E_2)$; (c) $P(E_1|E_2) = P(E_1)$.

Example 2.2.4 (See Example 2.2.3.) Consider the problem of drawing two cards from an ordinary deck of 52 playing cards. Between draws, the card on the first draw is replaced and the cards are reshuffled. To compute the probability that both cards are spades, notice that if E_1 is the event of a spade on the first draw and E_2 is the event of a spade on the second draw, E_1 and E_2 are independent events. Hence,

$$P(E_1E_2) = P(E_1)P(E_2) = \frac{1}{4} \times \frac{1}{4} = \frac{1}{16}$$

Usually, when dealing with probabilities defined on a finite sample space, the combining of mutually exclusive events can be a tedious procedure involving a great deal of counting. Often, this procedure can be simplified by introducing the concepts of permutations and combinations. To begin with, n! is defined as the product of the integers 1, 2, ..., n, that is, $n! = 1 \cdot 2 \cdots n$. Notice that $n! = n(n - 1)!$. Thus, in order to ensure a mathematically consistent definition, 0! is defined as 1.

Consider now a box with r different compartments, and suppose there are $n \geq r$ distinguishable balls available to fill these r compartments. How many ways can these r compartments be filled (exactly one ball per compartment) by these n balls? Clearly,

there are n ways to fill the first compartment since there are n
balls available initially. Given that the first compartment is
filled, there are n - 1 balls available to fill the second compartment. Hence, there are n - 1 ways to fill the second compartment.
Continuing in this way, given that the first r - 1 compartments are
filled, there are n - r + 1 balls available to fill the rth compartment and hence n - r + 1 ways to fill the rth compartment. Thus,
the number of ways of filling the box in the prescribed manner is
$n(n - 1) \cdots (n - r + 1) = n!/(n - r)!$. This is precisely the number
of permutations of n objects taken r at a time which is denoted by
the symbol $^n p_r$. This leads to the following definition.

Definition 2.2.10 $^n p_r$, the *number of permutations* of n objects
taken r at a time $n \geqslant r$ is $n!/(n - r)!$.

In the description of permutations the order of how the balls
are placed in the compartments is important. For example, if there
are two compartments and three balls numbered 1, 2, and 3, then
(1,3) and (3,1) represents distinct permutations. If the balls,
themselves, that are chosen to fill the r compartments are
indistinguishable; then $^n p_r$ must be divided by r!. Thus, if the
order is unimportant, the number of ways in which n balls can be
placed in r compartments is $^n p_r /r!$, which is the number of combinations of n objects taken r at a time. This leads to the following
definition.

Definition 2.2.11 The *number of combinations* of n objects taken r
at a time, denoted by $\binom{n}{r}$, is $^n p_r /r!$ That is, $\binom{n}{r} = \dfrac{n!}{r!(n - r)!}$.

Combinations are especially important in describing dichotomous
events such as the coin-tossing problem of Examples 2.2.1 and 2.2.2.
To understand this more clearly, suppose a coin is tossed n times
showing r heads and n - r tails. One such sequence is simply r
heads followed by n - r tails. However, any rearrangement of the
sequence still yields r heads and n - r tails. Thus, the number of
ways r heads can appear in n tosses of a coin is $\binom{n}{r}$ since the order
of heads is unimportant and is truly unobservable.

Example 2.2.5 Consider the coin-tossing problem of Example 2.2.1
and 2.2.2 in which an unbiased coin is tossed three times. The
sample space of eight mutually exclusive events can be partitioned
into $\binom{3}{0}$ – 1 set of no heads, $\binom{3}{1}$ = 3 sets of 1 heads, $\binom{3}{2}$ = 3 sets
of 2 heads, and $\binom{3}{3}$ = 1 set of 3 heads. More generally, the sample
space of two mutually exclusive events, generated by n tosses of a
coin (usually but not necessarily unbiased), can be partitioned
into the $\binom{n}{r}$ sets of r heads, r = 0, 1, 2, ..., n. This leads to
the useful formula

$$\sum_{i=0}^{n} \binom{n}{r} = 2^n$$

A more general formula than the above is known as the *binomial
theorem* is now given for further reference.

Theorem 2.2.4 (Binomial Theorem) Suppose a and b are any two real
numbers and n ≥ 1 is an integer. Then,

$$(a + b)^n = \sum_{r=0}^{n} \binom{n}{r} a^{n-r} b^r \tag{2.7}$$

The proof of this theorem is left as an exercise for the
reader. However, it can be proved fairly easily by mathematical
induction on n with the aid of the useful combinatorial equality

$$\binom{n}{r-1} + \binom{n}{r} = \binom{n+1}{r} \tag{2.8}$$

which is also left as an exercise for the reader.

Problems

1. Consider the simple experiment of tossing an unbiased coin five
    times. Describe the sample space. Find the probabilities that
    the following occur: (i) exactly two heads, (ii) exactly three
    heads, (iii) exactly four heads, (iv) at least one head, (v)
    at most two heads, and (vi) at most five heads.

2. Suppose a single card is drawn from an ordinary bridge deck.
    Find the probabilities that: (i) it is the king of hearts,
    (ii) it is either a king or a heart, (iii) it is either a king

or a heart but not the king of hearts, (iv) it is neither a
king nor a heart, and (v) it is either a king or a heart and it
is neither a king nor a heart.

3. If E_1 and E_2 are any two events, show that $P[(E_1 - E_2) \cup E_2] =$
    $P(E_1 - E_2) + P(E_2)$. Extend this to k events E_1, E_2, ..., E_k.

4. Two persons play a game of dice. The first person to toss a 7
    wins. If A_1 tosses first and A_2 tosses second, what are the
    respective probabilities that A_1 wins and A_2 wins? The game
    continues until there is a winner. Ans. $P(A_1) = 6/11$,
    $P(A_2) = 5/11$.

5. Suppose two urns contain 5 white and 5 black balls and 8 white
    and 2 black balls, respectively. A ball is transferred from the
    first urn to the second urn, and then a ball is drawn from the
    second urn. What is the probability a white ball is selected?

6. Three urns contain 4 red, 3 blue; 3 red, 4 blue; 2 red, 5 blue
    balls, respectively. A ball is selected at random from one of
    the urns and it is red. What is the probability it came from
    the second urn?

7. An individual tosses a single die and a coin simultaneously.
    Find the probabilities that: (i) a 6 and tails occur, (ii) an
    even number and heads occurs, and (iii) heads occurs irrespective
    of the value of the die.

8. Prove that $\begin{pmatrix} n \\ r - 1 \end{pmatrix} + \begin{pmatrix} n \\ r \end{pmatrix} = \begin{pmatrix} n + 1 \\ r \end{pmatrix}$

9. Using the result in the previous problem, prove the binomial
    theorem.

10. Suppose a single die is tossed three times. Find the probability
    that the sum of the total on the three tosses is 7 or greater.

2.3   PROBABILITY DISTRIBUTIONS IN A SINGLE VARIABLE

Instead of trying to describe various sample spaces and probabilities
over these sample spaces, it is both easier and more systematic to
introduce the concept of random variables and their distributions.
In this section a single random variable is considered.

Definition 2.3.1 Suppose S is a sample space. The variable X is called a *random variable* if X is any real-valued function defined over S. The value x is the realization of X, i.e., the set of values in the domain of X.

Example 2.3.1 Consider the sample space of Examples 2.2.1 and 2.2.2. Let X = 0 if all three coins are tails, event E_7; let X = 1 if one coin is heads and the other two coins are tails, events E_1, E_6, and E_8; let X = 2 if two coins are heads and the other coin is tails, events E_2, E_4, and E_5; and, finally let X = 3 if all three coins are heads, event E_3. Then X is a random variable defined on the sample space S.

Definition 2.3.2 Suppose X is a random variable whose domain is the discrete set N, either finite or countably infinite. (N is usually the set of nonnegative integers {0, 1, 2, ...} or a subset thereof.) Suppose, further, there is a function f(x) such that for each value x assumed by X,

$\quad$ (i) $f(x) = pr\{X = x\} \geqslant 0$ $\hfill$ (2.9)

and

$\quad$ (ii) $pr\{A\} = \sum_{x \in A} f(x)^{\dagger}$ $\hfill$ (2.10)

for any subset A of N. Then, f(x) is said to be a *probability function* of the *discrete random variable* X, provided $pr\{N\} = 1$.

$\quad$ Analogous to Definition 2.2.7 it is clear that if f(x) is a probability function of the discrete random variable X defined on N, then

$\quad$ (i) $pr\{A\} \geq 0$, for each $A \subset N$. $\hfill$ (2.11)

$\quad$ (ii) $pr\{N\} = \sum_{x \in N} f(x) = 1$. $\hfill$ (2.12)

and

$\quad$ (iii) If A_1, A_2, ... is any sequence (finite or infinite) of disjoint sets from N,

---

$^{\dagger}x \in A$ is the notation for the point x being contained in the set A; $x \notin A$ means x is not contained in A.

$$\text{pr}\{\cup_i A_i\} = \Sigma_i \text{ pr}\{A_i\} \tag{2.13}$$

Conversely, if $f(x)$ is any function of $x \in N$, satisfying (2.11) to (2.13), then $f(x)$ is a probability function of the discrete random variable X.

Example 2.3.2 Let X, a random variable, be defined on $N = \{0, 1, 2, \ldots, n\}$, where $n > 0$ is a fixed integer. Let $f(x)$ be the function

$$f(x) = \text{pr}\{X = x\} = \binom{n}{x} p^x q^{n-x} \tag{2.14}$$

for $x \in N$, and $f(x) = 0$, for $x \notin N$, where $0 < p < 1$, and $q = 1 - p$. Then, $f(x)$ is a probability function for X. To show that $f(x)$ is a probability function, it suffices to show $f(x) > 0$ for $x \in N$ and $\Sigma_{x=0}^{n} f(x) = 1$. The first part is trivial by the way $f(x)$ is defined. For the second part, note that

$$\sum_{x=0}^{n} f(x) = \sum_{x=0}^{n} \binom{n}{x} p^{n-x} q^{x} \tag{2.15}$$

But, by the binomial theorem,

$$\sum_{x=0}^{n} \binom{n}{x} p^x q^{n-x} = (p + q)^n = 1 \tag{2.16}$$

The distribution (2.14) is called the *binomial distribution*.

Example 2.3.3 Let the random variable X be defined on N, the set of all positive integers, and let $f(x)$ be the function

$$f(x) = \text{pr}\{X = x\} = \left(\frac{1}{2}\right)^x \tag{2.17}$$

$x = 1, 2, \ldots$, and $f(x) = 0$ otherwise. Clearly, $f(x) > 0$ for $x \in N$. Furthermore,

$$\sum_{x=1}^{\infty} f(x) = \sum_{x=1}^{\infty} \left(\frac{1}{2}\right)^x = \frac{1}{1 - 1/2} - 1 = 1 \tag{2.18}$$

Thus, $f(x)$ is a probability function for X.

Definition 2.3.3 Suppose A is the set of all random variables $X \leq x$.

Then

$$pr\{A\} = pr\{X \leq x\} \tag{2.19}$$

is called the *cumulative distribution function* of X, or the cdf of
X. The usual notation for the cdf of X is F(x).

Discrete random variables and their corresponding probability
functions describe a variety of sample spaces such as those generated by the outcomes of the tossing of a coin, the tossing of a pair
of dice, or the drawing of cards (with or without replacement) from
a deck of cards. However, discrete random variables in themselves
are not sufficient to describe all sample spaces which are of interest and which are important in applications. For example, in
measuring the length of life of electronic equipment a discrete
random variable cannot be used, since the random variable of interest is time which is continuous. The notion of a continuous random
variable is now introduced.

Definition 2.3.4 Suppose X is a random variable whose domain is the
real line $(-\infty,\infty)$ or a subset thereof. Suppose $f(x) \geq 0$ is a continuous function such that (i) for all $x \in [a,b]$,

$$pr\{a < X < b\} = \int_a^b f(x) \, dx \tag{2.20}$$

and

(ii) if A is any subset of intervals on the real line,

$$pr\{A\} = \int_{x \in A} f(x) \, dx \tag{2.21}$$

then f(x) is said to be a *probability density function* of the *continuous random variable* X[†], provided $\int_{-\infty}^{\infty} f(x) \, dx = 1$.

It follows that $pr\{a \leq X \leq b\} = pr\{a < X \leq b\} = pr\{a < X < b\}$
when X is continuous.

To prove the above assertion,

$$pr\{a \leq X \leq b\} = pr\{X = a\} + pr\{a < X \leq b\}$$

[†]f(x) may have a finite number of discontinuities but the definition
still holds.

However, $\text{pr}\{X = a\} = \int_a^a f(x) \, dx = 0$, by rules of integral calculus.
Similarly, $\text{pr}\{X = b\} = 0$.

As in the case of discrete random variables, if $f(x)$ is a probability density function (pdf) of the random variable X defined on
$(-\infty,\infty)$ with at most a finite number of discontinuities, then:

(i) $\text{pr}\{A\} \geq 0$, for each subset A on $(-\infty,\infty)$ (2.22)

(ii) $\text{pr}\{-\infty < X < \infty\} = \int_{-\infty}^{\infty} f(x) \, dx = 1$ (2.23)

and

(iii) If A_1, A_2, $\ldots$ is any sequence (finite or infinite) of
disjoint sets on $(-\infty,\infty)$,

$$\text{pr}\{\cup_i A\} = \Sigma_i \ \text{pr}\{A_i\}$$ (2.24)

Conversely, if $f(x)$ is any function with at most a finite number of
discontinuities such that (2.22) to (2.24) hold, then $f(x)$ is a pdf
of the random variable X.

Example 2.3.4   Let the random variable X be defined on $(-\infty,\infty)$.   Let
$f(x)$ be the function

$$f(x) = \frac{1}{\sqrt{2\pi}} e^{-x^2/2} \qquad -\infty < x < \infty$$ (2.25)

Then $f(x)$ is a pdf for X.   To show this, it is first clear that
since $f(x) \geqslant 0$, (2.22) is satisfied.   Furthermore, (2.24) is clearly
satisfied.   Thus, it remains to show that

$$\int_{-\infty}^{\infty} \frac{1}{\sqrt{2\pi}} e^{-x^2/2} \, dx = 1$$ (2.26)

This is basically an exercise in integration that involves a transformation from cartesian to polar coordinates.   The details are
left as an exercise for the reader. $f(x)$ is called the *standard
normal density function*.

Example 2.3.5   Let the random variable X be defined on $(-\infty,\infty)$.
Suppose $f(x)$ is a function defined as

$$f(x) = \begin{cases} 0 & -\infty < x < 0 \\ 1 & 0 \leqslant x \leqslant 1 \\ 0 & 1 < x < \infty \end{cases} \qquad (2.27)$$

Clearly, $f(x)$ is a pdf for the random variable X and is called the *uniform density function*.

The cumulative distribution given by Definition 2.3.3 is applicable to discrete and continuous random variables alike. It is clear that if X is a discrete random variable, defined over a subset of the nonnegative integers (including possibly the set of nonnegative integers), then

$$F(x) = \sum_{y=0}^{[x]} f(y) \qquad (2.28)$$

where $f(x)$ is the probability function of X and $[x]$ is the greatest integer less than or equal to x. On the other hand, if X is a continuous random variable defined on $(-\infty, \infty)$ and if $f(x)$ is a pdf of X, then

$$F(x) = \int_{-\infty}^{x} f(y) \, dy \qquad (2.29)$$

If x is a point for which $F(x)$ is not continuous, then $F(x)$ is, by definition,

$$F(x) = \lim_{\substack{y \downarrow x}} F(y) = \lim_{\substack{y \to x \\ y > x}} F(y) \qquad (2.30)$$

Example 2.3.6 Consider the probability functions and pdf's in Examples 2.3.2 to 2.3.5. Then the cdf's for these random variables are, respectively,

$$\text{(i)} \quad F(x) = \begin{cases} 0 & x < 0 \\ \sum_{y=0}^{[x]} \binom{n}{y} p^y (1-p)^{n-y} & 0 \leq x \leq n \\ 1 & x > n \end{cases} \qquad (2.31)$$

(ii) $F(x) = \begin{cases} 0 & x < 1 \\ 1 - (1/2)^{[x]} & x \geq 1 \end{cases}$ (2.32)

(iii) $F(x) = \displaystyle\int_{-\infty}^{x} \frac{1}{\sqrt{2\pi}}\, e^{-y^2/2}\, dy \qquad -\infty < x < \infty$ (2.33)

(iv) $F(x) = \begin{cases} 0 & x < 0 \\ x & 0 \leq x < 1 \\ 1 & x \geq 1 \end{cases}$ (2.34)

It is instructive to show the graphs of $F(x)$ in both the discrete
and continuous cases.   Figures 2.1 and 2.2 are the graphs of (2.32)
and (2.34), respectively.

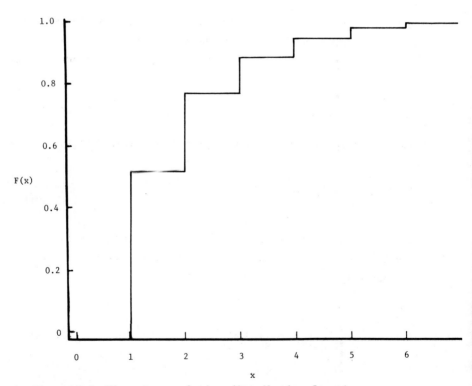

Figure 2.1 Discrete cumulative distribution function.
$F(x) = 1 - 1/2^{[x]}$; $x \geq 1$.

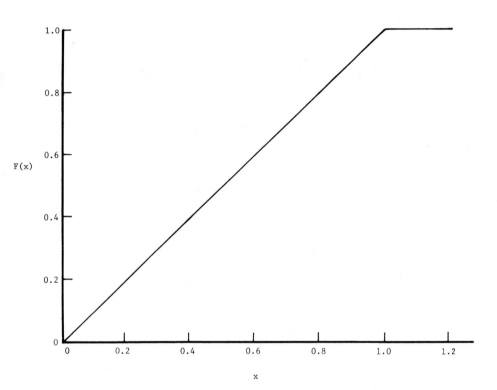

Figure 2.2 Continuous cumulative distribution function.
$F(x) = x$, $0 < x \leqslant 1$; $F(x) = 1$, $x > 1$.

Problems

1. Suppose $f(x) = e^{-\lambda}\lambda^x/x!$, $x = 0, 1, 2, \ldots$ and $f(x) = 0$, else-
    where. Show that $f(x)$ is a discrete probability function. $f(x)$
    is the Poisson model.

2. In problem 1, suppose $pr\{X = 1\} = pr\{X = 2\}$. Find $pr\{X > 1\}$.
    (Assume $\lambda > 0$.)

3. Show that $\int_{-\infty}^{\infty} \frac{1}{\sqrt{2\pi}} e^{-x^2/2}\,dx = 1$, and hence show that the stand-
    ard normal density function is a probability density function.

    Hint: $\left(\int_{-\infty}^{\infty} \frac{1}{\sqrt{2\pi}} e^{-x^2/2}\,dx\right)^2 = \int_{-\infty}^{\infty} \int_{-\infty}^{\infty} \frac{1}{2\pi} \exp\left[-\left(\frac{x^2 + y^2}{2}\right)\right]\,dy\,dx$

    Let $x = r \cos\theta$ and $y = r \sin\theta$ and transform x and y to polar
    coordinates. Thus,

    $\int_{-\infty}^{\infty} \int_{-\infty}^{\infty} \frac{1}{2\pi} \exp\left[-\left(\frac{x^2 + y^2}{2}\right)\right]\,dy\,dx = \frac{1}{2\pi} \int_0^{\infty} \int_0^{2\pi} re^{-r^2/2}\,d\theta\,dr$

4. Let $F(x) = 1 - p^{[x]}$, where $[x]$ is the largest integer $\leq x$,
    $x \geq 1$; $F(x) = 0$, $x < 1$, and $0 < p < 1$ is a real constant. Is
    $F(x)$ a cdf? If it is, what is the corresponding probability
    function $f(x)$, and what values does $f(x)$ take with positive
    probability?

5. Suppose $F(x)$ is the cdf of a continuous random variable X with
    at most a finite number of discontinuities. Show that the pdf
    $f(x) = dF(x)/dx$ at each point x for which $dF(x)/dx$ exists.

6. Graph the cdf's described by equation (2.31) and (2.33); for
    (2.31) assume $p = 1/2$; for (2.33) use Table A-1.

7. Suppose $f(x) = c/x^2$, if $x = 1, 2, \ldots$; $f(x) = 0$ otherwise.
    What is the value of c so that $f(x)$ is a probability function?
    Find, approximately, $pr\{X \geq 2\}$.

8. Suppose $f(x) = c/x^2$, if $x > 1$, and $f(x) = 0$ otherwise. Now
    find the value of c so that $f(x)$ is pdf. Find $pr\{X \geq 2\}$.

9. Should there be concern that the results in problems 7 and 8
    differ? Explain.

10. Suppose $f(x)$ is the probability function of a discrete random
    variable X defined over the set of nonnegative integers,

(0, 1, 2, ...). Show that if X = x, where x is an integer,

$f(x) = F(x) - F(x - 1)$

where $F(x)$ is the cdf of x.

2.4 MULTIVARIATE PROBABILITY DISTRIBUTIONS

It may occur in practice that a random variable has more than a
single component. For example, a demographer may be interested in
depicting a number of human characteristics such as height, weight,
ethnicity, age, sex, and geographic distribution. Each one of these
characteristics can, in turn, be considered random for a group of
individuals selected at random from the population. However, in
order to make inferences about the human population, all these characteristics are important and must be considered together. A random
variable that has more than one component is called a multivariate
random variable.

Definition 2.4.1 Suppose S is a sample space in k dimensions. Then
$(X_1,...,X_k)$ is called a *multivariate random variable* if X_i, i = 1,
2, ..., k, is a real-valued function defined on the ith dimension
of S.

Example 2.4.1 Consider the sample space generated by the tossing of
two coins. Let X_1 = 0 if the first coin is tails, X_1 = 1 if the
first coin is heads, X_2 = 0 if the second coin is tails, and X_2 = 1
if the second coin is heads. Then (X_1,X_2) is a random variable
defined on the two-dimensional sample space S.

Definition 2.4.2 Suppose $(X_1,X_2,...,X_k)$ is a (multivariate) random
variable whose ith component X_i has domain N_i, a discrete set,
either finite or countably infinite (N_i is usually the set of nonnegative integers or a subset thereof), i = 1, 2, ..., k. Suppose,
further, there is a function $f(x_1,...,x_k)$ such that for each value
x_i assumed by X_i, i = 1, 2, ..., k,

(i) $f(x_1,x_2,...,x_k) = pr\{X_1 = x_1,\ X_2 = x_2,\ ...,\ X_k = x_k\}$ (2.35)

and

(ii) $\text{pr}\{A\} = \sum_{(x_1,\ldots,x_k) \in A} f(x_1,x_2,\ldots,x_k)$ (2.36)

for any subset A of N, where N is the k-dimensional set whose ith component is N_i, i = 1, 2, ..., k. Then, $f(x_1,x_2,\ldots,x_k)$ is said to be a *probability function of the discrete multivariate random variable* $(X_1,\ldots,X_k)$, provided $\text{pr}\{N\} = 1$.

Analogous to Definition 2.2.7 it is clear that if $f(x_1,\ldots,x_k)$ is a probability function of the discrete multivariate random variable $(X_1,\ldots,X_k)$ defined on N, then

(i) $\text{pr}\{A\} \geq 0$, for each A $\in$ N   (2.37)

(ii) $\text{pr}\{N\} = \sum_{(x_1,\ldots,x_k) \in N} f(x_1,x_2,\ldots,x_k) = 1$ (2.38)

and

(iii) If A_1, A_2, ... is any sequence (finite or infinite) of disjoint sets from N,

$\text{pr}\{\cup_i A_i\} = \Sigma_i \ \text{pr}\{A_i\}$ (2.39)

Conversely, if $f(x_1,x_2,\ldots,x_k)$ is any function of $(x_1,x_2,\ldots,x_k) \in N$ satisfying (2.37) to (2.39), then $f(x_1,\ldots,x_k)$ is a probability function for the discrete multivariate random variable $(X_1,X_2,\ldots,X_k)$.

Convention: A multivariate random variable is termed a *random vector*.

Example 2.4.2 Let (X_1,X_2) be a random vector on (N_1,N_2) where $N_1 = \{0,1,\ldots,n\}$ and $N_2 = \{0, 1, \ldots, n - x_1\}$, n > 0 is a fixed integer, and x_1 is an integer between 0 and n. Let $f(x_1,x_2)$ be the function

$$f(x_1,x_2) = \text{pr}\{X_1 = x_1, \ X_2 = x_2\} = \frac{n!}{x_1!x_2!(n - x_1 - x_2)!} \ p_1^{x_1} p_2^{x_2}$$

$$\times \ (1 - p_1 - p_2)^{n-x_1-x_2}$$ (2.40)

for $(x_1,x_2) \in N$, where $N = (N_1,N_2)$ and $f(x_1,x_2) = 0$ for $(x_1,x_2) \notin N$, where $0 < p_1 < 1$, $0 < p_2 < 1$, and $p_1 + p_2 < 1$. Then $f(x_1,x_2)$ is a

multivariate probability function for (X_1, X_2). To show this, it
suffices to show $f(x_1, x_2) > 0$ for $(x_1, x_2) \in N$ and $\sum_{(x_1, x_2) \in N} f(x_1, x_2) =$
1. Clearly, $f(x_1, x_2) > 0$ for $(x_1, x_2) \in N$. For the second part note

$$\sum_{(x_1, x_2) \in N} f(x_1, x_2) = \sum_{x_1=0}^{n} \sum_{x_2=0}^{n-x_1} \frac{n!}{x_1! x_2! (n - x_1 - x_2)!} p_1^{x_1} p_2^{x_2}$$

$$\times (1 - p_1 - p_2)^{n - x_1 - x_2}$$

$$= \sum_{x_1=0}^{n} \frac{n!}{x_1!(n - x_1)!} p_1^{x_1} \sum_{x_2=0}^{n-x_1} \frac{(n-x_1)!}{x_2!(n - x_1 - x_2)!}$$

$$\times p_2^{x_2} (1 - p_1 - p_2)^{n-x_1-x_2}$$

Now, defining $n - x_1$ as n',

$$\sum_{x_2=0}^{n-x_1} \frac{(n - x_1)!}{x_2!(n - x_1 - x_2)!} p_2^{x_2} (1 - p_1 - p_2)^{n-x_1-x_2}$$

$$= \sum_{x_2=0}^{n'} \frac{n'!}{x_2!(n' - x_2)!} p_2^{x_2} (1 - p_1 - p_2)^{n'-x_2}$$

$$= (1 - p_1 - p_2 + p_2)^{n'} = (1 - p_1)^{n-x_1}$$

applying the binomial theorem. Thus,

$$\sum_{(x_1, x_2) \in N} f(x_1, x_2) = \sum_{x_1=0}^{n} \frac{n!}{x_1!(n - x_1)!} p_1^{x_1} (1 - p_1)^{n-x_1} = 1 \tag{2.41}$$

again applying the binomial theorem. $f(x_1, x_2)$ is called the *tri-nomial distribution*.

Example 2.4.3 (Extension of Example 2.4.2) Let $(X_1, X_2, \ldots, X_k)$ be a
random vector defined on $N = (N_1, N_2, \ldots, N_k)$, where $N_1 = \{0, 1, \ldots, n\}$,
$N_2 = \{0, 1, \ldots, n - x_1\}, \ldots, N_k = \{0, 1, \ldots, n - x_1 - \cdots - x_{k-1}\}$,
$n > 0$ is a fixed integer, and $x_1, x_2, \ldots, x_{k-1}$ are integers between 0 and n such that $\sum_{i=1}^{k-1} x_i < n$. Let $f(x_1, x_2, \ldots, x_{k-1})$ be the
function

$$f(x_1,x_2,\ldots,x_{k-1}) = \frac{n!}{x_1!x_2!\cdots(n - x_1 - x_2 - \cdots - x_{k-1})!}$$

$$\times\ p_1^{x_1} p_2^{x_2}\cdots(1 - p_1 - \cdots - p_{k-1})^{n-\sum_{i=1}^{k-1} x_i} \qquad (2.42)$$

for $(x_1,x_2,\ldots,x_k) \in N$, where $N = (N_1,\ldots,N_k)$ and $f(x_1,x_2,\ldots,x_k) = 0$, for $(x_1,x_2,\ldots,x_k) \notin N$, where $0 < p_1 < 1$, $0 < p_2 < 1,\ldots$, $0 < p_{k-1} < 1$, and $p_1 + p_2 + \cdots + p_{k-1} < 1$. Then $f(x_1,x_2,\ldots,x_{k-1})$ is a multivariate probability function. It is left as an exercise for the reader to show this. $f(x_1,x_2,\ldots,x_{k-1})$ is called the *multinomial distribution*.

Definition 2.4.3 Suppose A is the set of all random vectors $(X_1,X_2,\ldots,X_k)$ such that $X_i \leqslant x_i$, $i = 1, 2, \ldots, k$. Then

$$pr\{A\} = pr\{X_1 \leq x_1,\ X_2 \leq x_2,\ldots,\ X_k \leq x_k\} \qquad (2.43)$$

is called the *cumulative distribution function* (cdf) of $(X_1,X_2\ldots, X_k)$. The usual notation for the cdf of $(X_1,X_2,\ldots,X_k)$ is $F(x_1,x_2\ldots x_k)$. (Compare Definitions 2.3.3 and 2.4.3.)

Analogous to continuous random variables (see Definition 2.3.4) the probability density function of a continuous random vector $(X_1,X_2,\ldots,X_k)$ is now defined.

Definition 2.4.4 Suppose $(X_1,X_2,\ldots X_k)$ is a random vector whose ith component X_i has its domain the real line $(-\infty,\infty)$ or a subset thereof. Suppose $f(x_1,x_2,\ldots,x_k) > 0$ is a function such that for all $x_i \in [a_i,b_i]$, $i = 1, 2, \ldots, k,$

(i) $pr\{a_1 < X_1 < b_1,\ \ldots,\ a_k < X_k < b_k\}$

$$= \int_{a_k}^{b_k} \cdots \int_{a_1}^{b_1} f(x_1,\ldots,x_k)\ dx_1 \cdots dx_k \qquad (2.44)$$

and if A is any subset of k-dimensional intervals,

(ii) $pr\{A\} = \int \cdots \int_{(x_1,x_2,\ldots,x_k) \in A} f(x_1,x_2,\ldots,x_k)\ dx_1\ dx_2 \cdots dx_k$

$$\qquad (2.45)$$

then $f(x_1,x_2,\ldots,x_k)$ is said to be a joint *probability density function* of the *continuous random vector* $(X_1,X_2,\ldots,X_k)$, if the density integrates to one over its domain.

It can be shown that the probability of a $k - 1$ dimensional hyperplane is zero in a k-dimensional space, e.g., in two dimensions; $pr\{X = a, b_1 < y < b_2\} = 0$. (See the discussion following Definition 2.3.4 concerning single points on the real line.)

As in the case of discrete random vectors, if $f(x_1,x_2,\ldots,x_k)$ is a joint probability density function of $(X_1,X_2,\ldots,X_k)$ where X_i, the ith component, is defined on $(-\infty,\infty)$ with at most a finite number of discontinuities, then:

  (i) $pr\{A\} > 0$, for each subset A of k-dimensional intervals

  (ii) $pr\{-\infty < X_1 < \infty, \ldots, -\infty < X_k < \infty\}$

$$= \int_{-\infty}^{\infty} \cdots \int_{-\infty}^{\infty} f(x_1,\ldots,x_k) \, dx_1 \cdots dx_k = 1 \qquad (2.46)$$

and

  (iii) If A_1, A_2, $\ldots$ is any sequence (finite or infinite) of disjoint k-dimensional intervals,

$$pr\{U_i A_i\} = \Sigma_i \, pr\{A_i\} \qquad (2.47)$$

Conversely, if $f(x_1,x_2,\ldots,x_k)$ is any function with at most a finite number of discontinuities such that (i) to (iii) hold, then $f(x_1,x_2, \ldots,x_k)$ is a joint pdf of the random vector $(X_1,X_2,\ldots,X_k)$.

Example 2.4.4 Let the random vector (X,Y) be defined on the entire two-dimensional space $(-\infty,\infty) \times (-\infty,\infty)$. Let $f(x,y)$ be the function

$$f(x,y) = \frac{1}{\pi\sqrt{3}} e^{-(2/3)(x^2-xy+y^2)} \qquad -\infty < x < \infty, \; -\infty < y < \infty$$

$$(2.48)$$

Then $f(x,y)$ is a joint pdf for (X,Y). To show this, it is clear that (i) and (iii) above are both satisfied. Thus, it must be shown that

$$\int_{-\infty}^{\infty} \int_{-\infty}^{\infty} \frac{1}{\pi\sqrt{3}} e^{-(2/3)(x^2-xy+y^2)} \, dy \, dx = 1 \qquad (2.49)$$

given that

$$\int_{-\infty}^{\infty} \frac{1}{\sqrt{2\pi}} e^{-x^2/2} \, dx = 1$$

which is the result of Problem 3 and Section 2.3. This is done by completing the square. Again, the details are left as as exercise for the reader. The joint density function (2.48) is the bivariate normal density function.

Problems

1. Suppose $f(x,y) = (2\pi)^{-1} e^{-(x^2+y^2)/2}$, $-\infty < x < \infty$, $-\infty < y < \infty$. Show that $f(x,y)$ is a joint pdf in (X,Y).

2. Consider the joint density function $f(x,y)$ in problem 1. Using tables of the standard normal distribution find $\text{pr}\{X > 1, Y > 1\}$.

3. Consider the joint density function $f(x,y)$ defined by (2.48). Find $\text{pr}\{X > 1\}$.

4. Show that $f(x_1, x_2, \ldots x_k)$ as defined by (2.42) is a joint probability function in $(X_1, X_2, \ldots X_k)$.

5. Show that $f(x,y)$ as defined by (2.48) is a joint pdf in (X,Y).

6. Suppose $(X_1, \ldots X_k)$ is a continuous random vector whose pdf is $f(x_1, \ldots, x_k)$. Show that $F(x_1, \ldots, x_k)$, the cdf, can be written as

$$F(x_1, x_2, \ldots, x_k) = \int_{-\infty}^{x_1} \cdots \int_{-\infty}^{x_k} f(y_1, \ldots, y_k) \, dy_k \cdots dy_1$$

and, conversely,

$$f(x_1, \ldots x_k) = \frac{\partial^k F(x_1, \ldots, x_k)}{\partial x_1 \partial x_2 \cdots \partial x_k}$$

assuming the domain of X_i, the ith component of $(X_1, \ldots, X_k)$ is $(-\infty, \infty)$, $i = 1, 2, \ldots, k$.

7. Suppose $(X_1, \ldots, X_k)$ is a discrete random vector whose ith component X_i is defined over a subset of the nonnegative integers (including possibly the set of nonnegative integers). Show that $F(x_1, \ldots, x_k)$, the cdf, can be written as

$$F(x_1,\ldots,x_k) = \sum_{y_1=0}^{[x_1]} \cdots \sum_{y_k=0}^{[x_k]} f(y_1,\ldots,y_k)$$

where $f(x_1,\ldots,x_k)$ is the joint probability function of $(X_1,\ldots,X_k)$ and $[x_i]$ is the largest integer less than or equal to x_i, $i = 1, 2, \ldots, k$.

8. Suppose $(X_1,\ldots,X_k)$ is a discrete random vector defined as in the previous exercise. Show that if $(X_1,\ldots,X_k) = (x_1,\ldots,x_k)$, that is, $X_i = x_i$, for all i where each x_i is an integer, $i = 1, 2, \ldots, k$,

$$f(x_1,\ldots,x_k) = \Delta^k F(x_1,\ldots,x_k)$$

where $\Delta^2 F(x_1,x_2) = F(x_1,x_2) - F(x_1 - 1, x_2) - F(x_1, x_2 - 1) + F(x_1 - 1, x_2 - 1)$.

9. Suppose $f(x,y)$ is a function of the random vector (X,Y) defined as

$$f(x,y) = 3x^2 y + 3y^2 x \qquad 0 \leq x \leq 1,\ 0 \leq y \leq 1$$

show that $f(x,y)$ is a pdf for (X,Y). Find $F(x,y)$, the cdf of (X,Y).

10. Suppose $f(x,y)$ is a function of the random vector (X,Y) defined as

$$f(x,y) = k \qquad 0 \leq y \leq x,\ 0 \leq x \leq 1$$

Find the value of k such that $f(x,y)$ is a joint pdf in (X,Y), and find $F(x,y)$, the cdf of (X,Y).

2.5 MARGINAL AND CONDITIONAL DISTRIBUTIONS AND STOCHASTIC INDEPENDENCE FOR BIVARIATE DISTRIBUTIONS AND MOMENTS

In this section the important concepts of marginal and conditional distributions are discussed, and the extremely important concept of stochastic independence is introduced. That is, the question of how one random variable behaves in conjunction with another random variable is addressed.

Definition 2.5.1 (i) Suppose (X,Y) is a discrete random vector whose joint probability function is f(x,y), where X is defined over N_1 and Y is defined over N_2. The *marginal distribution* (probability function) of X, $f_1(x)$, is given as

$$f_1(x) = \sum_{y \in N_2} f(x,y) \tag{2.50}$$

(ii) If (X,Y) is a continuous random vector whose joint pdf is f(x,y), where the domain of each X and Y is $(-\infty, \infty)$ then the *marginal pdf of X*, $f_1(x)$, is

$$f_1(x) = \int_{-\infty}^{\infty} f(x,y)\, dy \tag{2.51}$$

Example 2.5.1 Let (X,Y) be a random vector defined on (N_1, N_2), where $N_1 = \{0,1,\ldots,n\}$ and $N_2 = \{0, 1, \ldots, n - x\}$, $n > 0$ is a fixed integer, and x is an integer between 0 and n. Let

$$f(x,y) = \frac{n!}{x!y!(n - x - y)!} p_1^x p_2^y (1 - p_1 - p_2)^{n-x-y} \tag{2.52}$$

for $(x,y) \in N$, where $N = (N_1, N_2)$ and f(x,y) = 0 for $(x,y) \notin N$, where $0 < p_1 < 1$, $0 < p_2 < 1$, and $p_1 + p_2 < 1$. In Example 2.4.2 this distribution, the trinomial distribution, was introduced.

$$f_1(x) = \sum_{y=0}^{n-x} \frac{n!}{x!y!(n - x - y)!} p_1^x p_2^y (1 - p_1 - p_2)^{n-x-y}$$

$$= \frac{n!}{x!(n - x)!} p_1^x \sum_{y=0}^{n-x} \frac{(n - x)!}{y!(n - x - y)!} p_2^y (1 - p_1 - p_2)^{n-x-y}$$

The above sum was encountered in Example 2.4.2. Thus, the marginal probability function is

$$f_1(x) = \frac{n!}{x!(n - x)!} p_1^x (1 - p_1)^{n-x} \qquad x = 0, 1, \ldots, n \tag{2.53}$$

Similarly, it can be shown that

$$f_2(y) = \frac{n!}{y!(n - y)!} p_2^y (1 - p_2)^{n-y} \qquad y = 0, 1, \ldots, n \tag{2.54}$$

Hence the marginal distributions for the trinomial are binomial.

Example 2.5.2 Let (X_1, X_2) be the continuous random vector, each of
whose components is defined on $(-\infty, \infty)$ and whose joint density function
tion is

$$f(x_1, x_2) = \frac{1}{2\pi} e^{-(x_1^2 + x_2^2)/2} \qquad -\infty < x_1 < \infty, \ -\infty < x_2 < \infty$$

$$(2.55)$$

The marginal densities $f_1(x_1)$ and $f_2(x_2)$, being symmetric, are given
as

$$f_i(x_i) = \frac{1}{\sqrt{2\pi}} e^{-x_i^2/2} \qquad -\infty < x_i < \infty; \ i = 1, 2 \qquad (2.56)$$

To see this, notice that

$$f_1(x_1) = \int_{-\infty}^{\infty} \frac{1}{2\pi} e^{-(x_1^2 + x_2^2)/2} \ dx_2 = \frac{1}{\sqrt{2\pi}} e^{-x_1^2/2}$$

$$\times \int_{-\infty}^{\infty} \frac{1}{\sqrt{2\pi}} e^{-x_2^2/2} \ dx_2$$

Thus, since $\int_{-\infty}^{\infty} \frac{1}{\sqrt{2\pi}} e^{-x_2^2/2} \ dx_2 = 1$, the result follows.

Definition 2.5.2 Suppose (X,Y) is a random vector (continuous or
discrete) whose joint distribution is $f(x,y)$. The *conditional distribution of Y given X = x (fixed)* is written $g(y|x)$ and is defined
as

$$g(y|x) = \frac{f(x,y)}{f_1(x)} \qquad (2.57)$$

where the domain of y may depend on x, where $f_1(x)$ is the marginal
distribution of X, and $f_1(x) > 0$.

Example 2.5.3 Let (X,Y) be the discrete random vector defined in
Example 2.5.1 (the trinomial). Then

$$g(y|x) = \frac{(n-x)!}{y!(n-x-y)!} \ \frac{p_2^y (1 - p_1 - p_2)^{n-x-y}}{(1 - p_1)^{n-x}}$$

Clearly, $(1 - p_1)^{n-x} \equiv (1 - p_1)^y (1 - p_1)^{n-x-y}$ is an identity for all p_1. Thus,

$$g(y|x) = \frac{(n - x)!}{y!(n - x - y)!} \, p_{2/1}^y \, (1 - p_{2/1})^{n-x-y}$$

$$y = 0, 1, \ldots, n - x \qquad (2.58)$$

where $p_{2/1} = p_2/(1 - p_1)$. Hence, the conditional distribution of a trinomial given a binomial is binomial. This can be extended to multinomial distributions.

Example 2.5.4 Let (X_1, X_2) be the continuous random vector defined in Example 2.5.2 Then

$$g(x_2|x_1) = \frac{1}{\sqrt{2\pi}} \, e^{-x_2^2/2} \qquad -\infty < x_2 < \infty \qquad (2.59)$$

Note that $g(x_2|x_1) = f_2(x_2)$. This concept is embodied in the following definition.

Definition 2.5.3 Suppose (X,Y) is a random vector whose joint distribution is $f(x,y)$. The random variables X and Y are said to be *stochastically independent*[†] if and only if

$$f(x,y) = f_1(x)f_2(y) \qquad (2.60)$$

over the entire domain of (X,Y) (i.e., for all x and y). Note that following from this definition, X and Y are independent if and only if $g(y|x) = f_2(y)$ over the entire domain of (X,Y). This is left as an exercise for the reader.

Example 2.5.5 Let (X_1, X_2) be the continuous random vector defined in Example 2.5.2. From the previous example it is clear that X_1 and X_2 are independent.

Example 2.5.6 Let (X,Y) be the discrete random variable defined in Example 2.5.1 (the trinomial). It was shown that

---

[†] The term "*independent*" is henceforth used instead of "*stochastically independent*" unless it causes confusion.

$$f_1(x) = \binom{n}{x} p_1^x (1 - p_1)^{n-x}$$

and

$$f_2(y) = \binom{n}{y} p_2^y (1 - p_2)^{n-y}$$

where the domain for both X and Y is 0, 1, ..., n. It is clear that $f(x,y) = f_1(x)f_2(y)$ does not hold for all X and Y. For example, $f(0,0) = (1 - p_1 - p_2)^n$ and $f_1(0) = (1 - p_1)^n$, $f_2(0) = (1 - p_2)^n$; hence $f(0,0) \neq f_1(0)f_2(0)$, and thus X and Y are not independent.

Definition 2.5.4 Suppose X is a random variable; then μ_r', the rth moment of X about the origin, is defined as

$$\mu_r' = \int_{-\infty}^{\infty} x^r f(x) \, dx \qquad (2.61)$$

if X is continuous, and

$$\mu_r' = \sum_{x \in N} x^r f(x) \qquad (2.61a)$$

if X is discrete and N is the domain of X, provided that for X continuous,

$$\int_{-\infty}^{\infty} |x^r| f(x) \, dx < \infty$$

and for X discrete,

$$\sum_{x \in N} |x^r| f(x) < \infty$$

An alternate notation for μ_r' is $E(x^r)$. This second symbolism leads to the following generalization.

Definition 2.5.5 Suppose X is a random variable and g(X) is any real-valued function of X; then

$$E(g(X)) = \int_{-\infty}^{\infty} g(x)f(x) \, dx \qquad (2.62)$$

if X is continuous, and

$$E(g(X)) = \sum_{x \in N} g(x)f(x) \tag{2.62a}$$

if X is discrete (N being the domain of X) provided $E(|g(X)|) < \infty$.
If $g(X) = X$, then $\mu_1' = E(X)$ is called the *mean of X* provided
$E(|X|) < \infty$. μ_1' is usually written as μ_x or μ when no confusion is
possible.   If $g(X) = X^2$, then $\mu_2' = E(X^2)$ is called the second moment
of X about the origin, provided $E(X^2) < \infty$.   If $g(X) = (X - \mu)^2$
(provided μ is finite), then $\sigma_x^2 = E[(X - \mu)^2]$ is termed the *variance*
of X.

Example 2.5.7   Let X be a discrete random variable having the binomial distribution (2.14). μ, the mean of X, is

$$\mu = \sum_{x=0}^{n} x \binom{n}{x} p^x q^{n-x} = \sum_{x=1}^{n} x \binom{n}{x} p^x q^{n-x}$$

$$= \sum_{x=1}^{n} \frac{n!}{(x-1)!(n-x)!} p^x q^{n-x} = np \sum_{x-1=0}^{n-1} \frac{(n-1)!}{(x-1)!(n-x)!}$$

$$\times p^{x-1} q^{n-x} \tag{2.63}$$

By the binomial theorem,

$$\sum_{x-1=0}^{n-1} \frac{(n-1)!}{(x-1)!(n-x)!} p^{x-1} q^{n-x} = 1$$

and hence $\mu = np$.   To compute σ^2, note that $E[(X - \mu)^2] = \mu_2' - \mu^2$.
(This is left as an exercise for the reader to verify.)   For the
binomial distribution

$$\mu_2' = \sum_{x=0}^{n} x^2 \binom{n}{x} p^x q^{n-x} = \sum_{x=2}^{n} x(x-1) \binom{n}{x} p^x q^{n-x} + \sum_{x=1}^{n} x \binom{n}{x} p^x q^{n-x}$$

$$\tag{2.64}$$

Now,

$$\sum_{x=1}^{n} x \binom{n}{x} p^x q^{n-x} = np$$

and

$$\sum_{x=2}^{n} x(x - 1) \frac{n!}{x!(n - x)!} p^x q^{n-x} = n(n - 1)p^2 \sum_{x-2=0}^{n-2} \binom{n-2}{x-2} p^{x-2} q^{n-x}$$

By the binomial theorem,

$$\sum_{x-2=0}^{n-2} \frac{(n - 2)!}{(x - 2)!(n - x)!} p^{x-2} q^{n-x} = 1$$

hence, $\mu_2' = n(n - 1)p^2 + np$ and $\sigma^2 = npq$.

Example 2.5.8 Let X be a continuous random variable having the standard normal density (2.25). μ, the mean of X, is

$$\mu = \int_{-\infty}^{\infty} \frac{1}{\sqrt{2\pi}} x e^{-x^2/2} \, dx \qquad (2.65)$$

Since $\mu = 0$,

$$\int_{-\infty}^{0} x e^{-x^2/2} \, dx = - \int_{0}^{\infty} x e^{-x^2/2} \, dx$$

Furthermore, since $\sigma^2 = \mu_2' - \mu^2$, and $\mu = 0$,

$$\sigma^2 = \int_{-\infty}^{\infty} \frac{x^2}{\sqrt{2\pi}} e^{-x^2/2} \, dx \qquad (2.66)$$

A simple integration by parts then shows that $\sigma^2 = 1$. Hence, if X has the standard normal density, then $\mu = 0$, $\sigma^2 = 1$. Furthermore, it is easy to show that if Y is a normally distributed variable related to X, a standard normal variable, by $Y = \mu + \sigma X$ then Y is normally distributed with mean μ and variance σ^2. Thus Y, an arbitrary normal random variable, is related to the standard normal random variable X by the equation $Y = \mu + \sigma X$, $\sigma > 0$, or equivalently, $X = (Y - \mu)/\sigma$.

The concept of moments can easily be extended to bivariate distributions. The following definition is the basis for this extension.

Definition 2.5.6   Suppose (X,Y) is a random vector whose joint
frequency function is f(x,y).   The (r,s)-st product moment of (X,Y)
about the origin is defined as

$$\mu'_{rs} = \int_{-\infty}^{\infty} \int_{-\infty}^{\infty} x^r y^s f(x,y) \; dy \; dx$$

if (X,Y) is continuous, and

$$\mu'_{rs} = \sum_{x \in N_x} \sum_{y \in N_y} x^r y^s f(x,y)$$

if (X,Y) is discrete and N_x and N_y are the respective domains of X
and Y, provided that for (X,Y) continuous,

$$\int_{-\infty}^{\infty} \int_{-\infty}^{\infty} |x^r y^s| f(x,y) \; dy \; dx < \infty$$

and for (X,Y) discrete,

$$\sum_{x \in N_x} \sum_{y \in N_y} |x^r y^s| f(x,y) < \infty$$

Just as Definition 2.5.6 is a generalization of Definition
2.5.4, so can Definition 2.5.5 be generalized to the bivariate case.
The details of this generalization are left as an exercise for the
reader.   In this regard, however, suppose that g(X,Y) is a real-
valued function of the random vector (X,Y) whose expectation
E[g(X,Y)] exists.   Further, if $g(X,Y) = (X - \mu_x)(Y - \mu_x)$, then
$E(X - \mu_x)(Y - \mu_y)$ is defined as the covariance of X and Y, which is
a measure of association between the two variables.   The alternate
notation of $E(X - \mu_x)(Y - \mu_y)$ is μ_{xy}.   Unfortunately, μ_{xy} fails as
a measure of association of X and Y since its value depends on the
units of measurement and hence cannot be compared with other distributions on a relative scale.   The following measure is independent of the measurement scale.

Definition 2.5.7   The *correlation coefficient* of the random vector
(X,Y) termed ρ_{xy} (or ρ when there is no possible confusion), is

given as

$$\rho = \frac{\mu_{xy}}{\sqrt{\sigma_x^2 \sigma_x^2}} \qquad (2.67)$$

provided X and Y both have finite second moments.

In accord with this definition the following theorem describes the range of ρ.

Theorem 2.5.1 $-1 \leq \rho \leq 1$.

Proof. Without loss of generality take $\mu_x = \mu_y = 0$. Then, by definition, $\rho = E(XY)/\sqrt{E(X^2)E(Y^2)}$. Thus it suffices to show that $\rho^2 \leq 1$, or equivalently, $[E(XY)]^2 \leq E(X^2)E(Y^2)$. This can be shown by appealing to the Schwarz-Cauchy inequality. However, a more elementary approach is used. Consider the function $(X - \alpha Y)^2 \geq 0$, where α is any real constant. Then, since $(X - \alpha Y)^2 \geq 0$ for all X and Y, $E(X - \alpha Y)^2 \geq 0$, where, by definition,

$$E(X - \alpha Y)^2 = \int_{-\infty}^{\infty} \int_{-\infty}^{\infty} (x - \alpha y)^2 f(x,y) \, dy \, dx$$

for X and Y both continuous (for example). Expanding this quadratic and using the fact that $E(ag(X) + bh(Y)) = aE(g(X)) + bE(h(Y))$ (Why?), it then follows that $E(X^2) - 2\alpha E(XY) + \alpha^2 E(Y^2) > 0$. Setting $\alpha = E(XY)/E(Y^2)$ yields the required conclusion.

Theorem 2.5.2 If X and Y are independent random variables, then $\rho = 0$.

Proof. $\mu_{xy} = E(XY) - \mu_x \mu_y$. Under the assumption of independence, however, $E(XY) = \mu_x \mu_y$. Hence $\mu_{xy} = 0$, implying $\rho = 0$. Generally, the converse of Theorem 2.5.2 is not true as the following example shows.

Example 2.5.9 Suppose (X,Y) is the discrete random vector whose probability function $f(x,y)$ is given in tabular form as

		X		
f(x,y):	-1	0	1	$f_2(y)$
-1	0.05	0.10	0.05	0.20
Y 0	0.05	0.10	0.05	0.20
1	0.10	0.40	0.10	0.60
$f_1(x)$	0.20	0.60	0.20	1.00

Clearly, X and Y are not independent since, for example, $f(0,0) = 0.10$, whereas $f_1(0)f_2(0) = 0.12$. However, $\mu_x = 0$. Thus, $\mu_x\mu_y = 0$. Furthermore,

$$E(XY) = (-1)(-1)(0.05) + (1)(-1)(0.05) + (-1)(1)(0.10)$$
$$+ (1)(1)(0.10) = 0$$

Hence, $\rho = 0$, and so X and Y are uncorrelated.

Since the moments of probability distributions are often important to obtain, especially the first two moments, methodology for obtaining moments is extremely useful. Thus, the following definition is important.

Definition 2.5.8 Suppose $(X_1,\ldots,X_k)$ is a k-dimensional random vector whose probability distribution $f(x_1,\ldots,x_k)$ admits moments of all orders. That is, $E(|X_1^{r_1}X_2^{r_2}\cdots X_k^{r_k}|) < \infty$ for all nonnegative integers r_i, i = 1, 2, ..., k. The moment-generating function of the random vector $(X_1,\ldots,X_k)$ is defined as

$$m_{\theta_1,\ldots,\theta_k}(X_1,\ldots,X_k) = E(e^{\theta_1 X_1 + \cdots + \theta_k X_k}) \tag{2.68}$$

where $\theta_1, \ldots, \theta_k$ are so chosen that the above expression is finite
To show that $m_{\theta_1,\ldots,\theta_k}(X_1,\ldots,X_k)$ generates moments, note that

$$m_{\theta_1,\ldots,\theta_k}(X_1,\ldots,X_k) = E\left[\sum_{r_1=0}^{\infty} \cdots \sum_{r_k=0}^{\infty} \frac{(\theta_1 x_1)^{r_1}}{r_1!} \cdots \frac{(\theta_k x_k)^{r_k}}{r_k!}\right]$$

Since each series is absolutely convergent, the expected valued E
and the summations may be interchanged. Thus,

$$m_{\theta_1,\ldots,\theta_k}(X_1,\ldots,X_k) = \sum_{r_1=0}^{\infty} \cdots \sum_{r_k=0}^{\infty} E\left[\frac{(\theta_1 x_1)^{r_1}}{r_1!} \cdots \frac{(\theta_k x_k)^{r_k}}{r_k!}\right]$$

It is then not difficult to see that differentiating $m_{\theta_1,\ldots,\theta_k}$
$(X_1,\ldots,X_k)$ r_1 times with respect to θ_1, r_2 times with respect to
θ_2, ..., r_k times with respect to θ_k and then setting
$\theta_1 = \theta_2 = \cdots = \theta_k = 0$ yields the product moment
$E(X_1^{r_1} X_2^{r_2} \cdots X_k^{r_k})$.

Example 2.5.10 Suppose X is the binomial random variable defined
in Example 2.3.2. The moment-generating function $m_\theta(X)$ is then

$$m_\theta(X) = \sum_{x=0}^{n} \binom{n}{x} (pe^\theta)^x q^{n-x} = (pe^\theta + q)^n \qquad (2.69)$$

applying the binomial theorem. It then follows that

$$\frac{dm_\theta(X)}{d\theta} = n(pe^\theta + q)^{n-1} pe^\theta$$

and

$$\frac{d^2 m_\theta(X)}{d\theta^2} = n(pe^\theta + q)^{n-1} pe^\theta + n(n-1)(pe^\theta + q)^{n-2} p^2 e^{2\theta}$$

Thus,

$$E(X) = \left.\frac{dm_\theta(X)}{d\theta}\right|_{\theta = 0} = np \qquad (2.70)$$

and

$$E(X^2) = \left.\frac{d^2 m_\theta(X)}{d\theta^2}\right|_{\theta = 0} = npq + n^2 p^2 \qquad (2.71)$$

in agreement with the results in Example 2.5.7.

Example 2.5.11   Suppose X is the normal random variable defined as

$$f(x;\mu,\sigma^2) = \frac{1}{\sqrt{2\pi}\,\sigma}\, e^{(-\frac{1}{2})\,[(x-\mu)/\sigma]^2} \qquad (2.72)$$

The moment-generating function of X is then

$$m_\theta(X) = \frac{1}{\sqrt{2\pi}\,\sigma} \int_{-\infty}^{\infty} e^{-\{(\frac{1}{2})\,[(x-\mu)/\sigma]^2 - \theta x\}}\, dx$$

Making the substitution $t = (x - \mu)/\sigma$ and completing the square,

$$m_\theta(X) = e^{\theta\mu+\theta^2\sigma^2/2}\, \frac{1}{\sqrt{2\pi}} \int_{-\infty}^{\infty} e^{-(t-\theta\sigma)^2/2}\, dt$$

$$= e^{\theta\mu+\theta^2\sigma^2/2} \qquad (2.73)$$

Now,

$$E(X) = \frac{dm_\theta(X)}{d\theta}\bigg|_{\theta=0} = \mu \qquad (2.74)$$

and

$$E(X^2) = \frac{d^2 m_\theta(X)}{d\theta^2}\bigg|_{\theta=0} = \mu^2 + \sigma^2 \qquad (2.75)$$

Thus, as in Example 2.5.8, the mean and variance of X are μ and σ^2, respectively.

Example 2.5.12   Suppose (X,Y) has the trinomial distribution

$$f(x,y) = \frac{n!}{x!\,y!\,(n-x-y)!}\, p_1^x p_2^y (1 - p_1 - p_2)^{n-x-y} \qquad (2.76)$$

$y = 0, 1, \ldots, n - x;\ x = 0, 1, \ldots, n$. The moment-generating function of (X,Y) is

$$m_{\theta_1,\theta_2}(X,Y) = (1 - p_1 - p_2 + e^{\theta_1} p_1 + e^{\theta_2} p_2)^n \qquad (2.77)$$

which can be shown by applying the binomial theorem twice. The details are left to the reader as an exercise. Furthermore,

$$E(X) = \left. \frac{\partial m_{\theta_1, \theta_2}(X,Y)}{\partial \theta_1} \right|_{\theta_1 = \theta_2 = 0} = np_1$$

$$E(Y) = \left. \frac{\partial m_{\theta_1, \theta_2}(X,Y)}{\partial \theta_2} \right|_{\theta_1 = \theta_2 = 0} = np_2$$

(2.78)

$$E(X^2) = \left. \frac{\partial^2 m_{\theta_1, \theta_2}(X,Y)}{\partial \theta_1^2} \right|_{\theta_1 = \theta_2 = 0} = np_1 q_1 + n^2 p_1^2$$

$$E(Y^2) = \left. \frac{\partial^2 m_{\theta_1, \theta_2}(X,Y)}{\partial \theta_2^2} \right|_{\theta_1 = \theta_2 = 0} = np_2 q_2 + n^2 p_2^2$$

(2.79)

where $q_i = 1 - p_i$, $i = 1, 2$; and

$$E(XY) = \left. \frac{\partial^2 m_{\theta_1, \theta_2}(X,Y)}{\partial \theta_1 \, \partial \theta_2} \right|_{\theta_1 = \theta_2 = 0} = n(n-1)p_1 p_2 \qquad (2.80)$$

$$\rho = \frac{E(XY) - E(X)E(Y)}{\sqrt{\sigma_x^2 \sigma_y^2}} = \frac{-p_1 p_2}{\sqrt{p_1 q_1 p_2 q_2}} \qquad (2.81)$$

Problems

1. Suppose

$$f(x,y) = \begin{cases} x + y & 0 < x < 1, \ 0 < y < 1 \\ 0 & \text{elsewhere} \end{cases}$$

Find the marginal pdf's of the random variable X and Y, and determine whether X and Y are independent.

2. Suppose (X,Y) is the discrete random vector whose probability function in tabular form is given as:

$$X$$

$f(x,y)$:		1	2	3
	1	1/72	1/36	5/72
Y	2	1/36	1/18	5/36
	3	1/12	1/6	5/12

Find the marginal probability functions of X and Y, and determine whether X and Y are independent.

3. Find the conditional distributions of Y given X = x for the bivariate random vectors in problems 1 and 2.

4. Consider the joint density function (2.48) in Example 2.4.4 of the random vector (X,Y). Find the marginal pdf's of X and Y. Are X and Y independent?

5. (a) Suppose X is a discrete random variable whose probability function is

$$f(x) = \begin{cases} \dfrac{6}{(\pi x)^2} & x = 1,\ 2,\ \ldots \\ 0 & \text{elsewhere} \end{cases}$$

Does E(X) exist? Explain.

(b) Suppose X is a continuous random variable whose pdf is

$$f(x) = \frac{1}{\pi} \left(\frac{1}{1 + x^2} \right) \qquad -\infty < x < \infty$$

Does E(X) exist? Explain. This is called the Cauchy distribution.

6. Suppose X is a random variable that has at least a finite second moment, i.e., $E(X^2) < \infty$. Show that $\sigma^2 \equiv E(X - \mu)^2 = E(X^2) - \mu^2$.

7. Suppose X is a continuous random variable defined on $0 < X < \infty$. Suppose, further, that $\mu'_r = r!$, r an integer. Show that $m_\theta(X) = (1 - \theta)^{-1}$ for $|\theta| < 1$, and hence that $f(x) = e^{-x}$, $x \geq 0$, and $f(x) = 0$, $x < 0$.

8. Suppose (X,Y) is a random vector whose joint probability function is

$$f(x,y) = \begin{cases} \dfrac{e^{-(\lambda_1 + \lambda_2)} \lambda_1^x \lambda_2^{y-x}}{x!\,(y - x)!} & x = 0,\ 1,\ \ldots;\ y = x,\ x + 1,\ \ldots \\ 0 & \text{elsewhere} \end{cases}$$

Find the correlation coefficient ρ.

9. Suppose that $(X_1, X_2, \ldots, X_k)$ is a random vector such that there
    exist functions $h_i(X_i)$ for which $E(|h_i(X_i)|) < \infty$, $i = 1, 2, \ldots,$
    k.   Suppose, further, $a_1, a_2, \ldots, a_k$ are any k constants.
    Show that

$$E\left(\sum_{i=1}^{k} a_i h_i(X_i)\right) = \sum_{i=1}^{k} a_i E\{h_i(X_i)\}$$

In mathematical terminology this shows that E is a linear
functional operator.

10. Verify in detail (2.77) and generalize the result for the k - 1
    variate multinomial (2.42).

11. Find the mean and variance of the Poisson distribution defined
    in problem 1 of Section 2.3.

2.6 PRINCIPLES OF STATISTICAL ESTIMATION

In the following two sections the problem under consideration is:
Given a frequency function $f(x; \theta_1, \ldots, \theta_k)$, where $\theta_1, \ldots, \theta_k$ are k
parameters (which when specified describe a particular statistical
model that can often be easily graphed), how, in the presence of
observations from $f(x; \theta_1, \ldots, \theta_k)$, can the parameters $\theta_1, \theta_2, \ldots, \theta_k$
be estimated so there is reasonable assurance that the model based
on these estimators fits the observed data within acceptable limits
according to some rational scheme? To describe one such rational
procedure, consider the following examples.

Example 2.6.1 Suppose X is a random variable which has the binomial probability distribution, i.e.,

$$f(x; p) = \binom{n}{x} p^x q^{n-x} \qquad (2.82)$$

x = 0, 1, $\ldots$, n. [See Equation (2.14).]   It was shown in Example
2.5.7 that E(X) = np. (Using moment-generating functions, this
fact was again demonstrated in Example 2.5.10.) Thus, E(X/n) = p,
and hence the random variable Y = X/n has the property that its

expected value is p. That is, if X is the number of successes in n
binomial trials, with p the probability success on a single trial,
then Y = X/n, which is the *proportion of successes*, has p as its
mean.

Example 2.6.2 Suppose X is normally distributed with mean μ and
variance σ^2. Thus, $m_\theta(X) = e^{\theta\mu+\theta^2\sigma^2/2}$. (See Example 2.5.11.) Now
if $X_1, \ldots, X_n$ are *any* independent identically distributed random
variables each of which has $m_\theta(X)$ as its moment-generating function, it can be shown $Y = \Sigma_{i=1}^{n} X_i$ has as its moment-generating function the value $[m_\theta(X)]^n$. Furthermore, if a is any constant,
$m_{a\theta}(X) = E(e^{a\theta X}) = E(e^{\theta aX}) = m_\theta(aX)$. If these rules pertaining to
moment-generating functions are applied, it then follows that

$$m_\theta(Y) = e^{n[\theta\mu+\theta^2\mu^2/2]} \qquad \text{and} \qquad m_\theta(\bar{X}) = e^{\theta\mu+\theta^2\sigma^2/2n}$$

where $\bar{X} = Y/n$. Since $m_\theta(\bar{X})$ is of the same form as $m_\theta(X)$, it follows
the $\bar{X}$ is normally distributed with mean μ and variance σ^2/n. These
latter two results are actually somewhat more general. If X has
mean μ and variance σ^2 and if $X_1, \ldots, X_n$ is a random sample of
size n drawn from the same population as X, then

$$E(\bar{X}) = \frac{1}{n} \sum_{i=1}^{n} E(X_i) = \frac{1}{n} n\mu = \mu$$

and

$$\sigma_x^2 = \text{var } \bar{X} = n^{-2} \text{var} \sum_{i=1}^{n} X_i = \frac{1}{n} \text{var } X_1 = \frac{\sigma_x^2}{n}$$

The latter result follows from the fact that the variance of the sum
of random variables is the sum of their variances when the random
variables are pairwise uncorrelated. One can also conclude by a
similar procedure as used in Example 2.6.2 that a linear combination of independent, normally distributed random variable is normally distributed.

Example 2.6.1 and 2.6.2 describe statistics Y and $\bar{X}$ which have the property that each of their expected values equals the associated population parameter. This idea leads to the following definition.

Definition 2.6.1 Suppose X_1, X_2, ..., X_n is a random sample of size n from a population whose frequency function is $f(x;\theta)$. Suppose $T(X_1,X_2,...,X_n)$ is a statistic formed from X_1, ..., X_n [e.g., $T(X_1,X_2,...,X_n) = \bar{X}$]. $T(X_1,X_2,...,X_n)$ is an *unbiased estimator* of θ provided $E\{T(X_1,X_2,...,X_n)\} = \theta$.

Example 2.6.3 In the previous examples of this section $T(X_1,X_2,..., X_n)$ was a linear function of the sample values X_1, X_2, ..., X_n. In this example suppose X_1, X_2, ..., X_n is a random sample from a normal distribution with mean μ and variance σ^2 (μ and σ^2 unknown). Consider the statistic $S^2 = \sum_{i=1}^{n}(X_i - \bar{X})^2/(n - 1) = (\sum_{i=1}^{n} X_i^2 - n\bar{X}^2)/(n - 1)$. $E(X_i^2) = \sigma^2 + \mu^2$. Thus, $E(\sum_{i=1}^{n} X_i^2) = n(\sigma^2 + \mu^2)$. Furthermore, $E(\bar{X}^2) = \sigma^2/n + \mu^2$, since $\sigma_{\bar{X}}^2 = \sigma^2/n$; and thus $E(S^2) = [n(\sigma^2 - \sigma^2/n)]/(n - 1) = \sigma^2$. This shows that S^2, which is not a linear function of X_1, ..., X_n, is an unbiased estimator of σ^2.

Unbiased estimators, while having the desirable property that on the average, over the long run, the estimator will equal its parameter, nothing in the definition reveals how quickly this is occurring. It is evident that the smaller the variance of an unbiased estimator, the closer the estimator is to its parameter value, other things being equal. The following definition thus is important.

Definition 2.6.2 Suppose $T(X_1,X_2,...,X_n) \equiv \sum_{i=1}^{n} a_iX_i$ is a *linear unbiased estimator* of a parameter θ, where X_1, ..., X_n is a random sample of size n from a frequency function $f(x;\theta)$, and a_1, ..., a_n are constants. $T(X_1,X_2,...,X_n)$ is called a *best linear unbiased estimator* of θ (b.l.u.e. estimator) provided $\text{var}\{T(X_1,X_2,...,X_n)\} \leq \text{var}\{T'(X_1,X_2,...,X_n)\}$, where $T'(X_1,X_2,...,X_n)$ is any other linear unbiased estimator of θ.

Example 2.6.4 Suppose $\bar{X}$ is the mean of a sample of size n from a
frequency function whose distribution mean is μ and variance is
$\sigma^2 < \infty$. The statistic $\bar{X}$ is a b.l.u.e. of μ. To show this, suppose
$T(X_1, X_2, \ldots, X_n) = \Sigma_{i=1}^{n} a_i X_i$. In order that $T(X_1, X_2, \ldots, X_n)$ be
unbiased, $\Sigma_{i=1}^{n} a_i = 1$, and conversely. To minimize var$\{T(X_1, X_2, \ldots, X_n)\}$, simple calculus methods are used. First, var $\{T(X_1, X_2, \ldots, X_n)\} = (\Sigma_{i=1}^{n} a_i^2)\sigma^2$, which can be written as

$$\text{var}\{T(X_1, \ldots, X_n)\} = \left[\sum_{i=1}^{n-1} a_i^2 + (1 - \sum_{i=1}^{n-1} a_i)^2\right]\sigma^2$$

since $\Sigma_{i=1}^{n} a_i = 1$. Then

$$\frac{\partial \text{var } T(X_1, \ldots, X_n)}{\partial a_i} = \left[2a_i - 2\left(1 - \sum_{i=1}^{n-1} a_i\right)\right]\sigma^2$$

$$= [2a_i - 2a_n]\sigma^2 \qquad i = 1, 2, \ldots, n - 1$$

The minimum value is then found by setting $\partial\text{var } [T(X_1, \ldots, X_n)]/\partial a_i = 0$, $i = 1, 2, \ldots, n - 1$, and solving for $a_1, \ldots, a_{n-1}$ and hence
a_n. Thus, $a_i = a_n$, $i = 1, \ldots, n - 1$, and since $\Sigma_{i=1}^{n} a_i = 1$, it
follows $a_i = 1/n$, $i = 1, 2, \ldots, n$.

A useful estimation procedure is the method of maximum likelihood which is known as maximum likelihood estimation (m.l.e.). This
idea is contained in the following definition.

Definition 2.6.3 Suppose X_1, X_2, $\ldots$, X_n is a random sample of
size n from a frequency function $f(x;\theta)$. Suppose $\hat{\theta} \equiv \hat{\theta}(X_1, \ldots, X_n)$
is an estimator of θ based on the random sample. $\hat{\theta}$ is said to be
the *maximum likelihood estimator* (m.l.e.) of θ if

$$\prod_{i=1}^{n} f(x_i; \hat{\theta}) \geq \prod_{i=1}^{n} f(x_i; \overset{\vee}{\theta}) \qquad\qquad (2.83)$$

where $\overset{\vee}{\theta} \equiv \overset{\vee}{\theta}(X_1, \ldots, X_n)$ is any other estimator of θ based on the
random sample. The function $\prod_{i=1}^{n} f(x_i; \theta)$ is called the *likelihood*

function. The reader at this point should think conceptually as to
the meaning of the m.l.e. One should note that for a given sample,
the m.l.e. is a "best fit" to the parameter in the sense that the
choice of the m.l.e. as the parameter estimator maximizes likelihood
at the n-dimensional point $(X_1, \ldots, X_n)$--the representation of the
sample in n-space.

There is a rather general procedure for obtaining m.l.e.'s
provided that certain regularity conditions [such as outlined in
Wasan (1970)] are met. These conditions on the parameters of most
common distributions are satisfied. Briefly, then, suppose $L(\theta; x_1,
x_2, \ldots, x_n)$ is, by definition, the likelihood function $\Pi^n_{i=1} f(x_i; \theta)$.
Assuming the regularity conditions hold the m.l.e. $\hat{\theta}$ of θ is
found as the solution to the equation

$$\frac{d \ln L(\theta; x_1, \ldots, x_n)}{d\theta} \Bigg|_{\theta = \hat{\theta}} = 0 \qquad\qquad (2.84)$$

Ordinarily, the equation in question will be simple and have but a
single solution. However, if multiple solutions exist and can be
found, then that solution that yields the largest likelihood is the
m.l.e., provided of course the regularity conditions are satisfied.
Finally, the reader should note that the concept of maximum likelihood can be extended to more than a single parameter.

Example 2.6.5 Suppose X_1, X_2, $\ldots$, X_n is a random sample of size n
from the population whose frequency is the two-point binomial, i.e.,

$$f(x; p) = p^x (1 - p)^{1-x} \qquad x = 0, 1 \qquad\qquad (2.85)$$

The m.l.e. of p is required. The likelihood is given by

$$L(p; x_1, x_2, \ldots, x_n) = \prod_{i=1}^{n} p^{x_i} (1 - p)^{1-x_i} \qquad\qquad (2.86)$$

Then,

$$\frac{d \ln L(p; x_1, \ldots, x_n)}{dp} = \frac{\Sigma_{i=1}^{n} x_i}{p} - \frac{(n - \Sigma_{i=1}^{n} x_i)}{1 - p}$$

and so $\hat{p} = \Sigma_{i=1}^n (x_i/n)$. Note that $\hat{p}$ is the same estimator as the unbiased estimator of the binomial parameter which was obtained in Example 2.6.1.

Example 2.6.6 Suppose X_1, X_2, ..., X_n is a random sample of size n from a normal distribution with mean μ and variance σ^2. In this case the m.l.e.'s of both μ and σ^2 are to be obtained. The likelihood is given by

$$L(\mu,\sigma^2;x_1,\ldots,x_n) = \prod_{i=1}^n (2\pi\sigma^2)^{-1/2} \exp \frac{-(x_i - \mu)^2}{2\sigma^2} \qquad (2.87)$$

It then follows that

$$\frac{\partial \ln L}{\partial \mu} = \sum_{i=1}^n \frac{x_i - \mu}{\sigma^2}$$

and

$$\frac{\partial \ln L}{\partial \sigma^2} = \frac{-n}{2\sigma^2} + \sum_{i=1}^n \frac{x_i - \mu}{2(\sigma^2)^2}$$

The m.l.e.'s $\hat{\mu}$ and $\hat{\sigma}^2$ are the unique values that jointly satisfy the equations

$$\left.\frac{\partial \ln L}{\partial \mu}\right|_{\mu=\hat{\mu}, \; \sigma^2=\hat{\sigma}^2} = 0 \quad \text{and} \quad \left.\frac{\partial \ln L}{\partial \sigma^2}\right|_{\mu=\hat{\mu}, \; \sigma^2=\hat{\sigma}^2} = 0$$

That is, $\hat{\mu} = \bar{X}$ and $\hat{\sigma}^2 = \Sigma_{i=1}^n (X_i - \bar{X})^2/n$. Note the difference between $\hat{\sigma}^2$ and S^2, the unbiased estimator of σ^2. From this example it then follows the m.l.e.'s are not generally unbiased.

In addition to unbiased and maximum likelihood estimators, there are other estimators that are used in practice. They include method-of-moments estimators, estimators based on sufficient statistics, and estimators based on percentiles. The method of moments and percentiles will be discussed in Chapter 5; remaining techniques have specific applications and are not discussed further here. The interested reader can obtain more information on these

other estimators in Hogg and Craig (1970), David (1970), and
Cramer (1964).

Problems

1. Suppose $T(X_1,\ldots,X_n)$ is an unbiased estimator of a parameter θ.
    Is e^{-T} an unbiased estimator of $e^{-\theta}$? Explain.

2. Suppose X is a binomial random variable, i.e., $f(x;p) =$
    $\binom{n}{p}p^x(1 - p)^{n-x}$. Show that $\hat{p}(1 - \hat{p})/(n - 1)$, where $\hat{p} = X/n$
    is an unbiased estimator of $p(1 - p)$.

3. Suppose $S_1^2 = \Sigma_{i=1}^{n_1}(X_{i1} - \bar{X}_1)^2/(n_1 - 1)$ and $S_2^2 = \Sigma_{i=1}^{n_2}(X_{i2} - \bar{X}_2)^2/$
    $(n_2 - 1)$ are both unbiased estimators of σ^2 based on random
    samples of size n_1 and n_2, respectively. How can S_1^2 and S_2^2 be
    combined to yield an unbiased estimator of σ^2 with smallest
    variance? Assume that var S_1^2 and var S_2^2 are finite and that
    var $S_1^2/ S_2^2 = (n_2 - 1)/(n_1 - 1)$.

4. Suppose $T(X_1,\ldots,X_n) = \Sigma_{i=1}^{n}a_iX_i$ is a b.l.u.e. of a parameter θ.
    Let $U(X_1,\ldots X_n)$ be any other linear unbiased estimator of θ.
    Prove that $\rho_{T,U} = \sqrt{\text{var } T/\text{var } U}$. Hint: Show that $T(X_1,\ldots,X_n) =$
    $\bar{X}$, and var $T = \sigma^2/n$.

5. Suppose X_1, X_2, $\ldots$, X_n is a random sample of size n from a
    population whose frequency function is $f(x;\theta) = (1/\theta)e^{-x/\theta}$,
    $x > 0$, $\theta > 0$; and zero elsewhere. Find the m.l.e. of θ based
    on this sample. Is the m.l.e. also the b.l.u.e.? Explain.

6. Suppose X_1, X_2, $\ldots$, X_n is a random sample of size n from a
    population whose frequency function is $f(x;\lambda) = \lambda e^{-\lambda x}$, $x > 0$,
    $\lambda > 0$; and zero elsewhere. Find the m.l.e. of λ based on this
    sample. Is the m.l.e. also the b.l.u.e.? Hint: Is the m.l.e.
    even unbiased?

7. Suppose X_1, X_2, $\ldots$, X_n is a random sample of size n from a
    population whose frequency function is $f(x;\lambda) = e^{-\lambda} \lambda^x/x!$,
    $x = 0, 1, \ldots\ldots$, and zero elsewhere. Find the m.l.e. and
    b.l.u.e. of λ. Do they coincide?

8. Suppose X_1, X_2, $\ldots$, X_n is a random sample of size n from a
    population whose frequency function is $(1 + \theta) x^{\theta}$, $\theta > 0$,
    $0 < x < 1$. Find the m.l.e. of θ based on this sample.

9. Suppose X_1, X_2, ..., X_n is a random sample from the normally
    distributed population, $f(x;\theta) = (2\pi\theta)^{-1/2} e^{-x^2/2\theta}$,
    $-\infty < x < \infty$ $\theta > 0$. Find the m.l.e. of θ based on this sample.

10. Suppose $T(X_1,...,X_n)$ is an unbiased estimator of θ and that
    var $T > 0$. Prove that T^2 cannot be an unbiased estimator of θ^2.

2.7 CONFIDENCE INTERVALS AND TESTS OF STATISTICAL HYPOTHESES

In the previous section, two methods for estimating parameters were
presented. In this section, methods for assessing the accuracy of
these (parameter) estimators are explored. In addition, assumptions
about parameter values, called hypotheses, are also discussed, For
example, in determining whether one type of electronic calculator
is more reliable, i.e., lasts longer between breakdowns, than another
type of calculator, the procedure involves testing a hypothesis. To
understand more clearly the intent of this section, consider the
following example.

Example 2.7.1 Suppose X_1, X_2, ..., X_n is a random sample of size n
from a normal population with unknown mean μ and known variance σ_0^2.
From Example 2.6.2 it follows that $\bar{X}$ is normally distributed with
mean μ and variance σ_0^2/n. Furthermore, from Example 2.5.8 it
follows that if $Z = \sqrt{n}(\bar{X} - \mu)/\sigma_0$, Z has the standard normal distribution [see, for example Hoel (1962)]. Since pr$\{-1.96 < Z < 1.96\}$ =
0.95, it follows that pr$\{-1.96 < \sqrt{n}(\bar{X} - \mu)/\sigma_0 < 1.96\}$ = 0.95. Hence,
by elementary algebraic manipulations, pr$\{\bar{X} - 1.96 \; \sigma_0/\sqrt{n} < \mu <$
$\bar{X} + 1.96 \; \sigma_0/\sqrt{n}\}$ = 0.95. The interval $\bar{X} \pm 1.96\sigma_0/\sqrt{n}$ is called a 95%
confidence interval for the mean (parameter) μ of the normal distri-
bution. For a more precise definition of a confidence interval see
Hogg and Craig (1970, pp. 193-195).

Under more general conditions the sample mean $\bar{X}$ of random sample
of size n from an arbitrary distribution has an approximate normal
distribution provided the first two moments of the parent (original)
distribution exist, and n, the sample size, is large. This is stated
more precisely as the *central limit theorem*.

Theorem 2.7.1   (Central Limit Theorem)   Let X_1, ..., X_n be a random sample if size n from a population whose distribution has finite mean and variance μ and σ^2, respectively.   The random variable $Z = \sqrt{n}(X - \mu)/\sigma$ has as its limiting distribution as n→∞ the standard normal distribution.   The proof of this form of the central limit theorem is not elementary and is omitted.

Example 2.7.2   Let X_1, X_2, ..., X_n be a random sample of size n ≥ 30 from a Poisson distribution with mean λ.   (See problem 1 of Section 2.3 and problem 11 of Section 2.5.)   Since the variance of the Poisson distribution is also λ, it follows from the central limit theorm that $\sqrt{n}(\bar{X} - \lambda)/\sqrt{\lambda}$ has an approximate standard normal distribution. An approximate 95% confidence interval for λ is then found by solving, in terms of λ, the quadratic equation $\lambda^2 - (2\bar{X} + 3.84/n) \lambda + \bar{X}^2 = 0$.   The smaller of the two roots denoted as λ_ℓ yields the lower limit of the confidence interval, whereas the larger root, λ_u provides its upper limit.   (See problem 1 at the end of the section.) As a numerical example, suppose $\bar{X} = 3.2$ and n = 100.   Then a 95% confidence interval for λ is (2.86,3.85).   However one should note that once the observations have been collected and the confidence interval has been calculated either λ is covered by it or it is not. The meaning of the confidence interval is that for all such confidence intervals, generated under identical conditions, 95% will include and 5% will exclude the parameter.

The purpose of a confidence interval is to obtain a set of values that, under sampling variation, yields a reasonable range of values for the parameter based on the data.   A confidence interval is then an *interval estimator* of the parameter.   There is another context in which inferences made on the parameter (s) are an important consideration.   This is termed a test of *hypothesis*.

Definition 2.7.1 A *statistical hypothesis* is an assumption about the distribution (frequency function) of a random variable.   If the frequency function involves parameters, then the hypothesis can be an assumption concerning the parameters.

Example 2.7.3 Suppose X is normally distributed with mean μ and variance unity. $\mu = 3$ is a hypothesis.

Definition 2.7.2 A test of hypothesis is a rule by which the sample space is divided into two regions--the region in which the hypothesis is accepted and the region in which the hypothesis is rejected.

A statistical hypothesis that is being tested is termed the *null hypothesis* and its symbol is H_0. The *alternative hypothesis* is the complement of the null hypothesis in general. However, the term also has a wider interpretation. For example, if H_0: $\mu = 3$ is the null hypothesis referred to in Example 2.7.3, then alternative hypothesis H_a is simply $\mu \neq 3$. However, a specific *alternative hypothesis* to H_0: $\mu = 3$ is H_1: $\mu = 4$.

Definition 2.7.3 Rejection of the null hypothesis H_0 when it is true is called *type I error*. The size of type I error is the probability of rejecting H_0 when it is true. Acceptance of H_0 when it is false is called *type II error*. The size of type II error is the probability of accepting H_0 when it is false. The size of type I and type II errors are usually termed α and β, respectively.

Definition 2.7.4 The *power* of a test is $1 - \beta$.

Testing statistical hypotheses and developing criteria for optimum procedures in performing statistical tests are not covered in this book.   However, the concept of power is extremely important in statistical inference. For the purposes in this book it suffices to state that if one has a choice between (among) two or more procedures in the testing of a given null hypothesis, one chooses the test with the greater power, i.e., the smaller value of β (if one exists). In many cases it turns out that the power of a test depends on the value of the specific alternative when H_0 is not true. In these cases no "most powerful" test exists. Tests that are "most powerful" against all alternatives are called *uniformly most powerful tests*. The interested reader can obtain more information about hypothesis testing in Hoel (1962) and Hogg and Craig (1970).

Problems

1. Suppose $X_1, \ldots, X_n$ is a random sample of size n from a Poisson distribution with mean λ. Assuming n is large, show that approximate 95% confidence interval for λ is $pr\{\lambda_\ell < \lambda < \lambda_u\} = 0.95$, where λ_ℓ and λ_u are the solutions to the quadratic equation $\lambda^2 - (2\bar{X} + 3.84/n)\lambda + \bar{X}^2 = 0$. Hint: If Z has a standard normal distribution, then $pr\{-1.96 < Z < 1.96\} = pr\{Z^2 < 3.84\}$, where $Z = \sqrt{n}(\bar{X} - \lambda)/\sqrt{\lambda}$.

2. Suppose $X_1, X_2, \ldots, X_n$ is a random sample of size n from a unit binomial distribution with success probability p, i.e., $f(x) = p^x(1 - p)^{1-x}$, $x = 0, 1, 0 < p < 1$. Let $\hat{p} = \bar{X}$. Show that if n is large, an approximate 95% confidence interval for p is $pr\{p_\ell < p < p_u\} = 0.95$ where p_ℓ and p_u are the solutions to the quadratic equation $(1 + 3.84/n)p^2 - (2\hat{p} + 3.84/n)p + \hat{p}^2 = 0$.

3. If p is defined as in the previous exercise, then $\sigma_{\hat{p}} = p(1 - p)/n$. (See Example 2.5.7.) Hence, show $\sigma_{\hat{p}} \leq 1/4n$. Using this result show that if $X_1, \ldots, X_n$ is a random sample from a unit binomial distribution, an approximate, conservative 95% confidence interval for p is $pr\{\hat{p} - 0.98/\sqrt{n} < p < \hat{p} + 0.98/\sqrt{n} \geq 0$.

4. Suppose a coin is tossed 100 times showing 56 heads and 44 tails. Obtain a 95% confidence interval for p, the probability of heads on a single toss, using the two methods in problems 2 and 3. Compare the results.

5. Verify the numerical confidence interval for λ, the Poisson parameter, in Example 2.7.2.

6. Suppose X is exponentially distributed, i.e., $f(x;\lambda) = \lambda e^{-\lambda x}$, $x \geq 0$, $\lambda > 0$, and zero elsewhere. Find $pr\{X \geq 0.5 \mid \lambda = 1\}$, $pr\{X \geq 1 \mid \lambda = 1\}$, and $pr\{X \geq 2 \mid \lambda = 1\}$.

7. Suppose X is normally distributed with mean μ and variance 1. Find $pr\{X > 1 \mid \mu = 0\}$, $pr\{X > 2 \mid \mu = 0\}$, $pr\{X > 1 \mid \mu = 1\}$ and $pr\{X > 2 \mid \mu = 1\}$.

8. The critical region of a test of hypothesis is the region of the sample space for which H_0, the null hypothesis, is rejected. Let H_0: $\mu = 0$ be the null hypothesis for X normally distributed with variance 1, and if $|X| > 2$ is the critical region of the test find α, the size of the type I error.

9. In the previous exercise suppose H_0: $\mu = 0$ is false and H_1: $\mu = 1$ is true. What is the value of β, the size of the type II error, assuming $|X| > 2$ is the critical region? Note: If $|X| > 2$ is the critical region or rejection region, $|X| \leq 2$ is the non-critical or acceptance region of the test.

10. Suppose X is uniformly distributed on $(0, \theta)$; i.e., $f(x) = 1/\theta$, $0 < x < \theta$, and $f(x) = 0$ elsewhere. Suppose, further, that H_0: $\mu = 1$. If $X > 0.9$ is the critical region, find α. If H_0 is false and H_1: $\mu = 2$ is true, find β.

REFERENCES

Bennett, C., and Franklin, N. (1954). *Statistical Analysis in Chemistry and the Chemical Industry*, John Wiley & Sons, Inc. New York.

Cramér, H. (1946). *Mathematical Methods of Statistics*, Princeton University Press, Princeton, N.J.

David, H. A. (1970). *Order Statistics*, John Wiley & Sons, Inc., New York.

Feller, W. (1957). *An Introduction to Probability Theory and Its Applications*, 2nd ed., Vol. 1, John Wiley & Sons, Inc., New York.

Hahn, G., and Shapiro, S. S. (1967). *Statistical Models in Engineering*, John Wiley & Sons, Inc., New York.

Hoel, P. G. (1967). *Introduction to Mathematical Statistics*, John Wiley & Sons, Inc., New York.

Hogg, R. V., and Craig , A. T. (1970). *Introduction to Mathematical Statistics*, Macmillan Publishing Company, Inc., New York.

Johnson, N. L., and Leone, F. C. (1964). *Statistics and Experimental Design in Engineering and the Physical Sciences*, John Wiley & Sons, Inc., New York.

Miller, I., and Freund, J. E. (1965). *Probability and Statistics for Engineers*, Prentice-Hall, Inc., Englewood Cliffs, N.J.

Wasan, M. T. (1970). *Parametric Estimation*, McGraw-Hill Book Company, New York.

Chapter 3

MODELS FOR MEASUREMENT: CONTINUOUS CASE

3.1   INTRODUCTION

In this chapter various continuous statistical models are presented.
The emphasis is on demonstrating how these models arise in nature
and their use in data analysis, especially in time-to-response situations, or in more general terms as a model for the output of a system.   For example, for purposes of advertising, a radio manufacturer
may require a study on the expected useful life of his product.
Another example may involve testing the mean tensile strength of a
new synthetic fiber.   Both of these situations typify the application of response distributional models.

The models presented in this chapter include the exponential,
the Weibull, the gamma, the extreme value, the lognormal, the beta,
and the logistic models.   These models are discussed in other con-
texts in the following references: Johnson and Kotz (1970), Hahn
and Shapiro (1967), Mann et al. (1974), and Gross and Clark (1975).
An interesting generalization of the exponential model is discussed
by Aroian and Robison (1966).   Their model is not explicitly discussed in this chapter.

3.2   THE EXPONENTIAL MODEL

Consider an event such as a failure of some mechanism.   If failure
is due to some random phenomenon such as accidental damage where
probability of occurrence is proportional to the length of the

interval and whose rate of occurrence is constant over time and if
there is no other cause of failure (such as deterioration of the
mechanism due to age), then clearly the probability of failure in
a given time interval is proportional to the length of interval.
This is true because the longer the interval, the greater the chance
of the occurrence of the random phenomenon. On the other hand, the
shorter the interval the less the chance of occurrence. To describe
this phenomenon mathematically, suppose the cdf for failure of the
mechanism is F(t) where by definition

$$F(t) = pr\{T \leq t\} \tag{3.1}$$

That is, F(t) is the probability that by time t the mechanism has
failed. It is assumed that $F(0) = 0$, i.e., when the mechanism is
brand-new it cannot be failed. Given that the mechanism has operated
until time t, the probability it fails in the time interval (of
length Δt) (t, t + Δt) is $\lambda \Delta t + o(\Delta t)$, where λ is constant with
respect to time and $o(\Delta t)$ is a function of Δt that goes to 0 faster
than Δt, i.e., $\lim_{\Delta t \to 0} o(\Delta t)/\Delta t = 0$. [For a discussion of the o
function in the context of probability the reader should see
Munroe (1951) or Feller (1957).] However, the probability the
mechanism fails in the interval (t, t + Δt) given it has operated
to t must be equal to $[F(t + \Delta t) - F(t)]/[1 - F(t)]$. Thus,

$$\frac{F(t + \Delta t) - F(t)}{1 - F(t)} = \lambda \Delta t + o(\Delta t) \tag{3.2}$$

If one divides through by Δt and taking the limit as $\Delta t \to 0$, (3.2)
then becomes

$$\frac{f(t)}{1 - F(t)} = \lambda$$

where $f(t) = dF(t)/dt$, or $-d[1 - F(t)]/dt$. The solution of this
differential equation using the initial condition $F(0) = 0$ is

$$1 - F(t) = e^{-\lambda t} \tag{3.3}$$

where $t \geq 0$, $\lambda > 0$. The constant λ is the *failure rate* and the
failure density f(t) is given as

$$f(t) = \begin{cases} \lambda e^{-\lambda t} & \lambda > 0, \ t \geq 0 \\ \\ 0 & \text{elsewhere} \end{cases} \qquad (3.4)$$

Another derivation of the exponential model based on the same
physical principles is as follows [see Feller (1957)]. Suppose
that the time interval $(0,t)$ can be divided into n equal intervals
each of length Δt so that $t = n\ \Delta t$, where n is large and Δt is small.
Suppose, further, that the probability the mechanism fails at any
of the points $k\ \Delta t$, $k = 1, 2, \ldots, n$ is constant, independent of
the other intervals and equals p. Thus, if T is the failure time of
the mechanism, then due to the independence property,

$$\text{pr}\{T > n\ \Delta t\} = (1 - p)^{n}$$

and the corresponding discrete probability function is geometric.
(See Section 4.5.) That is, if $f(k\ \Delta t)$ is the probability the mechanism fails at $k\ \Delta t$, then

$$f(k\ \Delta t) = (1 - p)^{k-1} p \qquad k = 1, 2, \ldots \qquad (3.5)$$

Thus, the expected time-to-failure is $\Delta t/p$. If the number of
intervals $n \to \infty$ and $\Delta t \to 0$ so that $\Delta t/p = \lambda$ and $n\ \Delta t = t$ remain
constant, it follows that

$$\text{pr}\{T > t\} = \lim_{\Delta t \to \infty} \left(1 - \frac{\Delta t}{\lambda}\right)^{t/\Delta t} = e^{-\lambda t} \qquad (3.6)$$

Equation (3.6) is equivalent to (3.3) and is a second derivation
of the exponential model.

The exponential model has been used extensively in applications
for three principal reasons:

(i)   Many time-response data, specifically failure data, can be
      fitted quite well by an exponential model.

(ii) Inferences concerning the one (unknown) parameter
      (equations 3.3 and 3.4) are relatively easy. That is, the
      theoretical aspects of the exponential model are not diffi-
      cult.

(iii)   Much has been written about the exponential model and so
        literature pertaining to this model is readily available.
        Some of the more important references appear in the biblio-
        graphy.

For these reasons, it is important to study this model in further
detail.

Suppose T is the time-to-failure of an electronic component
that has the exponential time-to-failure model. If μ_r' is the rth
moment about the origin of T, by direct integration it follows that

$$\mu_r' = \int_0^\infty \lambda t^r e^{-\lambda t} \, dt = \frac{r!}{\lambda^r} \tag{3.7}$$

Thus, the mean and variance of T are λ^{-1} and λ^{-2}, respectively, It
then follows that if λ is the failure rate of the component, its
mean time-to-failure is λ^{-1}, when then failure model is exponential.

Example 3.2.1 Suppose a component has a constant failure rate of 2
units per unit time. Then, the failure model is the exponential
with mean 1/2 and the corresponding failure density and survival
probability are, respectively,

$$f(t) = \begin{cases} 2e^{-2t} & t \geq 0 \\ 0 & t < 0 \end{cases} \tag{3.8}$$

and

$$1 - F(t) = e^{-2t} \qquad t \geq 0 \tag{3.9}$$

The plot of (3.8) and (3.9) is shown in Figure 3.1.   The mean
time-to-failure is one component per one-half unit of time, which
is the same as the standard deviation.

Besides the moments of the exponential failure model (in
particular, the mean and variance), the percentage points of the
cdf are quite useful and important. For example, it is usually
important to determine the time at which the probability is 0.5 (the

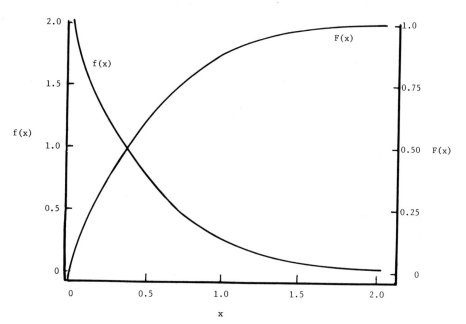

Figure 3.1 Failure density and cumulative distribution function of the exponential model ($\lambda = 2$).

median) the component is operating. Thus, if $t_{0.5}$ denotes this value, then

$$1 - F(t_{0.5}) = e^{-\lambda t_{0.5}} = 0.5$$

Thus, $t_{0.5} = \lambda^{-1} \ln 2$. That is, the *median failure time* for a component whose failure model is exponential with failure rate λ is $\lambda^{-1} \ln 2$. Other percentage points are readily calculated. In general, if t_p is the time at which the probability is p the component has not failed, then $t_p = \lambda^{-1} \ln p^{-1}$. The reader can easily find t_p for a variety of values of p and λ. For example, one can show quite easily that for the exponential model with mean 1/2 the median survival time is 0.35 time unit, whereas the mean survival time is 1/2 time unit, a somewhat larger value. It should not be too surprising that the mean and median survival times are not equal. In fact, only those continuous models that are symmetric about their means have the property of the mean and median being equal.

Suppose t_1, t_2, ..., t_n are failure times whose failure probability density function is (3.4). If $\bar{t}$ is the sample mean failure time, it can be shown that the m.l.e. of λ is given by

$$\hat{\lambda} = \bar{t}^{-1} = \frac{n}{\sum_{i=1}^{n} t_i} \tag{3.10}$$

In addition, $2n\lambda\bar{t}$ has the chi-square distribution with 2n degrees of freedom. [See Epstein and Sobel (1953).] The chi-square distribution is defined by (3.35a) in the next section. If $\chi^2_{2n,\alpha/2}$ and $\chi^2_{2n,1-\alpha/2}$ are, respectively, the $100\alpha/2$ and $100(1 - \alpha/2)$ percentage points of the chi-square distribution with 2n degrees of freedom, it follows that

$$pr\{\chi^2_{2n,\alpha/2} < 2n\lambda\bar{t} < \chi^2_{2n,1-\alpha/2}\} = 1 - \alpha \tag{3.11}$$

Hence a $(1 - \alpha)100\%$ confidence interval for λ is the interval (u_1, u_2), where $u_1 = \chi^2_{2n,\alpha/2}/2n\bar{t}$, and $u_2 = \chi^2_{2n,1-\alpha/2}/2n\bar{t}$.

It can be further shown that for large samples $\bar{t}$ is approximately normally distributed with mean $\mu = \lambda^{-1}$ and variance μ^2/n. (See Theorem 2.7.1.)   Hence, the statistic

$$Z = \frac{\sqrt{n}\ (\bar{t} - \mu)}{\mu} \qquad\qquad (3.12)$$

has an approximate standard normal distribution for sample sizes $n \geq 25$; and an approximate $(1 - \alpha)$ confidence interval for μ is given as

$$\mathrm{pr}\{\frac{\bar{t}}{1 + z_{1-\alpha/2}/\sqrt{n}} < \mu < \frac{\bar{t}}{1 - z_{1-\alpha/2}/\sqrt{n}}\} = 1 - \alpha \qquad\qquad (3.13)$$

where $z_{1-\alpha/2}$ is the $1 - \alpha/2$ percentage point of the standard normal distribution.   The interval (3.13) is obtained by noting $\mathrm{pr}\{-z_{1-\alpha/2} < z < z_{1-\alpha/2}\} = 1 - \alpha$, where $z_{1-\alpha/2}$ is the upper $1 - \alpha/2$ percentage point of the standard normal distribution, and Z is given by (3.12).   Furthermore, a test of the hypothesis H_0: $\mu = \mu_0$ vs. H_1: $\mu \neq \mu_0$ can be developed as follows.

Suppose the hypothesis H_0: $\mu = \mu_0$ vs. H_1: $\mu \neq \mu_0$ is under test and a sample of $n \geq 25$ failure times is collected from a population whose failure distribution is

$$f(t;\mu) = \frac{1}{\mu}\ e^{-t/\mu} \qquad t \geqslant 0,\ \mu > 0 \qquad\qquad (3.14)$$

The statistic $Z = \sqrt{n}(\bar{t} - \mu_0)/\mu_0$ has an approximate standard normal distribution when H_0: $\mu = \mu_0$ is true.   Thus, a test of H_0, the null hypothesis, is to reject H_0 at level α only if $|Z| > z_{1-\alpha/2}$ where $z_{1-\alpha/2}$ is the $100(1 - \alpha/2)$ percentage point of the standard normal distribution.   The formulation of this test of hypothesis H_0: $\mu = \mu_0$ vs. H_1: $\mu \neq \mu_0$ using (3.12) is left as an exercise to the reader.

Example 3.2.2   Suppose a certain transistor has a mean lifetime of 4.5 yrs.   A random sample of 50 transistors is taken from a production lot.   The test shows an accumulated time of operation of 200 yrs.   Is there any reason to believe the manufacturer should be concerned that a change in mean failure has occurred for this lot?

Use a 0.05 type I error.

Solution: $\bar{t}$ = 4 yr. and z = $\sqrt{50}$(4 - 4.5)/4.5 = -0.785. Given
α = 0.05, $Z_{0.975}$ = 1.96. Since -1.96 < -0.785 < 1.96, there is no
reason the manufacturer should be concerned that there has been a
change in mean failure time of the transistors in this lot.

Often when dealing with failure times, the following question
is very important. A sample of n failure times is obtained. Let
$x_{(1)}$ and $x_{(n)}$ be the smallest and largest failure times recorded.
Given a new observation X (from the same population), can one find
the probability that $x_{(1)}$ < X < $x_{(n)}$ a proportion (1 - α) of the
time? In other words, one is concerned with computing

$$\text{pr}\left\{\int_{x_{(1)}}^{x_{(n)}} \lambda e^{-\lambda x}\, dx\right\} \geq 1 - \alpha$$

More generally, if X_L and X_U are any two statistics based on the
sample X_1, ..., X_n, where $X_L \leq X_U$, one is interested in computing

$$\text{pr}\left\{\int_{x_L}^{x_U} \lambda e^{-\lambda x}\, dx\right\} \geq 1 - \alpha$$

If this probability is 1 - γ (say), the two statistics X_L and X_U are
called α *tolerance limits* with probability level 1 - γ. [See
Birnbaum (1962).] It is not hard to show that if $X_L = X_{(1)}$ and
$X_U = X_{(n)}$, the random variable V defined as $\int_{x_{(1)}}^{x_{(n)}} \lambda e^{-\lambda x}\, dx$ has the
probability density

$$g(v) = \begin{cases} n(n - 1)v^{n-2}(1 - v) & 0 < v < 1 \\ 0 & \text{elsewhere} \end{cases} \qquad (3.15)$$

which is a beta model to be described subsequently. What is more
important is that if f(x) is the pdf for any continuous random
variable X then v^*, defined as $\int_{x_{(1)}}^{x_{(n)}} f(x)\, dx$, has the same density as
as V. Thus, without loss of generality, $V \equiv \int_{x_{(1)}}^{x_{(n)}} f(x)\, dx$.

Example 3.2.3 Suppose a sample of size 25 failures is recorded
from a population whose failure density is f(x). Find the probability that a (new) observation X will lie between $X_{(1)}$ and $X_{(n)}$
95% of the time.

Solution: Integrating (3.15), it follows that

$$pr\{V > 0.95\} = \int_{0.95}^{1} 600v^{23} (1 - v) \, dv = 0.43$$

Thus, the probability is 0.43 that a new observation X lies between
the smallest and largest values of the original sample 95% of the
time.

If certain tolerance limits are specified to include a specific
proportion of the population $1 - \alpha$ (say) with probability $1 - \gamma$,
the minimum sample required to meet these specifications can be
computed. To demonstrate this, suppose $pr\{V > 1 - \alpha\} = 1 - \gamma$,
where α and γ are given. This is equivalent to the equation

$$n(1 - \alpha)^{n-1} - (n - 1)(1 - \alpha)^{n} = \gamma \tag{3.16}$$

The required sample size (as a function of α and γ) is the smallest
integer larger than the (nonintegral) solution of (3.16). Tables
of n can be computed for a given values of α and γ using an iterative procedure such as the Newton-Raphson technique. Such a table
is available in Birnbaum (1962).

Example 3.2.4 In reference to the failure distribution in the previous example, how large a sample of failures is required so that
the probability that a future failure time lies 95% of the time
between the smallest and largest observed time is 0.90?

Solution: Using (3.16), $n(0.95)^{n-1} - (n - 1)(0.95)^{n} = 0.10$.
Simplifying, one then has $(0.95)^{n-1} = (0.5n + 9.5)^{-1}$. The value
n which is the smallest integer larger than the solution to the
above equation is 77. Table IV in Birnbaum (1962) can be used to
obtain n.

Often in life-testing situations, n items are placed on test simultaneously. That is, the items are tested simultaneously and their ordered failure times $t_{(1)} \leqslant t_{(2)} \leqslant \cdots \leqslant t_{(n)}$ are recorded. Suppose now instead of waiting until all n items on test have failed (complete testing), only the first k failure times $t_{(1)} \leqslant t_{(2)} \cdots \leqslant t_{(k)}$ are recorded (censored testing). On the basis of these k recorded failure times, the problem is to predict $t_{(r)}$ the rth ordered failure time, where $k < r \leqslant n$. To accomplish this, define $S_k = \Sigma_{i=1}^{k} t_{(i)} + (n - k)t_{(k)}$. Lawless (1971) then derives the distribution of

$$u_r \equiv \frac{t_{(r)} - t_{(k)}}{S_k} \tag{3.17}$$

He shows that

$$pr\{u_r \geqslant t\} = D(n,k,r) \sum_{i=0}^{r-k-1} \frac{\binom{r - k - 1}{i}(-1)^i}{(n - r + i + 1)[1 + (n - r + i + 1)t]^k} \tag{3.18}$$

where $D(n,k,r) \equiv (n - k)!/[(r - k - 1)!(n - r)!]$. Thus, if t_α is that value of t such that $pr\{u_r \geqslant t_\alpha\} = \alpha$, it then follows from (3.17)

$$pr\{t_{(r)} < t_{(k)} + t_\alpha S_k\} = 1 - \alpha \tag{3.19}$$

The principal difficulty with this procedure is finding t_α values. The most convenient way is with a computer using the Newton-Raphson procedure. [For example, see Gross and Clark (1975), Chapter 6, for a discussion of the Newton-Raphson procedure.] In the important special case wherein r = n, i.e., on the basis of the first k failure times, it is required to predict the largest failure time, it follows from (3.18) that

$$pr\{u_n > t\} = (n - k) \sum_{i=0}^{n-k-1} \frac{\binom{n - k - 1}{i}(-1)^i}{(i + 1)[1 + (i + 1)t]^k} \tag{3.20}$$

From this it follows that if $t_{\alpha,n}$ is that value of t such that $pr\{u_n < t_{\alpha,n}\} = 1 - \alpha$, then a one-sided $100(1 - \alpha)\%$ prediction interval for $t_{(n)}$ is $(0, t_{(k)} + t_{\alpha,n} S_k)$.

Example 3.2.5 Suppose 10 identical components are put on test simultaneously. The failure law of each component is exponential. Assume that the failure times of the first four components are at 10, 17, 27, and 36 hours, respectively. Find a 95% upper prediction limit for total elapsed test time.

Solution: An algebraic derivation shows that

$$pr\{u_n \leqslant t\} = \sum_{i=0}^{n-k} \binom{n-k}{i}(-1)^i(1 + it)^{-k} \qquad (3.20a)$$

Thus, for the case n = 10, k = 4, the equation to be solved for u_{10} is

$$\sum_{i=0}^{6} \binom{6}{i}(-1)^i(1 + iu_{10})^{-4} = 0.95$$

An application of the Newton–Raphson procedure or some comparable numerical technique [e.g., see Gross and Clark (1975, pp. 205–208) for details] then is used to show u_{10} = 2.10. Now u_{10} = $(t_{(10)}$ - $t_{(4)})/S_4$, where $t_{(4)}$ = 36 hrs. and S_4 = 306 hrs. Hence, $pr\{t_{(10)} \leqslant 679\}$ = 0.95. That is, one can be 95% confident that the total elapsed time of the test will not exceed 679 hrs.

Problems

1. Suppose T is distributed exponentially according to (3.4). Find the moment generating function of T and show that the rth moment about the origin is given by (3.7).

2. Suppose T is distributed exponentially according to (3.4). Further, suppose $\lambda > 0$ and $pr\{0 \leqslant t \leqslant 1\}$ = $pr\{1 \leq t \leq 2\}$. Find λ, μ, $t_{0.5}$ and σ^2.

3. Suppose $\bar{t}$ is the sample mean for a sample of size n > 25, for which each observation is exponentially distributed according to (3.14). Assuming $Z = \sqrt{n}(t - \mu_0)/\mu_0$ has an approximate standard normal distribution when H_0: $\mu = \mu_0$ is true, show that a test whose type I error is α rejects H_0 only when $|z| > z_{1-\alpha/2}$, where $z_{1-\alpha/2}$ is the 100$(1 - \alpha/2)$ percentage point of the

standard normal distribution.

4. Suppose light bulbs fail exponentially with a mean life of 11.3
    months. The manufacturer in testing the quality control of his
    product finds that in a sample of 400 light bulbs the mean life
    is 10.5 months. Should he be concerned that there has been a
    change (for the worse) in the mean failure time of his product?
    Use $\alpha = 0.05$ as the size of the type I error.

5. In the previous exercise construct a 95% confidence interval
    for μ, the true mean life of light bulbs based on the data pre-
    sented.

6. If T is exponentially distributed with mean $\mu = \lambda^{-1}$, show that
    for constants $t_1 > 0$, $t_2 > 0$, $\mathrm{pr}\{T > t_1 + t_2 \mid T > t_2\} =$
    $\mathrm{pr}\{T > t_1\}$. This is the so-called memoryless property of the
    exponential distribution.

7. Suppose a sample of 100 failures is recorded from a population
    whose failure density if $f(x)$. Find the probability that a
    future observation drawn randomly from this population lies
    between the minimum and maximum failure times in the sample 90%
    of the time.

8. Show that (3.18) becomes (3.20) for the special case $r = n$.

9. Show that (3.20) can be further simplified so that

$$\mathrm{pr}\{u_n \leqslant t\} = \sum_{i=0}^{n-k} \binom{n-k}{i} (-1)^i (1 + it)^{-k}$$

10. Suppose 10 identical television receivers are put into opera-
    tion simultaneously. The failure law of each receiver is
    exponential. Assume the failure times of the first 4 components
    are at 31, 71, 84, and 106 months, respectively. Find a 95%
    upper prediction limit for the time at which the last receiver
    fails.

3.3 THE WEIBULL MODEL

Consider a transformation of the time axis, $Z = T^\gamma$, where $\gamma > 0$ is
a constant. If failures occur randomly on $0 < z < \infty$, then following

the derivations in the previous section, one sees that

$$1 - F(z) = e^{-\lambda z} \qquad \lambda > 0, \, z \geqslant 0 \qquad\qquad (3.21)$$

Accordingly, the failure density function is

$$f(z) = \lambda e^{-\lambda z} \qquad\qquad\qquad (3.22)$$

However, $Z = T^\gamma$. If a change of variable technique is used [e.g., see Theorem 5.3.1 or Hoel (1962) for explicit details concerning the change of variable procedure], the change of variable $Z = T^\gamma$ yields

$$g(t) = f(z) \left| \frac{dz}{dt} \right| \qquad \cdot$$

If it is noted that $\left| dz/dt \right| = \gamma t^{\gamma-1}$, it follows that

$$g(t;\lambda,\gamma) = \begin{cases} \lambda \gamma t^{\gamma-1} e^{-\lambda t^\gamma} \\ 0 \text{ elsewhere} \end{cases} \qquad \lambda > 0, \, \gamma > 0, \, t \geqslant 0 \qquad (3.23)$$

The pdf (3.23) is called the Weibull model after W. Weibull, who in his article (1951) proposed its use as general failure model.

The Weibull model has been used rather extensively in applications in recent years because:

(i)   In many instances for which failure data cannot be fitted by an exponential, the Weibull model can be used.

(ii) Inferences concerning the two (unknown) parameters λ and γ (equation 3.23), while more difficult to make than for the exponential model have been dealt with in the last 8 to 10 yrs. Mann et al. (1974) and Gross and Clark (1975) treat these problems in some detail.

(iii) The Weibull is one of the most frequently used parametric models in failure data analysis. Thus, the literature concerning it is growing rapidly and is not difficult to find. Some of the more important references appear among the list at the end of this chapter.

For these reasons, this model is studied in some detail.

Suppose T the time-to-failure of a piece of mechanical equipment has the Weibull model. If $\mu_r^{'}$ is the rth moment about the origin of T, then $\mu_r^{'}$ is formally given as

$$\mu_r^{'} = \int_0^\infty \lambda\gamma t^{r+\gamma-1} e^{-\lambda t^\gamma} \, dt \qquad (3.24)$$

A transformation of variable $y = \lambda t^\gamma$ shows that

$$\mu_r^{'} = \frac{\Gamma(r/\gamma + 1)}{\lambda^{r/\gamma}} \qquad (3.25)$$

where the function $\Gamma(w)$, the gamma function is defined as $\Gamma(w) = (w - 1)!$ if w is a positive integer or in general

$$\Gamma(w) = \int_0^\infty y^{w-1} e^{-y} \, dy \qquad (3.26)$$

for $w > 0$. Thus, the mean and variance of T are, respectively, $\lambda^{-1/\gamma} \Gamma(1/\gamma + 1)$ and $\lambda^{-2/\gamma} [\Gamma(2/\gamma + 1) - \Gamma^2(1/\gamma + 1)]$.

The relationship between the failure rate and the mean time-between-failure is not as simple for the Weibull failure model as it is for the exponential failure model. The failure rate for the Weibull model, obtained using the formula $f(t)/[1 - F(t)] = \lambda(t)$ (see the derivations of 3.2 and 3.3), which is $\lambda\gamma t^{\gamma-1}$. Thus, for the Weibull failure model, the failure rate increases (decreases) if $\gamma > 1$ ($\gamma < 1$). Note that if $\gamma = 1$, the Weibull and exponential models are the same.

Example 3.3.1 The unit exponential failure model ($\lambda = 1$) described is a Weibull failure model with $\lambda = \gamma = 1$.

Example 3.3.2 Suppose a component fails according to the Weibull failure model

$$g(t) = \begin{cases} 2te^{-t^2} & t > 0 \\ 0 & \text{elsewhere} \end{cases} \qquad (3.27)$$

The failure rate and survival probability are, respectively, $\lambda(t) = 2t$ and

$$1 - F(t) = e^{-t^2} \qquad t \geq 0 \qquad\qquad\qquad (3.28)$$

where $\lambda(t)$ and $F(t)$ are the generic notations for failure rate and failure distribution function, respectively.   The plots of (3.27) and (3.28) are shown in Figure 3.2.

This particular Weibull variate is also known as the Raleigh model.

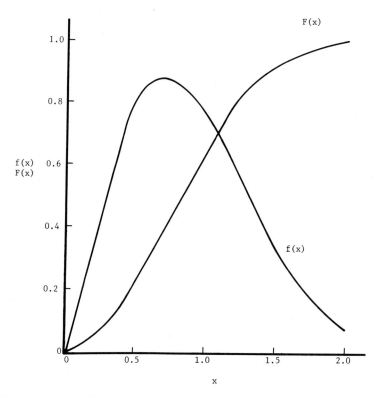

Figure 3.2 Failure density and cumulative distribution function of the Weibull model ($\gamma = 2$, $\lambda = 1$).

The percentage points of the Weibull cdf are useful and not difficult to obtain. Analogous to the method developed for the exponential model, the *median failure time* for the Weibull failure model given by (3.23) is $(\lambda^{-1} \ln 2)^{1/\gamma}$. More generally, if t_p is that time for which the probability is p the component has not failed, then $t_p = (\lambda^{-1} \ln p^{-1})^{1/\gamma}$. For example 3.3.2 the median time-to-failure is 0.83 hrs., whereas the mean failure time is 0.89 hrs. Although, they are not equal, these times are closer to each other. than in the unit exponential failure model, which is not surprising since the density function (3.27) is more nearly symmetric than the exponential.

The problem of inference for the two-parameter Weibull failure model, i.e., point and interval estimation and tests of hypothesis, has received a great deal of attention in recent years. Two recent texts, Mann et al. (1974) and Gross and Clark (1975) discuss the inference problems in considerable detail. It should be noted here that for samples of 25 or more times to failure (for a parti-cular component) maximum likelihood estimators of λ and γ are generally suitable. In small sample situations this is not true. A paper by Gross and Lurie (1977) compares maximum likelihood estimators with two other estimators proposed by Bain and Antle (1967), concluding that for small samples, i.e., less than 25, the Bain-Antle estimators are better in that their bias is smaller as well as their root-mean-squared error. A smaller bias and root-mean-square estimator is preferable [see Gross and Lurie (1977)].

The maximum likelihood estimators of λ and γ, $\hat{\lambda}$ and $\hat{\gamma}$ (say), are the solution of the equations

$$\hat{\lambda} = \frac{n}{\sum_{i=1}^{n} t_i^{\hat{\gamma}}} \tag{3.29}$$

and

$$\left(\frac{n}{\sum_{i=1}^{n} t_i^{\hat{\gamma}}} \right) \left(\sum_{i=1}^{n} t_i^{\hat{\gamma}} \ln t_i \right) = n\hat{\gamma}^{-1} + \sum_{i=1}^{n} \ln t_i \tag{3.29a}$$

To find $\hat{\gamma}$, it is necessary to solve (3.29a) iteratively, which can be accomplished by using the Newton-Raphson technique. This involves choosing an appropriate initial value for $\hat{\gamma}$ and then iterating until $\hat{\gamma}$ is approximated to within the desired degree of accuracy. Once $\hat{\gamma}$ has been determined, $\hat{\lambda}$ is obtained using (3.29). Details of this procedure are found in Gross and Clark (1975).

To obtain a $(1 - \alpha)100\%$ confidence interval for μ, the mean time-to-failure for the Weibull failure model, is fairly involved. The details of this procedure are given by Gross and Clark (1975). Here the (approximate) interval has as its limits $\hat{\mu} \pm \sqrt{\text{var } \hat{\mu}} \; z_{1-\alpha/2}$, where

$$\hat{\mu} = \hat{\lambda}^{-1/\hat{\gamma}} \; \Gamma\left(1 + \frac{1}{\hat{\gamma}}\right) \tag{3.30}$$

$$\text{var } \hat{\mu} \doteq \left(\frac{\hat{\mu}}{\hat{\gamma}\hat{\lambda}}\right)^2 \text{var } \hat{\lambda} + \left(\frac{\hat{\mu}^2}{\hat{\gamma}^4}\right)\left[\ln \hat{\lambda} + \Psi\left(1 + \frac{1}{\hat{\gamma}}\right)\right]^2 \text{var } \hat{\gamma}$$

$$- \frac{2\hat{\mu}^2}{\hat{\lambda}\hat{\gamma}^3}\left[\ln \hat{\lambda} + \Psi\left(1 + \frac{1}{\hat{\gamma}}\right)\right] \text{côv } (\hat{\gamma},\hat{\lambda}) \tag{3.31}$$

where $z_{1-\alpha/2}$ is the upper $(1 - \alpha/2)$ percentage point of the standard normal distribution,

$$\Psi(x) \equiv \frac{\Gamma'(x)}{\Gamma(x)} \qquad \text{for } x > 0 \tag{3.32}$$

which is called the digamma function, $\Gamma(x)$ is the gamma function (see 3.26), and the values $\text{var } \hat{\lambda}$, $\text{var } \hat{\gamma}$, and $\text{côv}(\hat{\lambda},\hat{\gamma})$ are obtained from the matrix equation

$$\begin{bmatrix} \text{var } \hat{\lambda} & \text{côv}(\hat{\lambda},\hat{\gamma}) \\ \text{côv}(\hat{\lambda},\hat{\gamma}) & \text{var } \hat{\gamma} \end{bmatrix}^{-1} = \begin{bmatrix} a_{11} & a_{12} \\ a_{12} & a_{22} \end{bmatrix} \tag{3.33}$$

with $a_{11} = n/\hat{\lambda}^2$, $a_{12} = n(\hat{\lambda}\hat{\gamma})^{-1}[\Psi(2) - \ln \hat{\lambda}]$, and

$$a_{22} = n(\hat{\lambda}\hat{\gamma}^2)^{-1}\{\Psi'(2) + [\Psi(2)]^2 - 2 \ln \hat{\lambda}\Psi(2) + (\ln \hat{\lambda})^2\}$$

Example 3.3.3[†]   In a study, 20 patients receiving an analgesic to
relieve headache pain had the following relief time (in hours):
1.1, 1.4, 1.3, 1.7, 1.9, 1.8, 1.6, 2.2, 1.7, 2.7, 4.1, 1.8, 1.5,
1.5, 1.2, 1.4, 3.0, 1.7, 2.3, 1.6, and 2.0.   A Weibull model was
fitted to these relief times.   From (3.29) and (3.29a), one uses a
Newton-Raphson iteration procedure with a starting value $\hat{\gamma}_0 = 0.5$
to obtain the final estimates $\hat{\gamma} = 2.79$ and $\hat{\lambda} = 0.12$.   Equation (3.30)
is used to obtain an estimate of the mean relief time.   Thus, $\hat{\mu} =$
$(0.12)^{-0.36}\Gamma(1.36)$.   With the aid of a gamma function table [Pearson
and Hartley (1966)] $\hat{\mu} = 1.89$ hrs.   To calculate $\hat{var}\ \hat{\mu}$, the following
quantities must be computed: (i) $(\hat{\mu}/\hat{\lambda}\hat{\gamma})^2$; (ii) $(\hat{\mu}^2/\hat{\gamma}^4)$; (iii) $\ln \hat{\lambda} +$
$\Psi(1 + \hat{\gamma}^{-1})$ and its square; and (iv) $2\hat{\mu}^2/\hat{\lambda}\hat{\gamma}^3$.   From these values a_{11},
a_{12}, and a_{22} may be calculated and hence $\hat{var}\ \hat{\lambda}$, $\hat{var}\ \hat{\gamma}$, and $\hat{cov}(\hat{\lambda},\hat{\gamma})$.
To facilitate these computations, tables of $\Psi(x)$, the digamma func-
tion, are necessary.   Tables for both $\Psi(x)$ and $\Psi'(x)$, the trigamma
function, are available in Abramowitz and Stegun (1964) for values
$1 \leq x \leq 2$.   If $x > 2$ or $x < 1$, the recurrence relations $\Psi(x + 1) =$
$\Psi(x) + x^{-1}$ and $\Psi'(x + 1) = \Psi'(x) - x^{-2}$ may be used with the tables.
See Bury (1975).   Now, $(\hat{\mu}/\hat{\lambda}\hat{\gamma})^2 = (1.89)/(0.12)(2.79)^2 = 31.868$,
$(\hat{\mu}^2/\hat{\gamma}^4) = 0.059$, $\ln \hat{\lambda} + \Psi(1 + \hat{\gamma}^{-1}) = \ln 0.12 + \Psi(1.358) = -2.120 -$
$0.105 = -2.225$, and $2\hat{\mu}^2/\hat{\lambda}\hat{\gamma}^3 = 2.741$.   Furthermore, $a_{11} = 20/(0.12)^2 =$
1388.889, $a_{12} = 20(0.423 - \ln 0.12)/(0.12)(2.79) = 151.927$, where
$\Psi(2) = 0.423$, $\Psi'(2) = 0.645$ and $a_{22} = 20[0.645 + (0.423)^2 -$
$(2 \ln 0.12)(0.423) + (\ln 0.12)^2]/(0.12)(2.79) = 30.712$.   The remainder
of the computation is left as an exercise to the reader.   One finds
that var $\mu = 0.023$ and hence an approximate 95% confidence interval
for μ has as upper and lower limits $1.89 \pm 1.96\sqrt{0.023}$ and so the
requisite interval is $(1.59, 2.19)$.

     Since tolerance limits are nonparametric in nature, the
methodology discussed in the previous section for applying tolerance

---

[†]This example is taken from Example 4.2.2 and 4.4.2 of Gross and
Clark (1975) with the kind permission of John Wiley and Sons, Inc.

limits applies to the Weibull distribution. Finally, other Weibull
estimation techniques including determination of prediction intervals
and linear estimation procedures are found in Mann et al. (1974).

Problems

1. Find the mean and variance of the Weibull variable T in Example
    3.3.2.

2. Show that the failure rate for the general two-parameter Weibull
    failure density given by (3.23) is $\lambda \gamma t^{\gamma-1}$.

3. The Weibull density is often written as $g(t;\theta,\gamma) = (\gamma/\theta)(t/\theta)^{\gamma-1}$
    $\exp[-(t/\theta)^{\gamma}]$, $\theta > 0$, $\gamma > 0$, $t \geq 0$. Show that this characteriza-
    tion of the Weibull density is equivalent to (3.23). Find a
    simple transformation involving θ and γ which transforms the
    Weibull variable to a unit exponential variable.

4. For the data in Example 3.3.3 verify that $\hat{\mu} = 1.89$ hr.,
    $\hat{\sigma}^2 = 0.023$ hr., and an approximate 95% confidence interval for
    μ is (1.59 hrs., 2.19 hrs.). Use the facts that $\Psi(2) = 0.4228$,
    and $\Psi'(2) = 0.6449$.

5. Find the equations whose solution provides the maximum likelihood
    estimators of λ and γ for the two-parameter Weibull density.
    Hence show that $\hat{\lambda}$ and $\hat{\gamma}$ are found by solving (3.29) and (3.29a.)

3.4   THE GAMMA MODEL

Consider again the event such as failure of some mechanism. If
failure is due to the accumulation of accidental damage, i.e., if it
takes n occurrences of damage before failure occurs, each occurrence
being random, then the time to failure of the mechanism $T = \Sigma_{i=1}^{n} X_i$,
where X_1 is the time between no damage and the first occurrence of
damage, X_2 is the time between the first and second occurrences of
damage, etc. Since each X_i is distributed exponentially with mean
λ^{-1} (say), it is necessary to obtain the distribution of T. This
is most conveniently accomplished using moment-generating functions.
The moment-generating function for X_i is $m_\theta(X)$ where one has

$m_\theta(X) = (1 - \theta/\lambda)^{-1}$, $|\theta| < \lambda$. (See Problem 1, Section 3.2.)
Since $X_1, \ldots, X_n$ are mutually independent, it follows that
$m_\theta(T) = (1 - \theta/\lambda)^{-n}$. But $(1 - \theta/\lambda)^{-n}$ is the moment-generating
function of the gamma density, i.e.,

$$f(t;\lambda,n) = \frac{\lambda^n t^{n-1} e^{-\lambda t}}{(n - 1)!} \qquad t \geq 0, \ \lambda \geq 0, \ n \geq 1 \qquad (3.34)$$

(see Problem 8, Section 3.4). Thus, T, the total survival time,
has a gamma density with scale parameter λ and shape parameter n.

More generally, the gamma failure model has the form

$$f(t;\lambda,\gamma) = \frac{\lambda^\gamma t^{\gamma-1} e^{-\lambda t}}{\Gamma(\gamma)} \qquad t \geq 0, \ \lambda > 0, \ \gamma > 0 \qquad (3.35)$$

where λ and γ are the scale and shape parameters respectively,
and γ is not necessarily an integer value. The gamma failure model
is a natural extension of the exponential failure model in the
sense that if an item is comprised of several components, each of
which fails according to the same exponential model, the (total)
time-to-failure of the item obeys the gamma failure model. For
this reason the gamma failure model has been used rather extensively
in reliability applications. Furthermore, some recent applications
in the biomedical area have involved gamma and gamma-like failure
models in multi-organ systems such as lung and kidneys. [See
Gross et al. (1971).] As in the case with the Weibull failure
model, a more extensive discussion of the parameter estimation
problems is found in Gross and Clark (1975) and Mann et al. (1974).

Suppose the time-to-failure of an item T has the gamma failure model. The rth moment about the origin, μ_r', can be obtained
in two ways: (i) by evaluating the derivative $d^r m_\theta/d\theta^r|_{\theta=0}$, or
(ii) directly, by evaluating the integral $[\lambda^\gamma/\Gamma(\gamma)]\int_0^\infty t^{r+\gamma-1}e^{-\lambda t}\, dt$.
In any event,

$$\mu_r' = \frac{\Gamma(r + \gamma)}{\Gamma(\gamma)} \left(\frac{1}{\lambda}\right)^r \qquad (3.36)$$

$$= (r - 1 + \gamma)(r - 2 + \gamma) \cdots \gamma/\lambda^r$$

Thus, μ and σ^2 for the gamma failure model are, respectively, γ/λ and γ/λ^2.

The failure rate for the gamma model is somewhat more compli-
cated than the failure rates for the exponential and the Weibull
failure models. Using the formula $\lambda(t) = f(t)/[1 - F(t)]$, the
failure rate for the gamma failure model is $t^{\gamma-1} e^{-\lambda t} / \int_t^\infty y^{\gamma-1} e^{-\lambda y} \, dy$.
It can be shown that the gamma failure rate increases if $\gamma > 1$,
decreases if $\gamma < 1$, and reduces to the constant failure rate if
$\gamma = 1$. The demonstration of this is left as an exercise to the
reader. Note, however, that if $\gamma \neq 1$, then

$$\lim_{t \to \infty} \frac{t^{\gamma-1} e^{-\lambda t}}{\int_t^\infty y^{\gamma-1} e^{-\lambda y} \, dy} = \lambda \qquad (3.37)$$

That is, the failure rate of the gamma failure model is bounded
(above if $\gamma > 1$, and below if $\gamma < 1$). Very often, a researcher may
be somewhat perplexed in not knowing whether to fit a Weibull or
gamma model to his failure data. A plot of the empirical hazard
rate [e.g., see Gross and Clark (1975)] should be helpful using
the gamma model if the failure rate appears to be bounded and the
Weibull model if the failure rate appears to be unbounded.

A special form of the gamma distribution is the chi-square
distribution where $\lambda = 1/2$ and $\gamma = k/2$, i.e.,

$$f(x;1/2,k) = \frac{1}{2^{k/2} \Gamma(k/2)} x^{k/2-1} e^{-x/2} \qquad x > 0 \qquad (3.35a)$$

Example 3.4.1 The unit exponential failure model is a gamma failure
model with $\lambda = \gamma = 1$.

Example 3.4.2 Suppose a component fails according to the gamma
failure model

$$f(t) = \begin{cases} te^{-t} & t \geq 0 \\ 0 & \text{elsewhere} \end{cases} \qquad (3.38)$$

The failure rate and survival probability are, respectively,
$\lambda(t) = t/(t + 1)$ and

$$1 - F(t) = (t + 1)e^{-t} \qquad t \geq 0 \qquad (3.39)$$

The plots of (3.38) and $F(x)$ are shown in Figure 3.3.

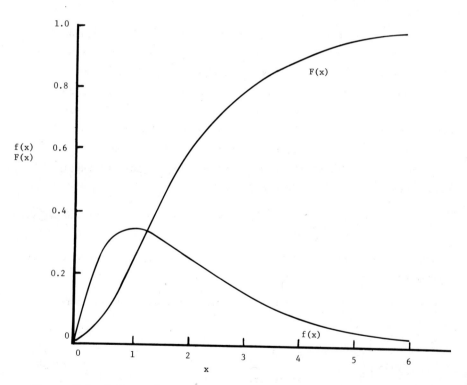

Figure 3.3 Failure density and cumulative distribution function of the gamma model $(\gamma = 2, \lambda = 1)$.

The gamma distribution percentage points are useful but not easily obtained, since the gamma density is not integrable in closed form. In general, if t_p is the time for which the probability is p that the component has not failed, then t_p is the solution of the equation

$$\frac{\lambda^{\gamma}}{\Gamma(\gamma)} \int_{t_p}^{\infty} t^{\gamma-1} e^{-\lambda t} \, dt = p \tag{3.40}$$

For the gamma density in Figure 3.3, the median time-to-failure is then the solution of the equation

$$(t_{0.5} + 1) e^{-t_{0.5}} = 0.5 \tag{3.41}$$

Using an iterative procedure such as the Newton-Raphson method, it is not hard to find that $t_{0.5} = 1.68$ hr. The mean failure time is 2 hr, since $\lambda = 1$ and $\gamma = 2$ for this model. The symmetry of the gamma model depends on γ, the shape parameter. As γ increases the gamma model approaches a symmetric distribution--the normal.

Inference problems for the two-parameter gamma failure model are discussed in some detail by Gross and Clark (1975) and Mann et al. (1974). Furthermore, these authors deal with the three-parameter gamma model considering a location parameter as well as a scale and shape parameter. As with the Weibull model, inference considerations for the gamma model are based on the maximum likelihood criteria.

The maximum likelihood estimators of λ and γ, $\hat{\lambda}$ and $\hat{\gamma}$ (say), are the solution of the equations

$$\frac{\hat{\lambda}}{\bar{t}} = \hat{\gamma} \tag{3.42}$$

and

$$\frac{\Gamma'(\hat{\gamma})}{\Gamma(\hat{\gamma})} - \ln \hat{\gamma} = \ln R \tag{3.43}$$

where $R = \Pi_{i=1}^{n} t_i^{1/n}/\bar{t}$ (i.e., R is the ratio of the geometric mean to the arithmetic mean) and

$$\Gamma'(\hat{\gamma}) = \int_0^\infty x^{\hat{\gamma}-1} \ln xe^{-x} dx \qquad (3.44)$$

To find $\hat{\gamma}$, a graphical method developed by Wilk et al. (1962) is recommended. This method describes the relationship between $(1 - R)^{-1}$ and $\hat{\gamma}$ and includes a table for values of $(1 - R)^{-1}$ between 0 and 50. The table is conveniently reproduced in Gross and Clark (1975).

As with the Weibull model, $(1 - \alpha)100\%$ confidence interval for λ, the mean of the gamma model, is somewhat involved. The details of this procedure appear in Gross and Clark (1975). The approximate confidence interval is

$$\hat{\mu} \pm \sqrt{\text{vâr } \hat{\mu}} \; z_{1-\alpha/2}$$

where

$$\hat{\mu} = \frac{\hat{\gamma}}{\hat{\lambda}} \qquad (3.45)$$

$$\text{vâr } \hat{\mu} = \left(\frac{\hat{\gamma}}{\hat{\lambda}2}\right)^2 \text{vâr } \hat{\lambda} + \left(\frac{1}{\hat{\lambda}}\right)^2 \text{vâr } \hat{\gamma} - \frac{2\hat{\gamma}}{\hat{\lambda}} \hat{\text{cov}}(\hat{\lambda},\hat{\gamma}) \qquad (3.46)$$

and the values vâr $\hat{\lambda}$, vâr $\hat{\gamma}$, côv$(\hat{\lambda},\hat{\gamma})$ are obtained from the matrix equation

$$\begin{bmatrix} \text{vâr } \hat{\lambda} & \hat{\text{cov}}(\hat{\lambda},\hat{\gamma}) \\ \hat{\text{cov}}(\hat{\lambda},\hat{\gamma}) & \text{vâr } \hat{\gamma} \end{bmatrix}^{-1} = \begin{bmatrix} a_{11} & a_{12} \\ a_{12} & a_{22} \end{bmatrix} \qquad (3,47)$$

with $a_{11} = n\hat{\gamma}/\hat{\lambda}$, $a_{12} = -n/\hat{\lambda}$, and $a_{22} = n\Psi'(\hat{\gamma})$ noting that $\Psi'(x) = d^2 \ln \Gamma(x)/dx^2 = \Sigma_{i=0}^{n} (x + i)^{-2}$ is the trigamma function.

Example 3.4.3   Choi and Wette (1969) took a random sample of size 200 from a gamma distribution with $\lambda = 1$ and $\gamma = 3$.   They found

$\bar{t} = 2.9047$ and $\Pi_{i=1}^{200} t_i^{1/200} = 2.4432$. Thus, $R = 0.8411$, and $(1 - R)^{-1} = 6.2940$. By linear interpolation in Table 1 in Gross and Clark (1975), $\hat{\gamma} = 3.0462$. Thus, $\hat{\lambda} = 1.0487$ and $\hat{\mu} = \hat{\gamma}/\hat{\lambda} = 2.9047$, and it was found with the aid of a computer that $\Psi'(3.0462) = 0.3874$. Equation (3.47) shows that $\hat{\text{var}} \hat{\lambda} = 0.0118$, $\hat{\text{var}} \hat{\lambda} = 0.0848$, $\hat{\text{cov}}(\hat{\lambda},\hat{\gamma}) = 0.0291$. Hence, $\hat{\text{var}} \hat{\mu} = 0.0139$, and an approximate 95% confidence interval for μ is $(2.6737, 3.1357)$.

Tolerance limits, being nonparametric in nature, are discussed in Section 3.2. Prediction intervals for the gamma model can be worked out in a manner similar to the exponential model. For details of the derivation the reader is directed to Lawless (1971).

Problems

1. Suppose a component's life distribution is governed by the gamma model $f(t) = 4t^2 e^{-2t}$, $t \geq 0$. Find the values of the shape and scale parameters, γ and λ, respectively. Find μ, σ^2, and $t_{0.50}$.

2. For the gamma failure model show that $\mu_r' = (r - 1 + \gamma)(r - 2 + \gamma)\ldots\gamma/\lambda^r$, using either method described in the text.

3. Suppose T_1, T_2, ..., T_n are n independent times-to-failure from the exponential failure model $f(t;\lambda) = \lambda e^{-\lambda t}$, $t > 0$. Let $Y = \Sigma_{i=1}^n T_i$. Show, by convolution or a transformation of variables, that Y follows the gamma failure model (3.34). Hint: See Gross and Clark (1975), pp. 15-16, or Hahn and Shapiro (1967), Chapter 5.

4. Verify that the maximum likelihood estimators for the gamma model are found by solving (3.43) for $\hat{\gamma}$ and then substituting into (3.42) to obtain $\hat{\lambda}$.

5. For the data in Example 3.4.3 verify that $\hat{\text{var}} \hat{\mu} = 0.0139$ and hence that an approximate 95% confidence interval for μ is $(2.6737, 3.1357)$.

6. If $\gamma > 1$, show that the hazard rate for the gamma model, $t^{\gamma-1} e^{-\lambda t} / \int_t^\infty y^{\gamma-1} e^{-\lambda y} \, dy$, increases, and hence show λ is an upper bound for the hazard rate. Derive an analogous result if $\gamma < 1$.

7. Suppose T_1 is a failure time from an exponential failure model
    with mean time-to-failure of 18 hr, and T_2 is a failure time
    from an exponential failure model with mean time-to-failure of
    24 hr. If T_1 and T_2 are independent and if $Y = T_1 + T_2$, is the
    failure model of Y gamma? Can one compute E(Y) and var Y
    easily? Explain.

8. Suppose $f(t;\lambda,n) = \lambda^n t^{n-1} e^{-\lambda t}/(n - 1)!$. Show that $m_\theta(T) =$
    $(1 - \theta/\lambda)^{-n}$.

3.5   OTHER CONTINUOUS MODELS FOR MEASUREMENT

Among the continuous models the exponential, Weibull, and gamma
all can be derived from the basic physical principles, as has been
demonstrated. Some of the more recent models were developed
empirically rather than physically. For example, Murthy et al.
(1973) describe a response model whose hazard rate is initially
decreasing, relatively constant for a long period of time, and
eventually becomes increasing. Such a model often applies to
equipment that has a high risk of failure initially, a relatively
long period of useful life, and a final period in which age takes
its toll. This is not unlike the survival pattern in human beings
as can be seen from a life-expectancy table for the total U.S.
population [cf. Gross and Clark (1975), pp. 27–30]. Unfortunately,
the model of Murthy et al. is quite complex, requiring four parameters to be estimated. There are other models that can be
derived from physical situations which are important in application. These include the lognormal and extreme value models. For
instance Kao (1965) shows that if failure can be regarded as the
breaking point in the accumulation of damage to an item that is
proportional to the previous time of damage occurrence, the
time-to-failure of the item follows a lognormal distribution. The
extreme value distributions arise quite naturally as the distribution of the maximum or minimum of a sample of independent, identically distributed random variables. Gumbell (1958) provides an
excellent treatise on extreme-value distributions.

The derivation of the lognormal model is now given. Suppose an item fails only when damage to it has reached a specified level. Further, the times at which the damage occurs are $T_1 < T_2 < \cdots < T_n$; T_n is the failure time. Let $T_i - T_{i-1} = S_i T_{i-1}$, $i = 1, 2, \ldots, n$, where S_i is a random variable and T_0 is the time of initial observation, i.e., the time between the ith occurrence and (i - 1)th occurrence of damage is randomly proportional to the time of the (i - 1)th occurrence of damage. It then follows that $\sum_{i=1}^{n} S_i = \sum_{i=1}^{n} (\Delta T_i / T_i)$, where $\Delta T_i = T_i - T_{i-1}$, $i = 1, \ldots, n$. The limiting value as $\Delta T_{i-1} \to 0$ and n large shows

$$\sum_{i=1}^{n} S_i = \int_{T_0}^{T_n} \frac{dT}{T} = \ln \frac{T_n}{T_0}$$

since, for ΔT_{i-1} small and n large, $\sum_{i=1}^{n} (\Delta T_i / T_i) \approx \int_{T_n}^{T_n} dT/T$. If $S_1, \ldots, S_n$ are independent but not necessarily identically distributed random variables such that (i) $E|S_i^{2+\varepsilon}| < \infty$, for each S_i, and

$$\text{(ii)} \quad \lim_{n \to \infty} \frac{1}{\sigma^{2+\varepsilon} \left(\sum\limits_{i=1}^{n} S_i \right)} \sum_{i=1}^{n} \mu_{2+\varepsilon}(i) = 0$$

where $\sigma^2 \left(\sum_{i=1}^{n} S_i \right) = \sum_{i=1}^{n} \text{var } S_i$, and $\mu_{2+\varepsilon}(i) = E|S_i - E(S_i)|^{2+\varepsilon}$, then $\sum_{i=1}^{n} S_i$ is asymptotically a normally distributed random variable. (This is the Lyapunov version of the central limit theorem.) Thus, for large n, $\ln T_n$ is normally distributed, implying T_n has the lognormal distribution. A similar derivation of this result is found in Mann et al. (1974).

To fix ideas suppose Y has a normal distribution with mean μ and variance σ^2. A transformation of variables $Y = \ln T$ shows that T has the lognormal distribution

$$f(t; \mu, \sigma^2) = \left(\sqrt{2\pi}\sigma t \right)^{-1} e^{-[(\ln t - \mu)]^2 / 2\sigma^2} \tag{3.48}$$

where $t > 0$, $\sigma > 0$. To find the kth moment about the origin of the lognormal model it should be noted that if T follows the lognormal model and $T = e^Y$, Y follows the normal model. Thus,

$$\mu_k^{'} = E(T^k) = E(e^{Yk}) = e^{k\mu+k^2\sigma^2/2} \qquad\qquad (3.49)$$

with an application of (2.73) (the moment-generating function of the normal model). Hence, the mean and variance of the lognormal model are, respectively, $e^{\mu+\sigma^2/2}$ and $e^{2\mu+\sigma^2}(e^{\sigma^2} - 1)$.

If t_p is the time for which the probability is p that the item (whose failure time is lognormally distributed) has not failed, then $t_p = e^{y_p}$, where y_p is the pth percentage point of the normally distributed variable Y whose mean is μ and variance is σ^2. In particular, $t_{0.50} = e^{\mu}$.

Example 3.5.1 Let an item fail according to the lognormal failure model.

$$f(t) = \frac{1}{\sqrt{2\pi}t} e^{-(1/2)[\ln t]^2} \qquad t > 0 \qquad\qquad (3.50)$$

The failure rate and survival probability are left as exercises for the reader to obtain. The plot of (3.50) is shown in Figure 3.4.

The lognormal model described by (3.50) is in standard form. It is clear that for the standard lognormal model

$$\mu_k^{'} = e^{k^2/2} \qquad\qquad (3.49')$$

and hence the mean and variance of the standard lognormal model are, respectively, $\sqrt{e}$, and $e(e - 1)$. It is also clear that the median $t_{0.50} = 1$.

Since the failure rate of the lognormal model first increases and then decreases over time and since the model is characterized by the random variable T, where $Y = \ln T$ is normally distributed thereby enabling one to appeal to widely known results concerning normality, the model has enjoyed some popularity in reliability and survival analysis.

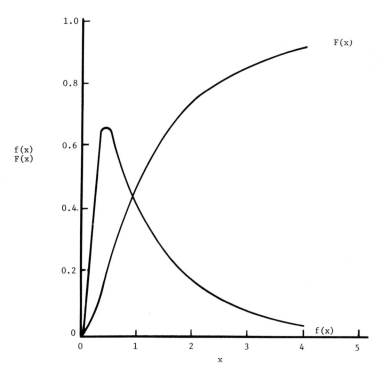

Figure 3.4 Failure density and cumulative distribution function of the lognormal model ($\sigma = 1$, $\mu = 0$).

Inference problems for the two-parameter lognormal distribution are discussed in some detail by Mann et al. (1974). They also discuss the three-parameter lognormal. Here, only the problem of obtaining the maximum likelihood estimators for μ and σ^2, the lognormal parameters, is considered. Suppose T has the lognormal distribution (3.48). Then Y = ln T is normally distributed with mean μ and variance σ^2. Thus, if T_1, $\ldots$, T_n is a random sample for the lognormal distribution (3.48), Y_1, $\ldots$, Y_n is a random sample from a normal distribution whose likelihood function is given by (2.87). It then follows that

$$\hat{\mu} = \bar{y} = \sum_{i=1}^{n} \frac{\ln t_i}{n} \tag{3.51}$$

and

$$\hat{\sigma}^2 = \frac{\sum_{i=1}^{n} (\ln t_i)^2 - n\left(\sum_{i=1}^{n} \ln t_i / n\right)^2}{n} \tag{3.52}$$

are the maximum likelihood estimators of μ and σ^2, respectively. Standard methods can then be used to obtain a $(1 - \alpha)100\%$ confidence interval for μ. If $\sigma = \sigma_0$ is known, a 95% confidence interval (say) for μ is $\bar{y} \pm 1.96\sigma_0/\sqrt{n}$. If σ is unknown and hence estimated from the data, a 95% confidence interval for μ is $\bar{y} \pm t_{n-1,0.975}\hat{\sigma}/\sqrt{n-1}$, where $t_{n-1,0.975}$ is the upper 2-1/2 percentage point of Student's t distribution with n - 1 df's. For unknown σ, it is known that $n\hat{\sigma}^2/\sigma^2$ has a chi-square distribution with n - 1 df's. Thus, a two-sided 95% confidence interval for σ^2 is

$$\frac{n\hat{\sigma}^2}{\chi^2_{0.975,n-1}} < \sigma^2 < \frac{n\hat{\sigma}^2}{\chi^2_{0.025,n-1}}$$

where $\chi^2_{0.025,n-1}$ and $\chi^2_{0.975,n-1}$ are the lower and upper 2-1/2 percentage points, respectively, of the chi-square distribution with n - 1 df's.

To obtain an approximate $(1 - \alpha)100\%$ confidence interval for the mean of the lognormal model, consider the following. Let $\tau = \mu + \sigma^2/2$, and $\hat{\tau} = \hat{\mu} + n\hat{\sigma}^2/2(n - 1)$, where $\hat{\mu}$ and $\hat{\sigma}^2$ are obtained from (3.51) and (3.52). In large samples $\hat{\tau}$ is approximately normally distributed with mean τ and variance $\sigma_{\hat{\tau}}^2$. Now

$$\sigma_{\hat{\tau}}^2 = \text{var } \hat{\mu} + \frac{\text{var}[n\hat{\sigma}^2/(n - 1)]}{4}$$

since $\hat{\mu}$ and $n\hat{\sigma}^2/2(n - 1)$ are independently distributed. [See Rao (1952), p.46.] But $\text{var}[n\hat{\sigma}^2/(n - 1)] = n^2\sigma^4/(n - 1)^3$, which is left as an exercise to the reader. Also var $\hat{\mu} = \sigma^2/n$. Thus, an approximate $(1 - \alpha)100\%$ confidence interval for $e^{\mu+\sigma^2/2}$, the mean of the lognormal model, is $(\exp(\hat{\tau} - z_{1-\alpha/2}\hat{\sigma}_{\hat{\tau}}), \exp(\hat{\tau} + z_{1-\alpha/2}\hat{\sigma}_{\hat{\tau}}))$ where

$$\hat{\sigma}_{\hat{\tau}}^2 = \frac{\hat{\sigma}^2}{n - 1} + \frac{n^2\hat{\sigma}^4}{4(n - 1)^3}$$

and $z_{1-\alpha/2}$ is the $(1 - \alpha/2)100\%$ point of the standard normal model.

Example 3.5.2 A component fails according to a lognormal model. A total of 100 failures is recorded showing $\hat{\mu} = 2.00$ and $\hat{\sigma}^2 = 0.99$; i.e., the geometric mean failure time of these 100 components is 7.39 hrs. This follows, since $\hat{\mu} = \Sigma_{i=1}^{n} \ln t_i/n$ and $e^{\hat{\mu}} = \exp(\Sigma_{i=1}^{n} \ln t_i/n) = \Pi_{i=1}^{n} t_i^{1/n}$, the geometric mean. Furthermore, $\hat{\tau} = 2.50$ and a simple calculation shows $\hat{\sigma}_{\hat{\tau}}^2 = 0.0125$. Thus, a 95% confidence interval for the mean time-to-failure of the component is $(e^{2.28}$ hrs., $e^{2.72}$ hrs.) or (9.78 hrs., 15.18 hrs.). Note that the estimate for the mean failure time of the component is 12.18 hrs., using $e^{\hat{\tau}}$, where $\hat{\tau} = \hat{\mu} + n\hat{\sigma}^2/2(n - 1) = 2.5$.

There are three types of so-called extreme-value distributions all arising as the limiting distributions of the smallest order statistic. The type III extreme-value distribution is actually a three-parameter Weibull model, i.e., if T is a random variable whose response model is the type III extreme-value model, then

$$f(t;\lambda,\gamma,\beta) = \lambda\gamma(t - \beta)^{\gamma-1}e^{-\lambda(t-\beta)^{\gamma}} \tag{3.53}$$

where $t > \beta$, $\gamma > 0$, $\lambda > 0$, is the probability density function of T. If $\beta = 0$, then (3.53) reduces to (3.23). β is a location parameter. Estimation procedures for this model are discussed in Mann et al. (1974).

The type II extreme-value model has as its pdf

$$f(t;\lambda,\gamma,\beta) = \gamma\lambda[\lambda(\beta - t)]^{-(\gamma+1)}e^{-[\lambda(\beta-t)]^{-\gamma}} \tag{3.54}$$

where $t < \beta$, $\gamma > 0$, $\gamma > 0$. Further discussion of this model can be found in Gumbel (1958).

The *type I extreme-value distribution* can be obtained from the Weibull model as follows. Suppose T has the Weibull distribution (3.23). The survival model $1 - G(t;\lambda,\gamma)$ is given by the equation

$$1 - G(t;\lambda,\gamma) = e^{-\lambda t^{\gamma}}$$

where $t \geqslant 0$, $\lambda > 0$, $\gamma > 0$. Let $Y = \ln T$ be survival model for Y; then $1 - F(y;\alpha,\beta)$ is written

$$1 - F(y;\alpha,\beta) = \exp(-e^{(y-\alpha)/\beta}) \tag{3.55}$$

where $-\infty < y < \infty$, $\alpha = (-\ln \lambda)/\gamma$, and $\beta = \gamma^{-1} > 0$. The density function for the type I extreme-value distribution is

$$f(y;\alpha,\beta) = \frac{1}{\beta}e^{(y-\alpha)/\beta}\exp(-e^{(y-\alpha)/\beta}) \tag{3.56}$$

where $-\infty < y < \infty$, $\alpha = (-\ln\lambda)/\gamma$, and $\beta = \gamma^{-1} > 0$. Another way in which the extreme-value model arises is as the limiting distribution of the smallest observation of a set of independent identically distributed normal random variables. [See Cramér (1946) for a derivation.] The limiting distribution of the largest observation of a set of independent, identical normal variables is

$$f(y;\alpha,\beta) = \frac{1}{\beta}e^{-(y-\alpha)/\beta}\exp(-e^{-(y-\alpha)/\beta}) \tag{3.56a}$$

The mean and variance of the type I extreme-value distribution
are obtained as follows. If $\alpha = 0$, $\beta = 1$, i.e., the standardized
extreme-value distribution, then μ_k', the kth moment about the
origin, is seen to be $\int_0^\infty (\ln x)^k e^{-x} dx$. For k = 1 and 2 with the
aid of Euler's formula[†] one finds $\mu = 0.5772$ and $\mu_2' = 1.6449 +$
$(0.5772)^2$. Thus, the mean and variance of the type I extreme-value
distribution for minimum (maximum) are $\alpha - 0.5772$ ($\alpha + 0.5772$) and
$1.6449\beta^2$, respectively. The failure rate for the extreme-value
model given by the formula $\lambda(t) = f(y)/[1 - F(y)]$ is $e^{(y-\alpha)/\beta}/\beta$.
Note that the failure rate increases as y increases. Finally, y_p,
the pth percentage point for the extreme-value model, is given as
$y_p = \alpha + \beta \ln \ln (1 - p)^{-1}$.

Example 3.5.3   The density function of the standard type I extreme-
value model for minimum is given as

$$f(y) = e^y \exp(-e^y) \qquad -\infty < y < \infty \qquad (3.57)$$

The survival probability $1 - F(y)$ is

$$1 - F(y) = \exp(-e^y) \qquad -\infty < y < \infty \qquad (3.58)$$

A plot of f(y) and F(y) is shown in Figure 3.5.

It should be noted that the extreme-value model is Figure 3.5
is a skewed-left model being the model represented by (3.56). That
is, it has a long tail to the left of the origin, whereas it tails
off sharply to the right of zero. The model represented by (3.56a)
is a skewed-right model. In the general case of type I extreme-
value models given by (3.56) and (3.56a) the parameters α and β
are location and scale parameters, respectively, Hence, the shape
of these models is invariant. Contrast this to the Weibull, gamma,
and lognormal models, noting that each has a shape parameter.

Estimation procedures for the type I extreme-value distribution usually involve linear estimation methods. Mann et al. (1974)
contain an excellent and thorough discussion of linear estimation

---

[†]Euler's formula: $\int_0^\infty \ln x e^{-x} dx = 0.5772$.

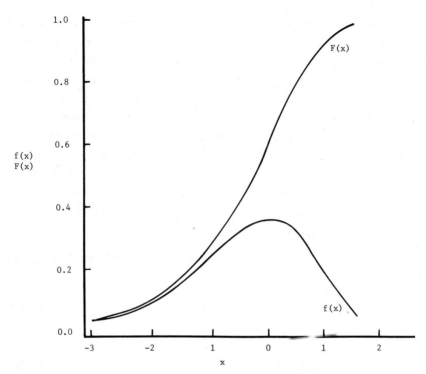

Figure 3.5 Failure density and cummulative distribution function
of the type I extreme-value model   ($\beta = 1$, $\alpha = 0$).

techniques for the Weibull and the type I extreme-value models.
To obtain the maximum likelihood estimators of α and β in complete
samples, notice that if y has the type I extreme-value distribution with the parameters α and β, $T = e^y$ has the Weibull distribution with parameters $\lambda = e^{-\alpha/\gamma}$ and $\gamma = \beta^{-1}$.   (See equation 3.5.5.)
Hence, the maximum likelihood estimators $\hat{\alpha}$ and $\hat{\beta}$ of α and β,
respectively, are $\hat{\alpha} = -\ln \hat{\lambda}/\hat{\gamma}$ and $\hat{\beta} = \hat{\gamma}^{-1}$, where $\hat{\lambda}$ and $\hat{\gamma}$ are
obtained from (3.29) and (3.29a).   This follows from a well-known
theorem that assures the likelihood function remains invariant

under one-one transformation of the parameters.

Other inference problems, such as obtaining a confidence interval for the mean of a type II extreme-value model, are not discussed further. These problems are dealt with in a straight-forward way.

Another specialized model that arises from both physical and empirical considerations is the logistic model. Bartlett (1955) shows that during an epidemic, if everyone in the population can be classified as either infected or susceptible and that if the rate of growth of infectives is proportional to the number already in-fected multiplied by the number still susceptible, the resulting model is the logistic model. If $x(t)$ is the number of infectives at time t and n is the total population size, the clearly

$$\frac{dx(t)}{dt} = \lambda x(t)[n - x(t)] \tag{3.59}$$

where λ is the constant of proportionality. Under the initial condition $x(0) = n/2$, this differential equation has the solution

$$\frac{x(t)}{n} = (1 + e^{-\lambda nt})^{-1} \tag{3.60}$$

Equation (3.60) is known as the *logistic growth model* or, more simply, the *logistic model*.

The logistic distribution also arises as the mixture[†] of type I extreme-value distributions. See Dubey (1969) or Johnson and Kotz (1970) for the details.

Suppose the random variable T follows the logistic distribution with parameters α and β; the density function of T is

$$f(t;\alpha,\beta) = \beta^{-1} e^{-(t-\alpha)/\beta} [1 + e^{-(t-\alpha)/\beta}]^{-2} \tag{3.61}$$

If $t - \alpha$ replaces t and β^{-1} replaces λn, (3.61) follows from (3.60)

[†]The random variable X has a distribution that is the mixture of distributions of F_1 and F_2 if there exists a p, $0 < p < 1$, such that $F(x) = pF_1(x) + (1 - p)F_2(x)$, for all x.

by differentiation. The domain of t is the entire real line, i.e.,
$-\infty < t < \infty$. The parameters α and β are the location and scale parameters, respectively, of the logistic model. There is no shape
parameter and the density function is clearly symmetric about α.
The standard logistic model is obtained when $\alpha = 0$ and $\beta = 1$.

Moments are obtained in terms of the moments for the standard
logistic distribution. First of all, it is clear that all odd
moments are zero since the standard logistic model is symmetric
about the origin. To compute the even moments note that if μ_{2r} is
the 2rth moment of the logistic model, where $r \geq 1$.

$$\mu_{2r} = \int_{-\infty}^{\infty} t^{2r} e^{-t} (1 + e^{-t})^{-2} \, dt = 2 \int_{0}^{\infty} t^{2r} e^{-t} (1 + e^{-t})^{-2} \, dt$$

Now $(1 + e^{-t})^{-2} = \sum_{i=1}^{\infty} (-1)^{i-1} i e^{-(i-1)t}$, which follows from standard
power series results. Interchanging the order of integration and
summation, one obtains

$$\mu_{2r} = 2 \sum_{i=1}^{\infty} (-1)^{(i-1)} i \int_{0}^{\infty} t^{2r} e^{-it} \, dt$$

Noting that the integral is a gamma function, it then follows that

$$\mu_{2r} = 2\Gamma(2r + 1) \sum_{i=1}^{\infty} \frac{(-1)^{i-1}}{i^{2r}}$$

Finally, it can be shown that

$$\sum_{i=1}^{\infty} \left[\frac{1}{i^{2r}} + \frac{(-1)^i}{i^{2r}} \right] = \frac{1}{2^{2r-1}} \sum_{i=1}^{\infty} \frac{1}{i^{2r}}$$

Hence,

$$\mu_{2r} = 2\Gamma(2r + 1)[1 - 2^{-(2r-1)}]\zeta(2r) \tag{3.62}$$

where $\zeta(k) = \sum_{i=1}^{\infty} i^{-k}$, $k > 1$, is the Riemann zeta function [see Johnson
and Kotz (1970) or Gross et al. (1971) for a discussion]. Noting
that $\sum_{i=1}^{\infty} i^{-2} = \pi^2/6$, it follows that the variance of the standard

logistic model is $\sigma^2 = \pi^2/3$. For a general logistic model with location and scale parameters α and β, respectively, $\mu = \alpha$ and $\sigma^2 = \beta^2\pi^2/3$. The failure rate for the logistic model is $\beta^{-1} e^{-(t-\alpha)/\beta}$, which increases as t increases and is bounded below and above by 0 and β^{-1}, respectively. Finally, the pth percentage point for the logistic model is $t_p = \alpha + \beta \ln p(1 - p)^{-1}$.

Example 3.5.4   The density function of the standard logistic model is

$$f(y) = e^{-y}(1 + e^{-y})^{-2} \tag{3.63}$$

where $-\infty < y < \infty$. The survival probability, $1 - F(y)$, is

$$1 - F(y) = e^{-y}(1 + e^{-y})^{-1} \tag{3.64}$$

A plot of $f(y)$ and $F(y)$ is shown in Figure 3.6.

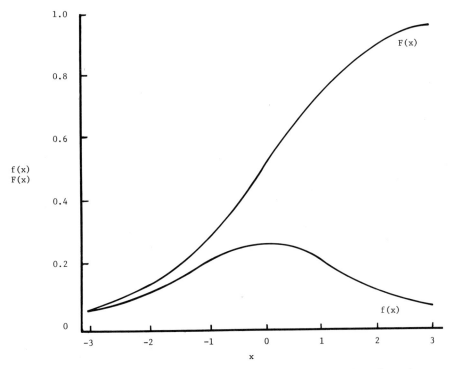

Figure 3.6 Failure density and cumulative distribution function of the logistic model   ($\alpha = 0$, $\beta = 1$).

There are two standard methods of estimating parameters of the logistic model; maximum likelihood and least squares. Both are described briefly.

It is easy to show that $\hat{\alpha}$ and $\hat{\beta}$, the maximum likelihood estimators of α and β, respectively, are the solution to the equations

$$\sum_{i=1}^{n} \left[1 + e^{-(t_i - \hat{\alpha})/\hat{\beta}} \right]^{-1} = \frac{n}{2} \tag{3.65}$$

and

$$\sum_{i=1}^{n} \left[\frac{(t_i - \hat{\alpha})}{\hat{\beta}} \; \frac{1 - e^{(t_i - \hat{\alpha})/\hat{\beta}}}{1 + e^{(t_i - \hat{\alpha})/\hat{\beta}}} \right] = n \tag{3.66}$$

To obtain the least-squares estimators $\tilde{\alpha}$ and $\tilde{\beta}$ (say), one makes makes use of the cdf $F(t)$. Specifically, since

$$F(t) = [1 + e^{-(\gamma + \xi t)}]^{-1}$$

where $\xi = \beta^{-1}$ and $\gamma = -\alpha\beta^{-1}$, a transformation is used to show that

$$\ln \frac{F(t)}{1 - F(t)} = \gamma + \xi t$$

If one uses in place of $F(t)$, $\hat{F}(t)$ = (no. of samples $\leq$ t)/n, where n is the sample size, and omitting the largest observation in the sample (in order that $\ln\{\hat{F}(t)/[1 - \hat{F}(t)]\}$ remain finite), the model

$$v = \gamma + \xi t + e$$

where $v = \ln\{\hat{F}(t)/[1 - \hat{F}(t)]\}$ and e is the error term, is a linear regression model. Thus, estimators of γ and ξ and hence α and β are the least-squares estimators. Unfortunately, the error terms are not independently distributed. Hence, the ordinary least-squares methods must be modified to be applied in this case. Discussion of this procedure in a related problem is found in Gross and Clark (1975). Finally, it should be noted that the least-squares estimators which are obtainable in closed form very often serve as

excellent first approximations to the maximum likelihood estimators.
Since the latter estimators have more desirable large sample properties, the recommended procedure is to use the least-squares
estimates and then use the Newton-Raphson procedure in two dimensions to obtain the maximum likelihood estimates. A discussion of
the two dimensional Newton-Raphson procedure is found in Gross and
Clark (1975).

Another specialized model that is considered in this section
is the beta model. One way to see how the beta model arises is to
consider the following problem: Suppose an item has a failure time
T_1 that is distributed according to the gamma model,

$$f(t_1;\lambda,\gamma_1) = \frac{\lambda^{\gamma_1} t_1^{\gamma_1-1} e^{-\lambda t_1}}{\Gamma(\gamma_1)} \qquad t_1 \geq 0, \ \gamma_1 > 0, \ \lambda > 0$$

and a second item (perhaps of the same kind) has a failure time T_2
that is distributed according to a second gamma model,

$$f(t_2;\lambda,\gamma_2) = \frac{\lambda^{\gamma_2} t_2^{\gamma_2-1} e^{-\lambda t_2}}{\Gamma(\gamma_2)} \qquad t_2 \geq 0, \ \gamma_2 > 0, \ \lambda > 0$$

A question that an engineer may have who is interested in the production of these items is what is the distribution of the fraction
of the failure time of the first item in proportion to the total
failure time of both items? Such a question is important if one is
concerned about allocating resources to both the manufacture and
maintenance of these items. Thus, statistically, the distribution
of the random variable $Y = T_1/(T_1 + T_2)$ is required, where T_1 and
T_2 are the times-to-failure of the items and are independent (see
Section 2.5). Thus,

$$f(t_1,t_2;\lambda,\gamma_1,\gamma_2) = \frac{\lambda^{\gamma_1+\gamma_2} t_1^{\gamma_1-1} t_2^{\gamma_2-1} e^{-\lambda(t_1+t_2)}}{\Gamma(\gamma_1)\Gamma(\gamma_2)}$$

A change of variable $Y = T_1/(T_1 + T_2)$ shows that $T_2 = T_1(1 - Y)/Y$, where the domain of Y is $0 < y < 1$. The Jacobian of this variable transformation is $|dt_2/dy| = t_1/y^2$. [See Hoel (1962), pp. 119–121.] Thus,

$$g(t_1,y;\lambda,\gamma_1,\gamma_2) = \frac{\lambda^{\gamma_1+\gamma_2} t_1^{\gamma_1+\gamma_2-1} y^{-(\gamma_2+1)} (1-y)^{\gamma_2-1} e^{-\lambda\{1+[(1-y)/y]\}t_1}}{\Gamma(\gamma_1)\Gamma(\gamma_2)}$$

To obtain the model for Y, unconditionally, the random variable T_1 is integrated between its limits 0 and ∞. If one notes that

$$\int_0^\infty t^{n-1} e^{-at} dt = \frac{\Gamma(n)}{a^n}$$

for any real numbers $n > 0$ and $a > 0$, it follows that

$$h(y;\gamma_1,\gamma_2) = \frac{\Gamma(\gamma_1 + \gamma_2)}{\Gamma(\gamma_1)\Gamma(\gamma_2)} y^{\gamma_1-1}(1 - y)^{\gamma_2-1} \qquad (3.67)$$

where $0 < y < 1$, $\gamma_1 > 0$, $\gamma_2 > 0$. The model described by (3.67) is called the *beta model*, and is used in Chapters 5–7. In these later chapters γ_1 and γ_2 are replaced by α and β, respectively.

A further specialized model is the linear hazard rate model. Recall that the exponential model is characterized by its constant hazard rate (constant as a function of time). In some applications, however, the constant hazard rate assumption does not hold. Boag (1949) and Kodlin (1967) consider survival times of 121 patients with breast cancer. Kodlin shows that their hazard rate can be approximated quite well with a linear function of time. Details of the procedure may be found in Kodlin's article or in Gross and Clark (1975). If $\lambda(t) = \lambda + \gamma t$ is the hazard rate of an item or, as in the breast cancer model, of an individual (noting that $\lambda(t) = f(t)/[1 - F(t)]$ is the equation that relates the density function, distribution function, and hazard rate for any model that represents the lifetime of a system), it follows that

$$\frac{-d[1 - F(t)]}{1 - F(t)} = \lambda + \gamma t$$

(See the discussion concerning the exponential model.) The
solution of this equation with the initial condition F(0) = 0 leads
to the equation

$$1 - F(t) = e^{-[\lambda t + (\gamma/2)t^2]}$$

The corresponding density function is

$$f(t;\lambda,\gamma) = (\lambda + \gamma t)e^{-[\lambda t + (\gamma/2)t^2]} \qquad t \geq 0 \tag{3.68}$$

This model is known as *Kodlin's model* or the *linear hazard rate
model*.

The final model that is considered in this chapter is the
Pareto model. The Pareto model is usually derived as follows:
Suppose X has the exponential with mean μ and that μ, in turn, has
a gamma model with parameters β (scale) and α (shape). The unconditional model of X is then written

$$f(x;\alpha,\beta) = \frac{\alpha\beta}{(\beta x + 1)^{\alpha+1}} \tag{3.69}$$

where x > 0, α > 0, β > 0. This model is discussed more fully in
Johnson and Kotz (1970).

Problems

1. For the lognormal model, explain carefully why $t_{0.50}$, the
    median, is e^{μ}, where μ and σ^2 are the parameters of the model.

2. Suppose T_1, T_2, ..., T_n is a random sample of size n with each
    T_i obeying the lognormal model. If $\hat{\sigma}^2$ is given by (3.52), show
    that var $n\hat{\sigma}^2/(n - 1) = n^2\sigma^4/(n - 1)^3$.

3. A component fails according to a lognormal model. A total of
    100 failures is recorded showing $\hat{\mu}$ = 5.00 ln hrs. and $\hat{\sigma}^2$ = 1.98.
    Find a 99% confidence interval for the mean time-to-failure of
    the component.

4. Show explicity that if T has a Weibull model with scale para-
    meter λ and shape parameter γ, Y = ln T has a type I extreme-
    value model with location parameter α = $(-\ln \lambda)/\gamma$ and shape
    parameter $\beta = \gamma^{-1}$ > 0.

5. Show that the mean and variance of the general extreme-value
    distribution with location and scale parameters α and β,
    respectively, are $\alpha - 0.5772\beta$ and $1.6449\beta^2$. Use the fact that
    $\int_0^\infty \ln x e^{-x}\, dx = -0.5772$ and $\int_0^\infty (\ln x)^2 e^{-x}\, dx = 1.6449 + (0.5772)^2$.

6. Consider the "mirror image" extreme-value distribution $f(y;\alpha,\beta) =$
    $1/\beta\ e^{-(y-\alpha)/\beta}\exp[-e^{-(y-\alpha)/\beta}]$, $-\infty < y < \infty$, $-\infty < \alpha < \infty$, $\beta > 0$. What
    are the mean, median, and variance of Y, the random variable
    that obeys this model? What is the failure rate and how does
    it differ from the failure rate of the extreme-value model
    (3.56)?

7. For the data in Example 3.3.3 consider a transformation $Y_i =$
    $\ln T_i$ for $i = 1, 2, \ldots, 20$. The data now come from a type I
    extreme-value model. Find the maximum likelihood estimates
    $\hat{\alpha}$ and $\hat{\beta}$ of α and β, respectively.

8. Show that the differential equation $dx(t)/dt = \lambda x(t)[n - x(t)]$,
    with $x(0) = n/2$, has the solution $x(t)/n = [1 + e^{-\lambda nt}]^{-1}$ and
    hence derive the logistic model from the given physical condi-
    tions. Hint: $\int dx/x(a + bx) = (1/a) \ln [x/(a + bx)] + C$, where
    C is a constant to be determined by the initial condition.

9. Show that the maximum likelihood estimators for the parameters
    of the logistic model are obtained by solving (3.65) and (3.66).

10. Show that

$$\sum_{i=1}^{\infty} \left[\frac{1}{i^{2r}} + \frac{(-1)^i}{i^{2r}} \right] = \frac{1}{2^{2r-1}} \sum_{i=1}^{\infty} \frac{1}{i^{2r}}$$

    and hence show how one derives (3.62).

11. The beta distribution is related to another distribution called
    the F distribution that has many applications in statistical
    inference. Show that if T is a beta-distributed random vari-
    able with parameters $\nu_1/2$ and $\nu_2/2$, then the transformation
    $T = (\nu_1/\nu_2)F(1 + \nu_1/\nu_2 F)^{-1}$ has the F distribution

$$P(F) = \frac{\nu_1^{\nu_1/2} \nu_2^{\nu_2/2} \Gamma[(\nu_1 + \nu_2)/2]}{\Gamma(\nu_1/2)\Gamma(\nu_2/2)} \frac{F^{\nu_1/2 - 1}}{(\nu_2 + \nu_1 F)^{(\nu_1 + \nu_2)/2}} \qquad (3.70)$$

REFERENCES

Abramowitz, M., and Stegun, J. A., Eds. (1964). *Handbook of Mathematical Functions*, National Bureau of Standards, U.S. Government Printing Office, Washington, D.C.

Aroian, L. A., and Robison, D. F. (1966). Sequential life tests for the exponential distribution with changing parameter, *Technometrics 8*, No. 2, pp. 217-227.

Bain, L. J., and Antle, C. E. (1967). Estimation of parameters in the Weibull distribution, *Technometrics 9*, No. 4, pp. 621-627.

Bartlett, M. S. (1955). *An Introduction to Stochastic Processes*, Cambridge University Press, Cambridge.

Birnbaum, Z. W. (1962). *An Introduction to Probability and Mathematical Statistics*, Harper & Brothers, Publishers, New York.

Boag, J. W. (1949). Maximum likelihood estimates of the porportion of patients cured by cancer therapy, *Roy. Stat. Soc. B11.*, pp. 15-53.

Bury, K. V. (1975). *Statistical Models in Applied Science*, John Wiley & Sons, Inc., New York.

Choi, S. C., and Wette, R. (1969). Maximum likelihood estimation of the parameters of the gamma distribution and their bias, *Technometrics 11*, No. 4, pp. 683-690.

Cramér, H. (1946). *Mathematical Methods of Statistics*, Princeton University Press, Princeton, N.J.

Dubey, S. D. (1969). A new derivation of the logistic distribution, *Naval Logistics Q. 16*, pp. 37-40.

Epstein, B., and Sobel, M. (1953). Life testing, *J. Amer. Stat. Assoc. 48*, No. 263, pp. 486-502.

Feller, W. (1957). *An Introduction to Probability Theory and Its Applications*, Vol 1 (2nd Edition), John Wiley & Sons, Inc., New York.

Gross, A. J., and Clark, V. A. (1975). *Survival Distributions: Reliability Applications in the Biomedical Sciences*, John Wiley & Sons, Inc., New York.

Gross, A. J., Clark, V. A., and Liu, V. (1971). Estimation of survival parameters when one of two organs must function for survival, *Biometrics 27*, No. 2, pp. 369-377.

Gross, A. J., and Lurie, D. (1977). Monte Carlo comparisons of parameter Weibull distributions, *IEEE Trans. on Reliability R-26*, No. 5, pp. 356-358.

Gumbel, E. J. (1958). *Statistics of Extremes*, Columbia University Press, New York.

Hahn, G., and Shapiro, S. S. (1967). *Statistical Models in Engineering*, John Wiley & Sons, Inc., New York.

Hoel, P. G. (1962). *Introduction to Mathematical Statistics*, 3rd ed., John Wiley & Sons, Inc., New York.

Johnson, N. L., and Kotz, S. (1970). *Distributions in Statistics: Continuous Univariate Distributions, Vols. 1 and 2*, Houghton Mifflin Co., Boston.

Kao, J. H. K. (1965). Statistical models in mechanical reliability, *Proc. 11th Nat. Symp. Reliability and Quality Control*, pp. 240–247.

Kodlin, D. (1967). A new response time distribution, *Biometrics 23*, No. 2, pp. 227–239.

Lawless, J. F. (1971). A prediction problem concerning samples from the exponential distribution with application in life testing, *Technometrics 13*, No. 4, pp. 725–730.

Mann, N. R., Schafer, R. E., and Singpurwalla, N. D. (1974). *Methods for Statistical Analysis of Reliability and Life Data*, John Wiley & Sons, Inc., New York.

Murthy, V. K., Swartz, G., and Yuen, K. (1973). *Realistic Models for Mortality Rates and Their Estimation, Vols. 1 and 2*, Department of Biomathematics Technical Report, University of California at Los Angeles.

Munroe, M. E. (1951). *Theory of Probability*, McGraw-Hill Book Company, New York.

Pearson, E. S., and Hartley, H. O. (1966). *Biometrika Tables for Statisticians, Vol. 1*, 3rd ed., Cambridge University Press, Cambridge, England.

Rao, C. R. (1952). *Advanced Statistical Methods in Biometric Research*, John Wiley & Sons, Inc., New York.

Weibull, W. (1951). A statistical distribution of wide applicability, *J. Appl. Mech. 18*, pp. 293–297.

Wilk, M. B., Gnanadesikan, R., and Huyett, M. J. (1962). Estimation of parameters of the gamma distribution using order statistics, *Biometrika 49*, pp. 525–545.

Chapter 4

MODELS FOR MEASUREMENT: DISCRETE CASE

4.1   INTRODUCTION

The focus of this chapter is the presentation of various discrete statistical models and how they arise physically and/or from experimental situations.   There are many discrete models in use.   For example, an auto dealer wishes to survey all customers who bought at least one automobile in the last three years.   The question asked very simply is "Are you satisfied with the automobile purchased at our dealership?"   Assuming all customers respond either yes or no, a simple model (the binomial model) can be developed to assess the level of customer satisfaction.   In another instance bacteria colonies are grown in a broth medium and are put on slides.   The number of bacteria colonies per slide is tabulated.   It is required to obtain the statistical model that determines, theoretically, the distribution of the number of bacteria colonies per slide.   Often, the model that is applied in this situation is the Poisson model.

The binomial and Poisson models are both derived from basic principles.   In fact, the Poisson model is obtained both as a limiting case of the binomial model and as the number of occurrences of an event in a specified time period, assuming that the event occurs according to the exponential model.   Other distributions that are also discussed include the hypergeometric, the geometric and negative binomial, the compound Poisson, and the power series models including the logarithmic series model.   An excellent overall discussion of the discrete probability models appears in Johnson and Kotz (1969).   Other accounts of discrete probability models can be

found in Feller (1957), Hahn and Shapiro (1967), Patil (1961, 1962,
1963, 1964a) (with particular reference to the generalized power
series distribution), Patil (1964b), and Haight (1967) (with
particular reference to the Poisson distribution).

4.2 THE BINOMIAL MODEL

Undoubtedly, the most common discrete probability model is the
binomial model. It has a long and honored history both in applications and theory. To demonstrate its applicability, the following
three examples describe situations in which one would use a binomial model to analyze experimental results.

Example 4.2.1 A physician treats 20 patients with elevated blood
pressures. The drug administered has a probability of 0.8 of
lowering the blood pressure in an individual patient. What is the
probability that at least 15 of the 20 patients respond successfully to treatment?

Example 4.2.2 An engineer designs a new electronic component.
After one year 30 out of 100 of these components have failed. Is
the engineer correct when he states "with probability 0.95 at
least 1/2 of all such components produced last one year or longer"?

Example 4.2.3 Two candidates D (Democrat) and R (Republican)
oppose each other in an election. A random survey of voters shows
53 favor R and 47 favor D. Is there reason to believe R will win
the election based on this survey if the chance of declaring R the
winner when he is not is 0.05?

At this time the reader should review the material on combinations and permutations as well as the material in Section 2.2 and
2.3, particularly Example 2.3.2.

With respect to the development of the binomial model, consider a sample of n light bulbs (from an infinite population of
light bulbs) for which the probability a light bulb will work is p
and the probability it is defective is q = 1 - p independent of the
other bulbs. To examine the probability that exactly x light bulbs

are good and $n - x$ are defective, consider one such arrangement of
the n light bulbs in which the first x are good and the remaining
$n - x$ are defective.   The probability of this occurrence is $p^x q^{n-x}$
as is shown below:

$$
\underbrace{pp \cdots p}_{x} \qquad \underbrace{qq \cdots q}_{n - x}
$$

This, however, is merely one arrangement of x good and $n - x$
defective light bulbs in a sample of size n.   In total, there are
$\binom{n}{x}$ such arrangements or combinations since the order of good and
defective light bulbs is unimportant.   Thus, the probability that
exactly x out of n light bulbs will be good and $n - x$ will be
defective is given by $f(x;p) \equiv pr\{X = x\}$, where

$$
pr\{X = x\} = \binom{n}{x} p^x q^{n-x} \tag{4.1}
$$

$x = 0, 1, 2, \ldots, n$. $f(x;n,p)$ is exactly the same probability model
as given by (2.14), i.e., the binomial probability model.

In Example 2.5.10, the moment-generating function $m_\theta(X)$ was
shown to be $(pe^\theta + q)^n$.   Hence $E(X) = np$ and var $X = npq$.   A
unique median does not exist.   If a unique median $X_{0.50}$ existed it
is the value of X such that

$$
\sum_{x=y+1}^{n} \binom{n}{x} p^x q^{n-x} = \sum_{x=0}^{y} \binom{n}{x} p^x q^{n-x} = 0.5 \tag{4.2}
$$

Hence, $x_{0.50} = (2y + 1)/2$ and $x_{0.50}$ is not attainable since it is
not an integer.

In large samples, appealing to the central limit theorem
(Theorem 2.7.1), X is normally distributed with mean np and variance npq.   Tables for the binomial model are available.   Those by
Pearson and Harley (1966) are among the more commonly used tables.
Note that these tables only contain values for $0 < p \leq 0.5$.   However, values for $0.5 < p \leq 1.0$ can be obtained by using $p' = 1 - p$
and $y = n - x$.

In general the mode of the binomial model is found by considering the ratio of $f(x + 1)/f(x) = (n - x)p/(x + 1)q$.   Now

$f(x + 1) \geq (\leq) \; f(x)$ and $(n + 1)p \geq (\leq)$ x.   Hence, mode (X) is the largest integer x_0 that is less than or equal $(n + 1)p$.   That is,

$$\text{mode}(X) = [(n + 1)p] \qquad (4.3)$$

where $[(n + 1)p]$ is the largest integer less than $(n + 1)p$.   If $(n + 1)p$ is itself an integer, then $f((n + 1)p) = f((n + 1)p - 1)$ is a double mode.

Estimation of the parameter p for n fixed, is accomplished by the standard maximum likelihood procedure.   The maximum likelihood estimator $\hat{p}$ of the parameter p was found in Example 2.6.5 and is x/n, where x is the number of successes in n binomial trials. In large samples $\hat{p}$ is approximately normally distributed with mean p and variance pq/n.   In order to obtain an approximate $(1 - \alpha)100\%$ confidence interval for p, the variance pq/n is replaced by $\hat{p}\hat{q}/n$. This is allowable in large sample theory [see, for example, Hogg and Craig (1970), p. 196].   Thus in sample sizes n > 30 an approximate $(1 - \alpha)100\%$ confidence interval for p is given as

$$\text{pr}\left\{\hat{p} - Z_{1-\alpha/2}\sqrt{\frac{\hat{p}\hat{q}}{n}} < p < \hat{p} + Z_{1-\alpha/2}\sqrt{\frac{\hat{p}\hat{q}}{n}}\right\} = 1 - \alpha \qquad (4.4)$$

where $Z_{1-\alpha/2}$ is the $1 - \alpha/2$ percentile of the standard normal distribution.   The limits for this interval are $\hat{p} \pm Z_{1-\alpha/2}\sqrt{\hat{p}\hat{q}/n}$. Often the interval is expressed in terms of its limits.   Alternate methods of obtaining a $(1 - \alpha)100\%$ confidence interval for p are considered in Problems 2 and 3 in Section 2.7.   The upper one-sided $(1 - \alpha)100\%$ confidence interval for p is $(\hat{p} + Z_{1-\alpha}\sqrt{\hat{p}\hat{q}/n}, \; 1)$ and the lower $(1 - \alpha)100\%$ confidence interval is $(0, \; \hat{p} - Z_{1-\alpha}\sqrt{\hat{p}\hat{q}/n})$, where $Z_{1-\alpha}$ is the $(1-\alpha)$ percentile of the standard normal distribution.

An application of the upper $(1 - \alpha)100\%$ confidence interval was seen in Example 4.2.2.

For small samples, i.e., n < 30, confidence intervals can be obtained using binomial tables such as are available in Pearson and Hartley (1966).   (See the discussion of Example 4.2.1 regarding these tables.)   The well-known identity involving binomial sums is

$$\sum_{x=x_0}^{n} \binom{n}{x} p^x q^{n-x} = \frac{n!}{(x_0 - 1)!(n - x_0)!} \int_0^p t^{x_0-1} (1 - t)^{n-x_0} \, dt$$

$$(4.5)$$

The right-hand side of (4.5) is an incomplete beta integral. Since the beta and F variable (see Problem 2, Section 3.5) are related by the transformation $t = (\nu_1/\nu_2)F(1 + \nu_1/\nu_2 F)^{-1}$, where t has a beta distribution with parameters $\nu_1/2$ and $\nu_2/2$ and F has an F distribution with ν_1 degrees of freedom in the numerator and ν_2 degrees of freedom in the denominator, the equations to obtain p_L and p_u, the lower and upper limits of the confidence interval, i.e.,

$$\sum_{x=x_0}^{n} \binom{n}{x} p_L^x q_L^{n-x} = \frac{\alpha}{2}$$

and

$$\sum_{x=0}^{x_0} \binom{n}{x} p_u^x q_u^{n-x} = \frac{\alpha}{2}$$

can be obtained from tables of the F distribution by means of the equation

$$P_L = 2x_0 F_{2x_0, 2(n-x_0+1), 1-\alpha/2} [2(n - x_0 + 1)$$

$$+ 2x_0 F_{2x_0, 2(n+x_0+1), \alpha/2}]^{-1} \qquad (4.6)$$

and

$$P_u = 2(x_0 + 1) F_{2(x_0+1), 2(n-x_0), 1-\alpha/2} [2(n - x_0) + 2(x_0 + 1)$$

$$\times F_{2(x_0+1), 2(n-x_0), \alpha/2}]^{-1} \qquad (4.7)$$

where $F_{2x_0, 2(n-x_0+1), \alpha/2}$ is the $\alpha/2$ percentage point of the F distribution with $2x_0$ and $2(n - x_0 + 1)$ degrees of freedom for the numerator and denominator, respectively, and a similar definition for $F_{2(x_0+1), 2(n-x_0), 1-\alpha/2}$. This discussion is similar to that in Johnson and Kotz (1969).

Example 4.2.4 Suppose in a sample of 30 binomial trials there were
11 successes and 19 failures. Here $x_0 = 11$, and n = 30. A table
of F values, e.g., Dixon and Massey (1969), is used to find
$F_{22,40,0.025} = 0.45$ and $F_{24,38,0.975} = 2.04$. Thus, from (4.6) and
(4.7) $p_L = 0.198$ and $p_u = 0.563$. Hence, a 95% confidence interval
for p, the probability of success, is (0.198,0.563).

Gross and Hosmer (1978) discuss a rather general procedure for
obtaining the sum of probabilities in the tails of many discrete
distributions.

The answers to the questions posed in Examples 4.2.1 - 4.2.3
can now be addressed. In example 4.2.1, it is required to compute
$\Sigma_{x=15}^{20} \binom{20}{x}(0.8)^x(0.2)^{20-x}$, noting that 0.8 is the probability the
blood pressure of any given patient is lowered and that one requires
the probability that at least 15 patients show a lowered blood
pressure. From the binomial table in Pearson and Hartley (1966),

$$\sum_{x=15}^{20} \binom{20}{x}(0.8)^x(0.2)^{20-x} = \sum_{y=0}^{5} \binom{20}{y}(0.2)^y(0.8)^{20-y} = 0.8042$$

Thus, the probability that at least 15 out of 20 patients exhibit
a lowered blood pressure is just about 4/5. In regard to Example
4.2.2, if p is the true probability of a component failure, a 95%
upper confidence limit for p, appealing to the central limit theorem,
is $\hat{p} + 1.645\sqrt{\hat{p}\hat{q}/n}$, where $\hat{p}$ is the observed proportion of failures
in the sample of size $n \geq 30$. Now $\hat{p} = 0.3$, and n = 100. Thus, the
upper 95% confidence limit for p is 0.376. That is, the engineer
is correct in his assessment that with probability 0.95 at least
1/2 of all components last one year or greater. In fact, since
$\hat{p} + 2.326\sqrt{\hat{p}\hat{q}/n}$ is the 99% upper confidence limit for p, which is
numerically 0.407, the engineer is actually somewhat conservative
about the reliability of his component. Finally, Example 4.2.3
deals with attempting to determine the winner, if possible, in an

election in which two candidates are entered. This problem can be
restated as a test of hypothesis problem. H_0, the null hypothesis,
is $p \leq 1/2$, versus H_1, the alternate hypothesis $p > 1/2$, where p
is the probability R will win the election. For the sample $\hat{p} = 0.53$
and $p_0 = 0.50$. Again, by the central limit theorem, the quantity
$z = (\hat{p} - p_0)/\sqrt{p_0 q_0/n}$ has an approximate standard normal distribution for $n \geq 30$. Thus, H_0: $p \leq p_0$ is rejected at the 0.05 level
of significance only if $z > 1.645$. Now, $z = (0.53 - 0.50)/$
$\sqrt{0.25/100} = 0.6$. Since the null hypothesis is not rejected, the
candidate R should not get overly excited about his chances of
winning the election--at least not on the basis of this relatively
small sample.

Figure 4.1 is a graph of the binomial distribution with
$p = 1/2$, $n = 10$. Notice that $f(x)$ is symmetric about $x = 5$. Such
symmetry exists only when $p = 1/2$. In other cases the binomial
model is skewed. It is skewed left if $p < 1/2$ and right if $p > 1/2$.
Furthermore, $E(X) = 5$ and var $X = 2.5$. No integer-valued median
exists. In fact, for most discrete models, a more widely used
measure of central tendency is the mode, or the most likely value.
In this example mode$(X) = 5$, coinciding with the mean.

Often discrete as well as continuous models are truncated.
That is a certain value (or values) of the domain over which the
model is defined is (are) omitted, and the model is redefined over
the restricted domain. For example, the binomial model is often
truncated at $X = 0$, i.e., the observations $X = 0$ are excluded. In
this case,

$$f^*(x;p|x > 0) = (1 - q^n)^{-1} \binom{n}{x} p^x q^{n-x} \tag{4.8}$$

$x = 1, 2, \ldots, n$ is the appropriate truncated binomial model.
Most of the results developed for the binomial model carry over
to the truncated binomial model in a straightforward manner. Con-
sequently, they are left as exercises for the reader to verify.
For example, if X has the truncated binomial distribution,

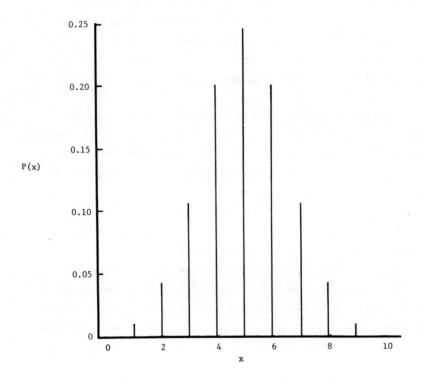

Figure 4.1   Binomial probability model   (p = 0.5, n = 10).

$$m_\theta(X|X > 0) = \frac{(pe^\theta + q)^n - q^n}{1 - q^n} \tag{4.9}$$

is the moment-generating function of X. Hence,

$$E(X|X > 0) = \frac{np}{1-q^n} \tag{4.10}$$

and

$$\text{var}(X|X > 0) = npq\left[\frac{1 - q^n - npq^{n-1}}{(1 - q^n)^2}\right] \tag{4.10a}$$

The maximum likelihood estimator of p for the truncated binomial model is, however, not as easily obtained as in the case of the ordinary binomial model. In this case $\hat{p}$, the maximum likelihood estimator of p, is found by solving, iteratively, the equation

$$\frac{x}{\hat{p}} - \frac{(n - x)/\hat{q} - n\hat{q}^{n-1}}{1 - \hat{q}^n} = 0 \tag{4.11}$$

where $\hat{q} = 1 - \hat{p}$.

An example of how the truncated binomial model may arise practically is the following. Suppose households of size n are examined for the occurrence of a certain noninfectious disease. The probability that an individual in the family has the disease is p and independent of whether other family members have the disease. Data are recorded for only those families with one or more ill members. The model that governs the probability of disease among these family members is then the truncated binomial model (4.8). (Note: A family with no ill members may have a probability p that a member is afflicted with the disease or may have a zero probability of affliction. There is no way to determine this until at least one family member has the disease.)

Finally, it should be noted that if X_1 and X_2 are independent binomial variables with probability functions

$$f(x_i;p) = \begin{pmatrix} n_i \\ x_i \end{pmatrix} p^{x_i} q^{n_i-x_i}$$

where $x_i = 0, 1, \ldots, n_i$, $i = 1, 2$, then the variable $Y = X_1 + X_2$
also is a binomial variable with model

$$f(y;p) = \begin{pmatrix} n_1 + n_2 \\ y \end{pmatrix} p^y q^{n_1+n_2-y} \qquad (4.12)$$

where $y = 0, 1, \ldots, n_1 + n_2$. This is easily proved using moment-
generating functions. First, $m_\theta(X_i) = (pe^\theta + q)^{n_i}$. Then
$m_\theta(X_1 + X_2) = m_\theta(X_1)m_\theta(X_2)$ for any pair of independent random
variables with moment-generating functions that exist. Finally,
since $m_\theta(y) \equiv m_\theta(X_1 + X_2) = (pe^\theta + q)^{n_1+n_2}$, and since this is the
moment-generating function for the binomial model (4.12), it
follows that Y has the binomial model (4.12). This result is easily
extended to the sum of k independent binomial variables provided
that p is the same for all k variables.

Problems

1. A device is tested 100 times. It fails 28 times and is success-
    ful 72 times. Find a 95% confidence interval for p, the prob-
    ability the device will be successful upon testing. If the
    device is tested 1000 times showing 720 successes and 280
    failures, how does the confidence interval change? Explain.

2. A used car dealer claims: "All my cars run trouble free for
    at least one year." To document this statement he offers a
    prospective customer a list of 10 customers each of whom has had
    had at least one year of trouble-free performance from his
    automobile. Would you believe the dealer? Explain.

3. In a new batch of light bulbs the actual proportion of defec-
    tives is unknown. How large a sample of light bulbs is neces-
    sary if we are to estimate the true proportion of defective
    light bulbs to within 10% with confidence 0.95? Hint:
    $pq \le 0.25$ for any p, $0 \le p \le 1$.

4. Redo Problem 3 if it is known that no more than 10% of all
    light bulbs are defective.

5. Without using moment-generating functions show that if X_1 and
    X_2 are independent binomially distributed random variables with
    sample sizes n_1 and n_2, respectively, but with the same success
    probability p, show that $Y = X_1 + X_2$ is binomially distributed
    with sample size $n_1 + n_2$ and success probability p. Hint:
    Use the identity

$$\sum_{x=0}^{y} \binom{n_1}{x}\binom{n_2}{y-x} = \binom{n_1 + n_2}{y}$$

6. Suppose in a sample of 25 patients with a certain disease 10 are
    cured. Find an exact 95% confidence interval for the true
    proportion of cures.

7. Suppose X is a unit binomial variable; i.e., X = 1 with prob-
    ability p and X = 0 with probability q. If n independent
    observations $X_1, \ldots, X_n$ are taken and r denotes the number of
    observations which is as large as X_1 show $pr\{r = n\} = 1 - p + p^n$.
    (Diploma in Statistics, Oxford University, 1951.)

8. Suppose X is normally distributed with mean 10 and variance 25.
    If a sample of five observations is drawn what is the prob-
    ability two are larger and three are smaller than 15?

9. Derive $m_\theta(X|X > 0)$ for the truncated binomial model. Hence, or
    otherwise, derive the formulas for $E(X|X > 0)$ and $var(X|X > 0)$.

10. Show that $\hat{p}$, the maximum likelihood estimator of p in the
    truncated binomial model, is given by (4.11).

4.3   THE POISSON MODEL

Suppose a slide is bombarded with bacterial colonies. Each colony
has a constant probability p of hitting the slide independent of
all other colonies. Suppose, further, that there are n colonies
altogether. The probability that x bacterial colonies hit the
slides is given by the binomial model, i.e.,

$$f(x;n,p) = \binom{n}{x} p^x q^{n-x}$$

where $x = 0, 1, \ldots, n$, $q = 1 - p$. If the number of bacterial colonies is very large, and the probability that an individual colony hitting the slide is small in such a way that np, the mean number of colonies that hit the slide, is constant, $f(x;p)$ can be approximated by the Poisson model,

$$P(x;m) = \frac{e^{-m} m^x}{x!} \tag{4.13}$$

where $x = 0, 1, 2, \ldots$ (see Problem 1, Section 2.3). In this approximation $m = np$. To see this more clearly, consider the following argument. Let

$$P(x;m) = \lim_{\substack{p \to 0 \\ n \to \infty \\ np=m}} \frac{n!}{(n-x)!x!} p^x q^{n-x}$$

If one observes that $p = m/n$ and $n!/(n-x)! = n(n-1) \cdots (n-x+1)$, it then follows after some elementary alegbra that

$$P(x;m) = \frac{m^x}{x!} \lim_{n \to \infty} \left(1 - \frac{1}{n}\right) \cdots \left(1 - \frac{x-1}{n}\right)\left(1 - \frac{m}{n}\right)^{n-x}$$

Since

$$\lim_{n \to \infty} \left(1 - \frac{1}{n}\right) \cdots \left(1 - \frac{x-1}{n}\right) = \lim_{n \to \infty} \left(1 - \frac{m}{n}\right)^{-x} = 1$$

and the product of limits is the limit of products, one obtains

$$P(x;m) = \frac{m^x}{x!} \lim_{n \to \infty} \left(1 - \frac{m}{n}\right)^n$$

However, a well-known calculus result shows that

$$\lim_{n \to \infty} \left(1 - \frac{m}{n}\right)^n = e^{-m}$$

Thus, $P(x;m)$ is given by (4.13). Much work has been done on how large an n and how small a p are required before one uses the Poisson approximation to the binomial model. In general, it is recommended that $p \leq 0.05$ and $n \geq 30$. One should also note that if p is large, i.e., near unity, a Poisson approximation based on q can be obtained merely by switching the roles of p and q.

To investigate the Poisson approximation to the binomial numerically, consider the following example.

Example 4.3.1 Suppose the sums $\sum_{x=0}^{3} \binom{100}{x} (0.01)^x (0.99)^{100-x}$ and $\sum_{x=0}^{3} (e^{-1}/x!)$ are compared noting that the second sum is the Poission approximation to the first sum. The first sum is 0.982, and the second sum is 0.981--an error of about 0.1%.

Another derivation of the Poisson model is obtained as follows: Suppose that telephone calls arrive randomly and uniformly throughout a finite time period $(0,t)$. Let Δt be a very small interval of time such that the probability a phone call arrives in the interval of length Δt is $\lambda \Delta t$ and the probability of more than one phone call in the interval $o(\Delta t)$ where $o(\Delta t)$ approaches zero faster than Δt; i.e., $\lim_{\Delta t \to 0} [o(\Delta t)/\Delta t] = 0$. Let $p_x(t)$ be the probability that x phone calls arrive during the time interval $(0,t)$. Then,

$$p_x(t + \Delta t) = \lambda p_{x-1}(t) \Delta t + (1 - \lambda \Delta t)p_x(t) + o(\Delta t) \qquad (4.14)$$

where $x = 0, 1, \ldots,$ and $p_{-1}(t)$ is defined as 0. Equation (4.14) follows from the fact that x calls occurring by time $t + \Delta t$ can happen in one of two ways (basically): $x - 1$ calls occurring by time t and an additional call in the interval Δt or x calls occurring by the time t with no additional calls in Δt. [Any other occurrences have the probability proportional to $o(\Delta t)$.] Since these two events are mutually exclusive with respective probabilities $\lambda p_{x-1}(t) \Delta t$ and $(1 - \lambda \Delta t)p_x(t)$, and any other probability can be assigned to $o(\Delta t)$, the justification of (4.14) is complete. If (4.14) is divided by Δt and $\Delta t \to 0$, then

$$p_x'(t) = \lambda p_{x-1}(t) - \lambda p_x(t) \tag{4.15}$$

where $p_x'(t) = dp_x(t)/dt$, $x = 0, 1, \ldots$.   The initial conditions
for this system of differential equations are $p_0(0) = 1$, $p_x(0) = 0$,
$x = 1, 2, \ldots$.   Use of the initial conditions shows how the system
of differential equations is solved recursively.   First of all,

$$p_0'(t) = \lambda p_0(t)$$

Therefore,

$$\ln p_0(t) = \lambda t + C_1$$

But $p_0(0) = 1$, and hence $C_1 = 0$.   Thus,

$$p_0(t) = e^{-\lambda t}$$

If one uses $e^{-\lambda t}$ in place of $p_0(t)$, it follows that

$$p_1'(t) + \lambda p_1(t) = \lambda e^{-\lambda t} \tag{4.15a}$$

which is a linear differential equation of the first order.   The
general solution of (4.15a) is

$$p_1(t)e^{\lambda t} = \lambda t + C_2$$

Since $p_1(0) = 0$,

$$p_1(t) = \lambda t e^{-\lambda t}$$

If this procedure is continued, it is then easy to verify that

$$p_x(t) = \frac{e^{-\lambda t}(\lambda t)^x}{x!} \tag{4.16}$$

where $x = 0, 1, \ldots$, which is the Poisson model with mean λt.
Thus, the probability that exactly x telephone calls arrive in
the time interval $(0,t)$ is obtained from the Poisson model with λt
as the expected number of calls during the interval.

Example 4.3.2 Phone calls arrive at a business randomly throughout the day according to the Poisson model at the rate of 0.5 calls

per minute. What is the probability that in a given five minute
period not more than three calls arrive? Here, $\lambda t = 2.5$. Thus,

$$pr\{X \leq 3\} = \sum_{x=0}^{3} \frac{e^{-2.5}(2.5)^x}{x!} = 0.758$$

A third and final derivation of the Poisson distribution that
is considered deals with counting the number of failures of a given
component during a given time interval $(0,t)$. Specifically,
suppose an item fails according to the exponential failure model

$$f(x;\lambda) = \lambda e^{-\lambda x}$$

where $x \geqslant 0$, $\lambda > 0$. Let $N(t)$ be the number of these components
that fail during the time interval $(0,t)$, and let $S_n = \sum_{i=1}^{n} X_i$.
S_n is distributed according to the gamma failure model (3.34).
Thus,

$$pr\{S_n \leq t\} = \frac{1 - e^{-\lambda t}[1 + \lambda t + \cdots + (\lambda t)^{n-1}]}{(n - 1)!}$$

The occurrence of the event $\{N(t) = n\}$ is equivalent to the joint
occurrence of the events $\{S_n \leq t\}$ and $\{S_{n+1} > t\}$ since S_n is the
total time it takes n component failures to occur. Thus,

$$pr\{N(t) = n\} = pr\{S_n \leq t \text{ and } S_{n+1} \geq t\}$$

But, $pr\{S_n \leq t \text{ and } S_{n+1} > t\} + pr\{S_n > t \text{ and } S_{n+1} > t\} =$
$pr\{S_{n+1} > t\}$. (For any two events A and B, $pr\{A \cap B\} + pr\{\bar{A} \cap B\} =$
$pr\{B\}$, where $\bar{A}$ is the complement of A.) However, if $S_n > t$, then
$S_{n+1} > t$. Hence $pr\{S_n > t \text{ and } S_{n+1} > t\} = pr\{S_n > t\}$. It then
follows that $pr\{N(t) = n\} = pr\{S_{n+1} > t\} - pr\{S_n > t\}$. Hence,

$$pr\{N(t) = n\} = \frac{e^{-\lambda t}(\lambda t)^n}{n!} \tag{4.17}$$

That is, the number of component failures follows a Poisson model
with λt as the expected number of failures during the interval
$(0,t)$.

There is a slight difference between the Poisson models (4.13) and (4.16) and (4.17).   The model (4.13) has no direct dependence on time, whereas (4.16) and (4.17) do.   In the latter situation, λ is actually a rate (λt being the mean).   Since there are, in essence, these two forms of the Poisson model, the model (4.13) is selected for examining its basic properties.

The moment-generating function $m_\theta(x)$ is

$$m_\theta(x) = \sum_{x=0}^{\infty} \frac{e^{\theta x}e^{-m}m^x}{x!} = e^{-m}\sum_{x=0}^{\infty} \frac{(me^\theta)^x}{x!} = \exp[m(e^\theta - 1)] \qquad (4.18)$$

From (4.18) or direct calculation, $E(X) = m$, and $\text{var}(X) = m$.   The verification is left as an exercise to the reader.   As with many other discrete probability models a unique median does not exist. If it does, then the median $X_{0.50}$ (an integer) is the value of x for which

$$\sum_{x=y+1}^{\infty} \frac{e^{-m}m^x}{x!} = \sum_{x=0}^{y} \frac{e^{-m}m^x}{x!} = 0.5 \qquad (4.19)$$

such that $x_{0.50} = (2y + 1)/2$.

The reader should now refer to Example 2.7.2.   In this example it was demonstrated that if X follows the Poisson model (4.13), then for random samples of size $n \geq 30$, $\bar{X}$ is approximately normally distributed with mean m and variance m/n.   Furthermore, if X is a Poisson variable, then X is approximately normally distributed with mean and variance both equal to m provided $m \geq 5$. [See Johnson and Kotz (1969).]

Tables of the Poisson distribution are readily available. The tables by Molina (1942) are perhaps the most widely used. These tables are somewhat more extensive, giving probabilities for values of m as large as 15 and cumulative probabilities for values of m as large as 60.

Figure 4.2 is a graph of the Poisson distribution with m = 5.

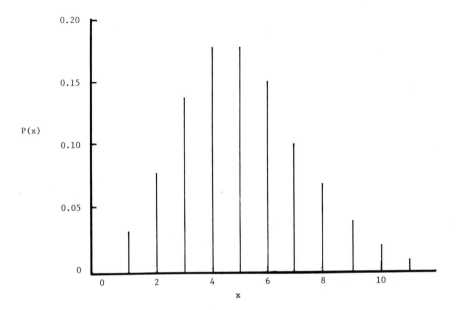

Figure 4.2 Poisson probability model   (m = 5).

Unlike the binomial, the Poisson model is never symmetric.
In fact, it is always skewed to the right. The mode of the Poisson
model is found exactly as is the mode of the binomial model.   The
mode is [m], where [m] is the largest integer less than or equal
to m.   If m itself is an integer (as in Figure 4.2), then the mode
occurs at both m - 1 and m.   (The values m = 4, 5 are the modal
values in Figure 4.2.)   The derivation of the Poisson mode is left
as an exercise to the reader.

     If X_1, ..., X_n is a random sample of size n from a Poisson
distribution, it is not difficult to show that $\hat{m}$, the maximum

likelihood estimator of the parameter m, is $\bar{X}$.   Since $\sqrt{n}(\bar{X} - m)/\sqrt{m}$
is approximately normally distributed with mean zero and variance
unity, an approximate $(1 - \alpha)100\%$ confidence interval for m is
obtained by solving the quadratic equation

$$m^2 - \left(2\bar{x} + \frac{\chi^2_{1,1-\alpha}}{n}\right) m + \bar{x}^2 = 0 \qquad (4.20)$$

for m, where $\chi^2_{1,1-\alpha}$ is the upper αth percentage point of the chi-
square distribution with one degree of freedom.   Equation 4.20 is
discussed in Example 2.7.2, initially for a 95% confidence interval.
The derivation of (4.20) is discussed in Problem 1, Section 2.7 for
the special case of a 95% confidence interval.

      Johnson and Kotz (1969) discuss another method for obtaining
an approximate $(1 - \alpha)100\%$ confidence interval for m that is
applicable quite generally.   This method is described as follows.
First of all, it is not difficult to show that

$$\text{pr}\{\chi^2_{2(x+1)} > 2m\} - \text{pr}\{\chi^2_{2x} > 2m\} = e^{-m}\frac{m^x}{x!} \qquad (4.21)$$

where χ^2_{2x} and $\chi^2_{2(x+1)}$ are chi-square variables with $2x$ and $2(x + 1)$
degrees of freedom, respectively.   This fact was used in the third
derivation of the Poisson model.   Suppose now m_L and m_U are values
of m such that

$$e^{-m_L} \sum_{y=x}^{\infty} \frac{m_L^y}{y!} = \frac{\alpha}{2} \qquad (4.22)$$

and

$$e^{-m_U} \sum_{y=0}^{x} \frac{m_U^y}{y!} = \frac{\alpha}{2} \qquad (4.23)$$

Johnson and Kotz (1969) then show that $m_L = (1/2)\chi^2_{2x,\alpha/2}$ and
$m_U = (1/2)\chi^2_{2(x+1),1-\alpha/2}$.   Thus, an approximate $1 - \alpha$ confidence
interval for m is the interval (m_L, m_U), the endpoints of which may

be obtained with the aid of chi-square tables.

Example 4.3.3 In a laboratory experiment the number of bacterial
colonies on slides follows the Poisson model. A total of 25 slides
is examined with $\bar{x}$ = 24.0 as the mean number of colonies per slide.
Find a 95% confidence interval for m, the true mean number of
bacterial colonies per slide. Using (4.20) an approximate 95%
confidence interval for m is obtained as the roots to the quadratic
equation

$$m^2 - 48.15m + 576 = 0$$

This yields (22.2,26.0) as the 95% confidence interval for m. If
(4.22) and (4.23) are used to obtain m_L and m_U, then the number of
bacterial colonies observed on all 25 slides also follows a Poisson
model. (The sum of Poisson distributed variables is also a Poisson
distributed variable. This is subsequently demonstrated.) If
x = 600 is the total number, then $m_L = 1/2 \ \chi^2_{1200,0.025} = 549.6$ and
$m_U = 1/2 \ \chi^2_{1202,0.975} = 654.0$. A per slide interval is obtained by
dividing each of these numbers by 25. Hence, (22.0,26.2) is the
95% confidence interval for m. This compares quite favorably to
the interval obtained using the normal approximation.

Other approximations to the Poisson model are found in
Johnson and Kotz (1969). Furthermore, they discuss bounds on
Poisson variables. For example, if X follows a Poisson model with
parameter m, then

$$\text{pr}\{X \leq m\} \geq e^{-1}$$

If m is an integer, then 1/2 can replace e^{-1}. This result is
derived by Teicher (1955).

A number of studies have been made concerning the Poisson
model that is truncated at X = 0. It is not difficult to show
that the Poisson model, truncated at X = 0, is written

$$P^*(x;m|x > 0) = \frac{(e^m - 1)^{-1} m^x}{x!} \tag{4.24}$$

where x = 1, 2, ... The following further results concerning $P^*(x;m|x > 0)$ are relatively easy to show and are left as exercises to the reader:

$$P^*(x;m|x > 0) = \lim_{\substack{n \to \infty \\ p \to 0 \\ np=m}} f^*(x;p|x > 0) \tag{4.25}$$

where $f^*(x;p|x > 0)$ is the truncated binomial model (4.8);

$$m_\theta(X|X > 0) = \frac{\exp(me^\theta) - 1}{e^m - 1} \tag{4.26}$$

$$E(X|X > 0) = \frac{m}{1 - e^{-m}} \tag{4.27}$$

and

$$\text{var}(X|X > 0) = \frac{m(1 - me^{-m} - e^{-m})}{(1 - e^{-m})^2} \tag{4.28}$$

The maximum likelihood estimator $\hat{m}$ of m, as is the case with the truncated binomial model, is found by solving, iteratively, the equation

$$\hat{m}(1 - e^{-\hat{m}}) - \bar{x} = 0 \tag{4.29}$$

where $\bar{x} = \Sigma_{i=1}^{n} x_i/n$ is the mean of a random sample of size n from a population whose probability model is given by (4.24).

Example 4.3.4 The following data refer to an epidemic of cholera in a village in India:

x	1	2	3	4	Total
n_x	32	16	6	1	55

where x is the number of cholera cases in a house and n_x represents the number of houses with x cholera cases. In addition to the 55 households having at least one cholera case, 168 households had no cases. Dahiya and Gross (1973) show how to obtain and estimate

the number of households that were infected but did not have any
active cases of cholera. The truncated Poisson model fits the data
of infected persons quite well, hence it can be used to estimate
the inactive infected cases and total infected cases. These data
were obtained by McKendrick (1926). Let n_0 be the (unknown) number
of households that were infected but did not have any active cases
of cholera. Let the total number of cases $N = \Sigma_{x=0}^{4} n_x$. An
estimator of $\hat{N}$ is $N = n/\hat{p}$, where $1 - \hat{p} = e^{-\hat{m}}$ and $\hat{m}$ is the maximum
likelihood estimator of m, the Poisson parameter, whose value is
obtained by solving (4.29). Dahiya and Gross show that $\hat{N}$ is
asymptotically normally distributed with mean N and variance
$Nq/(p - mq)$, where $q = e^{-m}$, m the Poisson parameter, and $p = 1 - q$.
Thus, a $100(1 - \alpha)\%$ confidence interval for N is the interval whose
limits are $\hat{N} \pm Z_{1-\alpha/2}\hat{\sigma}\sqrt{\hat{N}}$, where $\hat{\sigma} = \sqrt{\hat{q}/(\hat{p} - \hat{m}\hat{q})}$ and $Z_{1-\alpha/2}$ is the
upper $100(1 - \alpha/2)$ percentile of the standard normal variable.
Furthermore, if $n = \Sigma_{x=1}^{4} n_x$, then a $100(1 - \alpha)\%$ confidence interval
for n_0 is $(\hat{N} - n) \pm Z_{1-\alpha/2}\hat{\sigma}\sqrt{\hat{N}}$. Thus, a 95% confidence interval for
n_0, the unknown number of infected households showing no active
case of cholera, is found by obtaining $\hat{N}$, n, and $\hat{\sigma}$. $n = 55$,
$\hat{N} = 55/\hat{p} = 89$, where $\hat{m} = 0.97$ and $\hat{p} = 0.62$. Thus, $\hat{q} = e^{-\hat{m}} = 0.38$,
$\hat{\sigma}^2 = 0.38/[0.62 - (0.97)(0.38)] = 1.51$. Thus, an approximate 95%
confidence interval for n_0 is $34 \pm (1.96)(1.51)^{1/2}(9.43)$, or
$11 < n_0 < 57$.

Finally, if X_1 and X_2 are independent Poisson variables with
models

$$P(x_i; m_i) = \frac{e^{-m_i} m_i^{x_i}}{x_i!}$$

where $x_i = 0, 1, 2, \ldots$, $i = 1, 2$, then $Y = X_1 + X_2$ obeys the
Poisson model,

$$P(y; m_1 + m_2) = \frac{e^{-(m_1+m_2)} (m_1 + m_2)^y}{y!} \qquad (4.30)$$

This can be established using moment-generating functions as was
done in establishing a similar result for the binomial model. The
exact derivation is left as an exercise to the reader. As with the
binomial model, this result is easily extended to the sum of k
independent Poisson variables.

Problems

1. In a certain book there were 271 typographical errors out of
    322 pages. Assuming that errors are distributed randomly and
    in accordance to the Poisson model throughout the book, what is
    the probability that a page chosen at random contains no errors?
    If four pages are chosen at random, what is the probability
    that exactly two errors are found? At most two errors? At
    least two errors?

2. Phone calls obey a Poisson model with an arrival rate of m
    calls per minute. In a two-minute period the probability of
    exactly three calls is 0.19537. Find m. Hint: Try integer
    values for m.

3. As in Problem 2, phone call arrivals obey a Poisson model with
    an arrival rate of m calls per minute. In a three-minute
    period, the probability of at least one call arriving is 0.99.
    Find m.

4. Phone calls at three switchboards arrive according to Poisson
    models at the rates of one call per minute, three calls per
    minute, and five calls per minute. In a five-minute period,
    what is the probability of at least 40 phone calls arriving at
    all three switchboards?

5. Phone calls arrive at a switchboard at the rate of two calls
    per minute. Of those arriving calls, each call is put through
    to its party with probability 0.8. What is the arrival rate
    of those calls that are put through? Assume Poisson arrivals
    and that calls arrive independently of each other.

6. Prove that if X obeys the Poisson model with parameter m, that
    [m], the greatest integer less than or equal to m, is the mode.
    If m is an integer, show that m − 1 and m are a double mode.

7. If X obeys the Poisson model with parameter m, show that
   $E(X) = \text{var } X = m$.

8. Show that (4.24) is the appropriate truncated Poisson model
   when the observations are truncated at $X = 0$. Finally, show
   that $\hat{m}$ obtained from (4.29) is the maximum likelihood estimator
   of m.

9. If both $X = 0$ and 1 are truncated, what is the form of the
   Poisson model? What are the moment-generating function, the
   mean and the variance of X?

10. Show that if X_i obeys the Poisson model $P(x_i; m_i)$, $i = 1, 2, \ldots,$
    k, then $Y = \Sigma_{i=1}^{k} X_i$ obeys the Poisson model $P(y; \Sigma_{i=1}^{k} m_i)$.

4.4   THE HYPERGEOMETRIC MODEL

In many relevant applications, particularly in sampling, the
hypergeometric model plays an important role. For example, suppose
in surveying individuals in a given neighborhood as to whether
they are in favor of or against the development of a shopping mall
in their community, a random sample of persons is selected and
their responses, either yes, meaning for development, or no,
meaning against development, are recorded. If there are N persons
in the community, k of whom favor the shopping mall and N - k of
whom do not, and if the sample of n persons shows x favoring and
n - x not favoring the mall, then the probability model that
describes this outcome is

$$f(x; N, k, n) = \frac{\binom{k}{x}\binom{N-k}{n-x}}{\binom{N}{n}}$$ (4.31)

where $x = 0, 1, \ldots, k$, noting that both k and n (fixed) lie between 0 and N, and $\binom{N-k}{n-x}$ is defined to be zero if $n - x > N - k$.
The derivation of (4.31) follows from elementary combinatorial
considerations. Formally, the size of the sample space is $\binom{N}{n}$
since n individuals are selected from N individuals without
replacement. Furthermore, among the k individuals who favor the

shopping mall and N - k who oppose it, the total number of distinct
ways of selecting x and n - x who favor and oppose the shopping
mall, respectively, into the sample is $\binom{k}{n}\binom{N-k}{n-x}$. Thus, the probability that exactly X = x individuals in the sample favor development of the shopping mall and hence n - X = n - x oppose is given
by (4.31).

A second application of the hypergometric model is in the
field of sampling inspection. For example, suppose an electrical
firm manufactures N fuses a day k of which are good and N - k of
which are defective. A sample of size n is inspected daily, x of
which are good and n - x are defective. The probability that the
random variable X (number of defective fuses) assumes the value x
clearly is given by (4.31). That is, X obeys the hypergeometric
model.

Intuitively, if N, the population size, is large, and k is
large in such a way that k/N = p remains constant, then the
hypergeometric model should approximate the binomial model (4.1).
This is in fact the case as it is not difficult to show that

$$\lim_{\substack{n\to\infty \\ k\to\infty \\ k/N=p}} \frac{\binom{k}{x}\binom{N-k}{n-x}}{\binom{N}{n}} = \binom{n}{x} p^x q^{n-x} \qquad (4.32)$$

where x = 0, 1, ..., n. The verification of (4.32) is left as an
exercise to the reader. Now one can note the hierarchy that exists
among the three discrete models that have been discussed in this
chapter:

Hypergeometric	$\sim$	Binomial	$\sim$	Poisson
(4.31)		(4.1)		(4.13)
		N large		n large
		k large		p small
		k/N = p		np = m
		(constant)		(constant)

Here " $\sim$ " means "approximated by."

There is no easy closed form representation of the moment-generating function. However, the mean and variance of X, where X follows the hypergeometric model, are relatively easy to calculate, albeit somewhat tedious.

$$E(X) = \sum_{x=0}^{k} x \; \frac{\binom{k}{x}\binom{N-k}{n-x}}{\binom{N}{n}}$$

$$= \frac{nk}{N} \sum_{x^{'}=0}^{k} \binom{k^{'}}{x^{'}}\binom{N^{'}-k^{'}}{n^{'}-x^{'}} = \frac{nk}{N} \tag{4.33}$$

where $k^{'} = k - 1$, $N^{'} = N - 1$, $n^{'} = n - 1$, and $x^{'} = x - 1$. Similarly,

$$\text{var } X = \left(\frac{N-n}{N-1}\right) n \binom{k}{N}\left(1 - \frac{k}{N}\right) \tag{4.34}$$

The verification of (4.34) is left as an exercise to the reader. Again, it should be observed that if N and k are large, but $k/N = p$ is constant, then $E(X) = np$ and var $X = npq$, the mean and variance of the binomial model, respectively. Derivation of the median and mode are left as exercises for the reader. The mode of the hypergeometric model is given as the largest integer less than or equal to

$$\frac{(n + 1)(k + 1)}{N + 2} - 1$$

There are two maximum likelihood estimators to consider, the maximum likelihood estimators of k and N. Ordinarily k is unknown in applied situations, N may or may not be unknown. In fact, in the "capture-recapture" problem N is unknown. In this problem animals of a certain species are captured and tagged and then released. Shortly afterward (so that not enough time has elapsed to have caused a significant change in the animal population) animals of this species are "recaptured." The number with and without tags is recorded. From the collected data an estimate of total population N (say) is obtained. The maximum likelihood estimator is often used in this way.

The maximum likelihood estimator $\hat{k}$ is that value of k that maximizes $\binom{k}{x}\binom{N-k}{n-x}$. Writing (4.31) as a likelihood function L(k), $\hat{k}$ is that value of k such that $L(\hat{k}) \geq L(\hat{k} - 1)$ and $L(\hat{k}) > L(\hat{k} + 1)$, simultaneously. Thus, k is the greatest integer less than or equal to $(N + 1)x/n$. If $(N + 1)x/n - 1$ is an integer, then it and $(N + 1)x/n$ are both maximum likelihood estimators of k. Similarly, writing (4.31) as a likelihood function $L(\hat{N})$, $\hat{N}$ is that value of N such that $L(\hat{N}) \geq L(\hat{N} - 1)$ and $L(\hat{N}) > L(\hat{N} + 1)$, simultaneously. Hence, $\hat{N}$ is the greatest integer less than or equal to kn/x. If kn/x is an integer, kn/x and kn/x - 1 are both maximum likelihood estimators of N.

Figure 4.3 is a graph of the hypergeometric model with N = 10, k = 5, and n = 5. This model is symmetric and has a double mode at x = 2, 3.

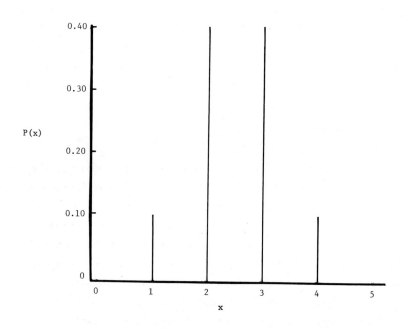

Figure 4.3 Hypergeometric probability model   (N = 10, k = 5, n = 5).

Tables for the hypergeometric model have been developed by Lieberman and Owen (1961). They give values for the individual and cumulative probabilities for a very wide range of values. The tables are to six decimal places.

In the capture-recapture problem, not only is the estimate of population size required but also an approximate confidence interval based on the estimate. However, var $\hat{N}$, where $\hat{N}$ is the maximum like-lihood estimator of N, is not finite ($\hat{N}$ has no finite moments). To remedy this situation in this and many other applied problems, the hypergeometric distribution with X = 0 truncated is used from which an approximate variance is calculated. The hypergeometric model truncated at X = 0 is easily shown to be

$$f(x;N,k,n) = \frac{\binom{k}{x}\binom{N-k}{n-x}}{\binom{N}{n} - \binom{N-k}{n}} \qquad (4.35)$$

where X = 1, 2, ..., k, noting that k and n (fixed) lie between 0 and N, and $\binom{N-k}{n-x}$ is defined to be zero if n - x > N - k. These are the same restrictions as for the hypergeometric model (4.31). Johnson and Kotz (1969) suggest other methods for approximating the variance of $\hat{N}$. The reader is referred to their book for a discussion. The capture-recapture problem is further treated in Example 4.4.2.

Example 4.4.1 In a motor parts factory 100 gaskets are manufactured daily, 95 of which are nondefective and 5 are defective. A sample of size 10 is chosen for inspection. Find the probability of at most 2 defective gaskets appearing in the sample. Using tables by Lieberman and Owen (1961), f(0;100,5,10) = 0.584, f(1;100,5,10) = 0.339, and f(2;100,5,10) = 0.070. Thus, the probability at most two defectives appear in the sample is the sum of these prob-abilities, which is 0.993. If one uses the binomial approximation to the hypergeometric with p = 0.95, n = 10, x = 0, 1, 2, the probabilities f(0;0.95) = 0.599, f(1;0.95) = 0.315, and f(2;0.95) = 0.075 are obtained. The relative errors in these

approximations are, respectively, 2.57%, -7.08%, and 7.14%. The
sum of the probabilities using the binomial approximation is 0.989
yielding a relative error of -0.40%. In this example, the binomial
approximation provides a reasonable result.

Example 4.4.2 (An Example of the Capture-Recapture Problem) In a
pond containing an unknown number of trout, 250 trout are netted,
tagged, and released back into the pond. One week later a sample
of 500 trout are netted with 75 tagged trout appearing in the sample.
In order to estimate N, the total number of trout in the pond, the
estimator $N* = (n + 1)(k + 1)/(x + 1) - 1$ is used. Its approximate variance is $N*^2$ $[N*/nk + 2(N*/nk)^2 + 6(N*/nk)^3]$ [see Johnson
and Kotz (1969, p. 147)]. In this application $N* = 1653$ and
$\sqrt{\text{var } N*} = 193$. Thus, a 95% confidence interval for N, the total
number of trout in the pond, is (1275,2031). This interval is
based on a normal approximation of $N*$.

Problems

1. Verify equation (4.32) and argue intuitively that for N and k
    large, with $p = k/N$, the binomial model is a good approximation
    to the hypergeometric model.

2. Derive the formula for the variance of the hypergeometric model.
    Show that if $N \to \infty$, $k \to \infty$, $p = k/N$ remains constant, var X =
    npq, the variance of the binomial model, provided n is fixed.
    Here $q = 1 - p$.

3. Find the mode of the hypergeometric model using the same method
    as was used to find the maximum likelihood estimators of k and
    N.

4. $E(1/X)$ is not finite for the hypergeometric model. Is
    $E[1/(X + 1)]$ finite? Explain. If it is finite can it be
    obtained in closed form? If it can, what is its value?

5. For the truncated hypergeometric model, $E(1/X)$ is finite. What
    difficulties do you find when you try to calculate it? Hint:
    Can you expand $1/x$ in a series that involves terms such as the
    following?

$$\frac{1}{x + 1} \, , \, \frac{1}{(x + 1)(x + 2)} \, , \, \cdots$$

6. In a motor parts factory 100 cylinders are manufactured daily
    90 of which are nondefective and 10 are defective. Find the
    probability that at most 3 defectives appear in a sample of 10
    cylinders that are randomly drawn for inspection.

7. Show that $\sum_{x=0}^{k} \binom{k}{x}\binom{N - k}{n - x} = \binom{N}{n}$ assuming that $n - x \le N - k$.
    Hint: Examine the coefficients on both sides of the identity
    $(1 + x)^{k}(1 + x)^{N-k} \equiv (1 + x)^{N}$.

8. In a valve company 10,000 valves are produced daily 100 of
    which are defective. Estimate the probability of finding no
    more than 3 defective valves if 50 valves are sampled daily.

9. Find the mode of the truncated hypergeometric model. How does
    this differ from the nontruncated case?

10. Find the maximum likelihood estimators for k and N for the
    truncated hypergeometric model. Are there difficulties in
    obtaining a closed-form solution?

11. A sampling plan calls for the inspection of 5 motors from a
    lot of 20. The lot is rejected when 1 or more defective motors
    are found in the sample. Find the probability that the lot is
    accepted if 3 of the 20 motors are defective. This calculation
    can be done for $k = 0, 1, \ldots, 20$. The resulting probabilities
    can be plotted as k/N. This curve is called an operating
    characteristic curve or OC curve and describes the sampling
    plan.

4.5   THE GEOMETRIC AND NEGATIVE BINOMIAL MODELS

Consider an event that can occur only at discrete points in time,
such as failure of a clock that records time in 1-second intervals.
Let p be the probability that the event occurs on the xth trial,
where p is constant from trial to trial. The probability that the
event occurs on the x-th trial *for the first time* is

$$f(x;p) = q^{x-1}p \tag{4.36}$$

where x = 1, 2, ..., and q = 1 - p. This follows because the probability that the event fails to occur on (x - 1) successive trials
is clearly q^{x-1}. Given the event has not occurred on (x - 1) trials,
the probability it occurs on the xth trial is p; hence the joint
probability that the event occurs on the xth trial *for the first
time* is given by (4.36). This model is the geometric model and is
the discrete analogue of the exponential model that was covered in
the previous chapter. In Exercise 6, Section 3.2, the reader was
asked to verify the memoryless property of the exponential model;
i.e., $\text{pr}\{X > t_1 + t_2 | X > t_1\} = \text{pr}\{X > t_2\}$, where t_1 and t_2 are any
two times. For the geometric model the analogous result holds;
i.e., if x_1 and x_2 are any two trials, then $\text{pr}\{X > x_1 + x_2 | X > x_1\} =$
$\text{pr}\{X > x_2\}$. This is also left as an exercise for the reader. It
was also shown in the previous chapter that if the time interval
(0,t) is divided into n equal intervals of length Δt so that t =
n Δt, where n is large and Δt is small; then as n → ∞ and Δt → 0 so
that $\Delta t/p$ and n Δt = t remain constant, the exponential model is
the limiting form of the geometric model under these conditions.

The moment-generating function $m_\theta(X)$ for the geometric model
is $pe^\theta(1 - e^\theta q)^{-1}$. Hence the mean and variance of the geometric
model are respectively p^{-1} and qp^{-2}. These are left as an exercise
for the reader to verify. It is clear that the mode of the geometric model always occurs at X = 1. Furthermore, percentage points
of this model are obtainable in closed form since

$$\text{pr}\{X \geq x_0\} = \sum_{x=x_0}^{\infty} q^{x-1}p = q^{x_0-1}$$

and hence

$$\text{pr}\{X < x_0\} = 1 - q^{x_0-1}$$

The maximum likelihood estimator $\hat{p}$ of p based on a sample of
size n observations is

$$p = \bar{x}^{-1}$$

where $\bar{x} = \Sigma_{i=1}^{n} x_i/n$, and $x_1, \ldots, x_n$ are the observations obeying the geometric model (4.36).   An approximate $(1 - \alpha)100\%$ confidence interval for p is obtained in the following way.   Since $\hat{p}^{-1} = \bar{x}$, $E(\hat{p}^{-1}) = p^{-1}$ and var $\hat{p}^{-1} = qp^{-2}/n$.   The expression $(\hat{p}^{-1} - p^{-1})/$ $\sqrt{\hat{q}\hat{p}}^{-2}/\sqrt{n}$ has an approximate standard normal distribution for $n > 30$. Hence, (p_L, p_U) is a $100(1 - \alpha)\%$ confidence interval for p, where $p_L = \hat{p}/(1 + z_{1-\alpha/2}\sqrt{\hat{q}/n})$ and $p_U = \hat{p}/(1 - z_{1-\alpha/2}\sqrt{\hat{q}/n})$.

Figure 4.4 is a graph of the geometric model with $p = 1/4$.

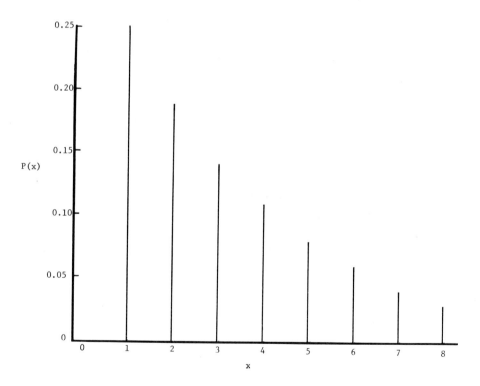

Figure 4.4   Geometric probability model $(p = 1/4)$.

Example 4.5.1 In exploring for oil, a large energy corporation has
been recording the average number of drilling excavations prior to
locating a productive oil well. In 50 experiences, it was found
that 4.2 drillings, on the average, were required before a well
was found. Find a 95% confidence interval for p, the probability
that a single drilling results in a productive well. To solve this
problem it is assumed that if p is the probability that a drilling
is productive, the number of drillings x necessary to find a productive well follows the geometric model (4.36). From the data
$\hat{p}$ = 1/4.2 = 0.238 is the estimated probability that a single drilling is productive. Applying the formula for obtaining a confidence
interval for p, one finds that (0.192,0.313) is the appropriate
95% confidence interval.

A natural extension of the geometric model is the negative
binomial model. Suppose that in drilling oil wells (Example 4.5.1)
a corporation will continue exploration in a given area until n
productive oil wells are located, where n is a predetermined value,
$n \geq 1$. Assuming that any well that is drilled has probability p
of being productive, and that all drillings are mutually independent, the probability it requires x - 1 drillings to achieve n - 1
productive wells is clearly the binomial model $\binom{x - 1}{n - 1} p^{n-1} q^{x-n}$ and
hence the probability it requires x drillings to achieve n pro-
ductive wells is

$$g(x;n,p) = \binom{x - 1}{n - 1} p^{n} q^{x-n} \tag{4.37}$$

where x = n, n + 1, ..., because the company will stop drilling
when they have found the nth productive well. (The last trial is
always a "success" where "success" is the event that occurs with
probability p.) The model (4.37) is called the *negative binomial*
model and is an extension of the geometric model (4.36) in that
if X_1, X_2, ..., X_n are identically distributed geometric random
variables each of which follows (4.36) and if $X = \sum_{i=1}^{n} X_i$, then X
obeys (4.37); i.e., X has the negative binomial model. The proof
of this result is left as an exercise to the reader. Thus, the

negative binomial model can in this sense be considered as the
discrete analogue to the gamma model (see Section 3.4), i.e., X
which follows the negative binomial model is $\Sigma_{i=1}^{n} X_i$, where X_i
follows the same geometric model.

The moment-generating function $m_\theta(X)$ is

$$m_\theta(X) = \sum_{x=n}^{\infty} \binom{x-1}{n-1} e^{\theta x} p^n q^{x-n}$$

$$= (pe^\theta)^n \sum_{y=0}^{\infty} \binom{n+y-1}{n-1} (e^\theta q)^y$$

Noting that if a is any constant such that $0 < a < 1$,
$\Sigma_{y=0}^{\infty} \binom{n+y-1}{n-1} a^y = (1-a)^{-n}$, it then follows that $m_\theta(X) =$
$(1 - e^\theta q)^n$. Hence, the mean and variance of the negative binomial
binomial model are, respectively, np^{-1} and nqp^{-2}. The mode of the
negative binomial model is $[(n-q)/(1-q)]$ where $[(n-q)/(1-q)]$
is the greatest integer less than or equal $(n-q)/(1-q)$. If
$(n-q)/(1-q)$ is an integer, then there is a double mode occurring
at $(n-q)/(1-q)$ and $(n-1)/(1-q)$.

The maximum likelihood estimator $\hat{p}$ or p is $\bar{x}^{-1}$, where
$\bar{x} = \Sigma_{i=1}^{k} x_i / \Sigma_{i=1}^{k} n_i$, and (x_i, n_i) $i = 1, \ldots, k$, are the numbers of
trials and success, respectively, in k negative binomial samples.
Exercises for the reader are to show $E(\hat{p}^{-1}) = p^{-1}$ and var $\hat{p} = qp^{-1}/$
$(\Sigma_{i=1}^{k} n_i)$. It then follows that an approximate $100(1-\alpha)\%$ confidence interval for p is (p_L, p_U) where $p_L = p/(1 + z_{1-\alpha/2}\sqrt{q/\Sigma_{i=1}^{k} n_i})$
and $p_U = p/(1 - z_{1-\alpha/2}\sqrt{q/\Sigma_{i=1}^{k} n_i})$. This is because $(\hat{p}^{-1} - p^{-1})/$
$\sqrt{\hat{q}\hat{p}^{-2}/\Sigma_{i=1}^{k} n_i}$ is approximately distributed as the standard normal
if $\Sigma_{i=1}^{k} n_i \geq 30$.

Example 4.5.2   In a military air operation it is required that 4
aircraft be able to fly each mission. Records have been kept for
the past 100 missions showing that an average of 6.5 planes were
prepared for each mission. That is, an average of 6.5 planes had
to be prepared in order that 4 fly. Find a 95% confidence interval

for p, the probability that an aircraft prepared for a mission will
fly the mission. Clearly, $\hat{p}$ = 4/6.5 = 0.615. Using the formula
for a confidence interval for p in the negative binomial case, it
follows that (0.579,0.655) is an appropriate 95% confidence interval
for p.

Problems

1. A clock has probability of failing at any second of time of
    2.5×10^{-8} independent of all times. Find the probability this
    clock lasts 1 month consisting of 30 days. Is this clock
    reliable? (That is, what is the probability the clock lasts
    365 days or more?)

2. Suppose X follows the geometric model (4.36). Show that if
    x_1 and x_2 are any two trials $pr\{X > x_1 + x_2 | X > x_1\}$ = $pr\{X > x_2\}$.

3. Suppose X follows the geometric model (4.36). Find $m_\theta(X)$,
    E(X), and var X.

4. Suppose X follows the geometric model (4.36) with p = 0.7.
    Find $pr\{2 < X < 6\}$.

5. Show that if $0 < a < 1$, $\sum_{y=0}^{\infty} \binom{n + y - 1}{n - 1} a^y = (1 - a)^{-n}$. Hint:
    Consider the series expansion for $(1 - a)^{-1}$ and take derivatives
    with respect to a of both sides.

6. Derive the confidence intervals for p for both the geometric
    and negative binomial models.

7. Obtain the mode of the negative binomial model and show that
    if $(n - q)/(1 - q)$ is an integer, a double mode occurs at
    $(n - q)/(1 - q)$ and $(n - 1)/(1 - q)$.

8. Show that the mean and variance of the negative binomial model
    are np^{-1} and nqp^{-2}, respectively.

9. A marine biologist wishes to estimate p the probability that a
    fish caught at random in a stream is a trout. He decides to
    fish until 10 trout are caught at which time he stops fishing.
    Find an estimate of p and an associated confidence interval if
    (i) 88 fish are caught before 10 trout are included; (ii) 198
    fish are caught before 10 trout are found.

10. Using moment-generating functions show that the sum of indepen-
    dently and identically distributed geometric random variables
    obeys the negative binomial model.

11. It is planned to send a man to the moon. Initially, four
    unmanned rockets will be fired and if they are all successful,
    then the manned launch will be attempted. Assuming all trials
    are independent and have the same success probability 0.90,
    find the probability the manned shot is unsuccessful.

4.6   THE MULTINOMIAL MODEL

The binomial model (equation 4.1) is designed to deal with questions
that are raised in Examples 4.2.1 through 4.2.3. That is, the
binomial model is the correct model to use when there are two categories of responses and all individuals in the sample appear in a
specified category with the same probability independently of each
other. Often, in applications, a model that initially is designed
as a two category of response model becomes a three (or more)
category model. For example, in Example 4.3.2 the random survey
of 100 voters may show 45 voters favor candidate R, 40 favor candidate D, and 15 voters are undecided. Hence, to the two categories
"R" and "D" is added yet a third category "undecided." In general,
a k-category of response situation may exist where all the categories are mutually exclusive such that each individual in the sample
appears in one of the k-categories with the same probabilities and
independently of each other.

   To fix ideas suppose that a manufactured item may appear in k
different conditions when it is for sale (e.g., an item of furniture
may be in first-class condition, i.e., no flaws, and also contain
various flaws that degrade its condition so that it may have to be
sold as a "second," "third," etc.). Let p_1, p_2, ..., p_k be the
respective probabilities that a given item appears in condition 1,
condition 2, ..., condition k. Then, if the first x_1 items appear
in condition 1, the next x_2 items appear in condition 2, ..., the

last x_k items appear in condition k, the overall probability of this occurrence is

$$p_1 \cdots p_1 \qquad p_2 \cdots p_2 \cdots p_k \cdots p_k$$

$$x_1 \text{ times} \qquad x_2 \text{ times} \qquad x_k \text{ times}$$

This, however, is but one arrangement of the n items in the sample, $n = \Sigma_{i=1}^{k} x_i$, according to the k conditions. Altogether there are $n!/(x_1! x_2! \cdots x_k!)$ such distinct arrangements. This is because all n items can be arranged in n! distinct ways. However, items in each of the k conditions cannot be distinguished among themselves; thus, the total number of distinct arrangements is $n!/(x_1! x_2! \cdots x_k!)$. Thus, the probability that exactly x_1 items appear in condition 1, x_2 items appear in condition 2, ..., x_k items appear in condition k, is given by $f(x_1,\ldots,x_k; p_1,\ldots,p_k) = \text{pr}\{X_1 = x_1, \ldots, X_k = x_k\}$, where

$$\text{pr}\{X_1 = x_1, \ldots, X_k = x_k\} = \frac{n!}{x_1! \cdots x_k!} p_1^{x_1} \cdots p_k^{x_k} \qquad (4.38)$$

with $x_i > 0$, $p_i > 0$, $i = 1, 2, \ldots, k$, $\Sigma_{i=1}^{k} x_i = n$ and $\Sigma_{i=1}^{k} p_i = 1$. This discrete model is the multinomial model which was discussed briefly in Chapter 2. (See Example 2.4.3 and equation 2.42.)

The joint moment-generating function of the k - 1 variables $X_1, X_2, \ldots, X_{k-1}$ is

$$m_{x_1,\ldots,x_{k-1}}(\theta_1,\ldots,\theta_{k-1})$$

$$= \sum_{x_1=0}^{n} \cdots \sum_{x_{k-1}=0}^{n-x_1-\cdots-x_{k-1}} \frac{n!}{x_1! \cdots x_k!} (p_1 e^{\theta_1})^{x_1} \cdots (p_{k-1} e^{\theta_{k-1}})^{x_{k-1}} p_k^{x_k}$$

where $p_k = 1 - p_1 - \cdots - p_{k-1}$ and $x_k = n - x_1 - \cdots - x_{k-1}$. It may be shown, using repeated summations, that

$$m_{x_1,\ldots,x_{k-1}}(\theta_1, \ldots, \theta_{k-1}) = (p_1 e^{\theta_1} + p_2 e^{\theta_2} + \cdots + p_k)^n$$

$$(4.39)$$

Note that x_k is uniquely determined by x_1, ..., x_{k-1} and p_k by
p_1, ..., p_{k-1}. The derivation of (4.39) is left as an exercise
for the reader. If (4.39) is used, one may obtain a vector of
means and matrix of variances and covariances (termed the variance-
covariance matrix) for the random vector $(X_1, X_2, ..., X_k)$. The
vector of means is $(np_1, np_2, ..., np_k)$, and the variance-covariance
matrix denoted by $\not{\mathrm{Z}}$ is

$$
\not{\mathrm{Z}} = \begin{bmatrix} np_1 q_1 & -np_1 p_2 & \cdots & -np_1 p_k \\ -np_1 p_2 & np_2 q_2 & \cdots & -np_2 p_k \\ \cdot\ \cdot\ \cdot & \cdot\ \cdot\ \cdot & \cdots & \cdot\ \cdot\ \cdot \\ -np_1 p_k & -np_2 p_k & & np_k q_k \end{bmatrix} \tag{4.40}
$$

where $q_i = 1 - p_i$, $i = 1, 2, ..., k$.

Modes and medians of multivariate distributions such as the
multinomial model are conceptually difficult to deal with and hence
their discussion is omitted.

If the various category probabilities p_i are small except the
kth category probability p_k (say) and if n is large in such a way
that $np_i = m_i$ is constant, $i = 1, 2, ..., k - 1$, then the approximate model that $(X_1, ..., X_{k-1})$ obeys is the multivariate Poisson
model

$$
P(x_1, ..., x_{k-1}; m_1, ..., m_{k-1}) = \prod_{i=1}^{k-1} \frac{m_i^{x_i}}{x_i!} e^{-(m_1+\cdots+m_{k-1})}
$$
$$\tag{4.41}$$

where $x_i = 0, 1, 2, ...$ for each $i = 1, 2, ..., k - 1$. The proof
of this result is a straightforward generalization of the univariate case which shows in an analogous way how the univariate Poisson
model approximates the binomial model in the case for which n is
large, p is small, but $np = m$ is finite. The proof of this result
is left as an exercise to the reader. It should be noted, further,

that $P(x_1, \ldots, x_{k-1}; m_1, \ldots, m_{k-1}) = \Pi_{i=1}^{k-1} P_i (x_i; m_i)$, where $P_i(x_i; m_i)$ is a univariate Poisson model; i.e., the variables X_1, $X_2, \ldots, X_{k-1}$ are mutually independent.

Estimation of $p_1, p_2, \ldots, p_k$ is accomplished by means of maximum likelihood. If the likelihood function of data obtained from the multinomial model is

$$L(p_1, \ldots, p_k; x_1, \ldots, x_k) = \frac{n!}{x_1! \ldots x_k!} p_1^{x_1} \cdots p_k^{x_k}$$

then

$$\ln L = C + \sum_{i=1}^{k} x_i \ln p_i$$

Now

$$\frac{\partial \ln L}{\partial p_i} = \frac{x_i}{p_i} - \frac{x_k}{p_k}$$

where $i = 1, 2, \ldots, k - 1$, because $p_k = 1 - p_1 - \cdots - p_{k-1}$. If one uses the constraints $\sum_{i=1}^{k} p_i = 1$ and $\sum_{i=1}^{k} x_i = n$, it then follows that $\hat{p}_i = x_i/n$ is the maximum likelihood estimator of p_i, $i = 1, 2, \ldots, k$.

A joint confidence interval, with confidence at least $(1 - \alpha)$, is obtained for all p_i's simultaneously, $i = 1, 2, \ldots, k$, by solving the quadratic equations $n(\hat{p}_i - p_i)^2 = z_{1-\alpha/k}^2 p_i q_i$ for p_i in terms of $\hat{p}_i$, n and $z_{1-\alpha/k}^2$, $i = 1, 2, \ldots, k$, where $z_{1-\alpha/k}$ is the upper α/k percentile of the standard normal distribution as long as $k > 2$. The two solutions $\hat{p}_{i,L}$ and $\hat{p}_{i,U}$ (say) are the endpoints of the interval for p_i, $i = 1, 2, \ldots, k$. Discussion of this procedure appears in Goodman (1965) or Johnson and Kotz (1969) and derived from a procedure discussed by Quesenberry and Hurst (1964).

Example 4.6.1 In an election survey 250 voters are asked their preference between two candidates, R and D. R is preferred by

135 voters, D is preferred by 90 voters, and 25 voters are
undecided. Find a joint 95% confidence region for the true proportions of voters in each category. Noting that $\hat{p}_1 = 0.54$,
$\hat{p}_2 = 0.36$, and $\hat{p}_3 = 0.10$ for the respective groups, and that
$n = 250$ and $z^2_{0.983} = 5.81$ ($\alpha = 0.05$ and $k = 3$), the quadratic
equations using Goodman's procedure are, respectively, $255.81p_1^2 -$
$275.81p_1 + 72.9 = 0$, $255.81p_2^2 - 185.81p_2 + 32.4 = 0$, and $255.81p_3^2 -$
$55.81p_3 + 2.5 = 0$, yielding as simultaneous confidence intervals
for p_1, p_2, and p_3, the respective intervals (0.46,0.61), (0.29,
0.44), and (0.06,0.16). Note that no candidate can be assured of
50% of the vote on the basis of this analysis; however, candidate
R is certainly the stronger candidate at this point.

There are other multivariate discrete models that are ana-
logues of the univariate models discussed in this chapter. However, these models do not appear often enough in applications to
warrant a full discussion here. The interested reader should see
Johnson and Kotz (1969), especially Chapter 11, for a further
discussion of some of these models. Finally, it should be noted
that the multinomial model is used extensively in the analyses of
categorical data, i.e., data that can be arranged into contingency
tables. Since this topic is covered very extensively by Bishop
et al. (1975), models for contingency tables are not discussed
here.

Problems

1. Verify, by use of repeated summations, equation (4.39).

2. Show that the vector of means for the multinomial model is
    $(np_1, np_2, \ldots, np_k)$ and that the variance-covariance matrix
    is given by (4.40).

3. Show that if $p_i \to 0$, $i = 1, 2, \ldots, k - 1$, and $n \to \infty$ such that
    $np_i = m_i$ is constant, $i = 1, 2, \ldots, k - 1$, the limiting prob-
    ability model for $(X_1, \ldots, X_{k-1})$ is given by (4.41).

4. Show, explicitly, that for the multinomial model $\hat{p}_i = x_i/n$,
    $i = 1, 2, \ldots, k$.

5. An inspector has 300 items to inspect. He finds 160 good, 80
    poor, and the remaining 60 items have not been inspected. Find
    simultaneous 95% confidence intervals for p_1, p_2, and p_3, the
    respective probabilities of good, poor, and not-inspected items.

6. Two persons toss a total of 25 coins each, recording the num-
    bers of heads and tails that are obtained. Write out the multi-
    nomial probability model that describes the numbers of heads
    and tails each person records (a) if all coins are unbiased;
    (b) if each coin the first person tosses has probability p_1
    of heads and if each coin the second person tosses has prob-
    ability p_2 of heads.

7. In an election three candidates are running, R the Republican,
    D the Democrat, and I the Independent. A recent poll shows
    $p_R = 0.3$, $p_D = 0.5$, and $p_I = 0.2$. If three people are chosen
    at random from all eligible voters, find the probability each
    will vote for a different candidate.

8. Suppose X is a continuous random variable whose pdf is $f(x) =$
    $2x$, $0 \leq x \leq 1$. Three observations are drawn from this distribu-
    tion at random. Find the probability that exactly one observa-
    tion is between 0 and 1/3, one observation between 1/3 and 2/3,
    and one observation is between 2/3 and 1.

9. Redo Problem 8 if $f(x) = 1$, $0 \leq x \leq 1$.

10. Given that an observation is drawn at random from either of the
    two distributions in the previous two problems, suppose that
    its value exceeds 2/3. What is the probability the observation
    was drawn from the distribution described in Problem 8?

4.7 COMPOUND MODELS

Often in a statistical experiment, the unknown parameter itself may
be considered as random and thus its behavior may be governed by a
probability model. For example, in the two-candidate problem
(Example 4.2.3) the probability p that the candidate R wins the
election may and in fact does vary at different times. As news

favorable (or unfavorable) to candidate R becomes available, the
value of p changes. If p itself can be considered to change
randomly, then p is a random variable whose probability density
function is g(p) (say). That is, $g(p) \geq 0$ for $0 \leq p \leq 1$ and
$\int_0^1 g(p) \, dp = 1$. [In the terminology of Bayesian statistics, $g(p)$
is a *prior density function* for p.]

In formulating the problem, suppose that $f(x;n,p)$ is the
binomial model for a fixed p in the interval $0 \leq p \leq 1$. That is,

$$f(x;n,p) = \binom{n}{x} p^x q^{n-x} \tag{4.42}$$

where $x = 0, 1, \ldots, n$, $q = 1 - p$. For example, in the voting
problem (Example 4.2.3) $f(x;n,p)$ is the probability that exactly
x out the n voters surveyed would vote for candidate R at that
specific time when the probability is p a given voter would vote
for R. It is assumed that P is a random variable whose density
function is $g(p)$, $0 \leq p \leq 1$, then the joint distribution of the
random vector (X,P) is

$$h(x,p;n) = \binom{n}{x} p^x q^{n-x} g(p) \tag{4.43}$$

where $x = 0, 1, \ldots, n$, $0 \leq p \leq 1$, $q = 1 - p$. Hence, the marginal
distribution of X is

$$f_1(x;n) = \int_0^1 \binom{n}{x} p^x q^{n-x} g(p) \, dp \tag{4.44}$$

where $x = 0, 1, \ldots, n$. The model described by (4.44) is called
a *compound model* for X, since it is "compounded" of two models,
the binomial model and the density function or model for p. The
compound model is not binomial, as the following example shows.

Example 4.7.1 Suppose in the voting problem $g(p) = 1$, $0 \leq p \leq 1$,
i.e., P is uniformly distributed on the unit interval. Thus,

$$f_1(x;n) = \binom{n}{x} \int_0^1 p^x (1 - p)^{n-x} \, dp$$

It is shown in Section 3.5 that if α and β are any positive real
numbers,

$$\int_0^1 p^{\alpha-1}(1 - p)^{\beta-1} \, dp = \frac{\Gamma(\alpha)\Gamma(\beta)}{\Gamma(\alpha + \beta)}$$

If one notes that x and n - x are integers, then

$$\int_0^1 p^x(1 - p)^{n-x} \, dp = \frac{x!\,(n - x)!}{(n + 1)!} = \frac{1}{(n + 1)} \frac{1}{\binom{n}{x}}$$

Hence, it follows that

$$f_1(x;n) = \frac{1}{n + 1} \tag{4.45}$$

where x = 0, 1, ..., n, which is the uniform model for the discrete
variable X. Thus, if the binomial model f(x;n,p) is compounded
with the continuous uniform model g(p) = 1, 0 < p < 1, the resulting
model for X is the discrete uniform model (4.45).

In conclusion, if the probability that an individual votes
for R is uniformly distributed between the values 0 and 1, the
number of votes (out of n) that candidate R receives is also
uniformly distributed at the values x = 0, 1, ..., n.

More generally, P will not be uniformly distributed since the
uniform assumption implies a value of P near zero is as likely as a
value near 1/2 or a value near unity. Often an experimenter or
survey inquirer has some idea as to the most likely value or values
P assumes. When this is the case, then a distribution or model for
P that has wide applicability is the *beta model*. This model is
defined for 0 < p < 1 and is written

$$b(p;\alpha,\beta) = \frac{\Gamma(\alpha + \beta)}{\Gamma(\alpha)\Gamma(\beta)} p^{\alpha-1}(1 - p)^{\beta-1} \tag{4.46}$$

0 < p < 1, α > 1, β > 1. The parameters α and β determine the
shape of the beta model. If b(p;α,β) replaces g(p) in (4.44), it
then follows that

$$f_1(x;\alpha,\beta,n) = \binom{n}{x}\frac{\Gamma(\alpha + \beta)}{\Gamma(\alpha)\Gamma(\beta)} \int_0^1 p^{x+\alpha-1}(1 - p)^{n-x+\beta-1} \, dp$$

Note that

$$\int_0^1 p^{x+\alpha-1}(1 - p)^{n-x+\beta-1} \, dp = \frac{\Gamma(\alpha + x)\Gamma[\beta + (n - x)]}{\Gamma(\alpha + \beta + n)}$$

This yields the marginal distribution of X as

$$f_1(x;\alpha,\beta,n) = \binom{n}{x}\frac{\Gamma(\alpha + \beta)}{\Gamma(\alpha)\Gamma(\beta)} \frac{\Gamma(\alpha + x)\Gamma[\beta + (n - x)]}{\Gamma(\alpha + \beta + n)} \qquad (4.47)$$

This model is termed the *binomial-beta* model. A discussion of the relationship between and among the factorials and gamma functions for this model is found in Raiffa and Schlaifer (1961). The first two moments are not difficult to obtain. In fact, the following are well-known results which are left as exercises to the reader:

$$E(X) = E_p E_x(X|P) \qquad (4.48)$$

and

$$\text{var } X = \text{var}_p E_x(X|P) + E_p \text{var}_x(X|P) \qquad (4.49)$$

The mean and variance for the beta model are $\alpha/(\alpha + \beta)$ and $\alpha\beta/[(\alpha + \beta)^2(\alpha + \beta + 1)]$, respectively. Hence,

$$E(X) = \frac{n\alpha}{\alpha + \beta} \qquad (4.48a)$$

and

$$\text{var } X = \frac{n\alpha\beta(n + \alpha + \beta)}{(\alpha + \beta)^2(\alpha + \beta + 1)} \qquad (4.49a)$$

These verifications are left as an exercise to the reader.

Direct estimation of α and β is difficult from a computational standpoint. However, a Bayesian method was developed by Fox (1966). The mode of the beta model (4.46) is $(\alpha - 1)/(\alpha + \beta - 2)$. Suppose that $\hat{p}$ is a subjective estimate of p. (For example, in the voting problem, Example 4.2.3, the subjective estimate of p may be 0.5.)

Furthermore, it is believed that the odds that the true value of
p lies in the interval $(\hat{p} - k\hat{p}, \hat{p} + k\hat{p})$ where $0 < k < 1$ is some
prechosen constant, e.g., $k = 0.05$, are a to b. That is, the
subjective probability the true value of p lies below $\hat{p} - k\hat{p}$ or
above $\hat{p} + k\hat{p}$ is $1 - a/(a + b)$ or $b/(a + b)$. It then follows that
estimates of α and β, $\hat{\alpha}$ and $\hat{\beta}$ (say), are found by solving, jointly,
the equations

$$\frac{\hat{\alpha} - 1}{\hat{\alpha} + \hat{\beta} - 2} = \hat{p} \tag{4.50}$$

and

$$\int_{\hat{p} - k\hat{p}}^{\hat{p} + k\hat{p}} b(p;\hat{\alpha},\hat{\beta}) \, dp = \frac{a}{a + b} \tag{4.51}$$

where $b(p;\hat{\alpha},\hat{\beta})$ is given by (4.46). Solution of these equations
cannot be obtained in closed form. Fox (1966) does give a table
for finding $\hat{\alpha}$ and $\hat{\beta}$ when $\hat{p}$, k, and $a/(a + b)$ are specified. Gross
(1971) modifies Fox's procedure to some degree in order to take
advantage of existing incomplete beta function tables such as the
tables by Harter (1964) and Pearson and Hartley (1966). Gross uses
the mean of the beta model as his subjective estimate of p. Then,
assuming $0 < k < 1$ is a prechosen value, he assesses the subjective
probability that p lies in the interval $(0,k\hat{p})$ where $\hat{p} = \hat{\alpha}/(\hat{\alpha} + \hat{\beta})$
is the subjective estimate of p. From these considerations, the
following equations are solved in terms of $\hat{\alpha}$ and $\hat{\beta}$:

$$\frac{\hat{\alpha}}{\hat{\alpha} + \hat{\beta}} = \hat{p} \tag{4.52}$$

and

$$\int_0^{k\hat{p}} b(p;\hat{\alpha},\hat{\beta}) \, dp = \frac{b}{a + b} \tag{4.53}$$

where $b(\hat{p};\hat{\alpha},\hat{\beta})$ is given by (4.46) and $b/(a + b)$ is the subjective
probability that p lies in the interval $(0,k\hat{p})$.

Example 4.7.2 Suppose in the voting problem g(p) is given by (4.46);
i.e., the density function of p is the beta model with parameters
α and β. Prior to sampling the voters, a group of opinion research
people subjectively assess candidate R's chances of winning to be
0.6. Furthermore, they agree that the probability is 0.90 that his
chances of winning are at least 0.4. Therefore, $k\hat{p} = 0.4$, and
hence k = 2/3. Applying equations (4.52) and (4.53) with the aid
of tables of the incomplete beta function [cf. Pearson and Hartley
(1966)], $\hat{α} = 6.0$ and $\hat{β} = 4.0$. If now 10 voters are sampled at
random, the marginal probability model $f_1(x)$, given by (4.47), is

$$f_1(x) = \binom{10}{x} \frac{9!}{5!3!} \frac{(5 + x)!(13 - x)!}{19!}$$

where x = 0, 1, ..., 10. Table 4.1 is the tabular form of $f_1(x)$
for the values x = 0, 1, ..., 10.

Table 4.1 Numerical Values of $f_1(x)$

x	$f_1(x)$
0	0.00310
1	0.01429
2	0.03751
3	0.07274
4	0.11457
5	0.15276
6	0.17504
7	0.17147
8	0.13932
9	0.08669
10	0.03251

A second compound model involving the binomial model is formulated as follows. Suppose a disease attacks members of household
according to the Poisson model

$$P(x;m) = \frac{e^{-m} m^x}{x!}$$

where x = 0, 1, 2, ... Furthermore, if any person is affected
with the disease, the probability he/she recovers is p, and hence
the probability of nonrecovery is q = 1 - p. It is assumed that
if x persons in a household are affected, their recovery probabilities are mutually independent and are equal. Hence, the conditional probability distribution that exactly Y individuals out
of the x ill persons in the family recover is

$$f(y;p|x) = \binom{x}{y} p^y q^{x-y}$$

where y = 0, 1, 2, ..., x. The problem now is to find the marginal
distribution of Y, the number of persons who are cured of a disease
within any family, irrespective of the number of family members
who are affected with the disease. The joint distribution of X
and Y is simply $f(y;p|x)P(x;m)$. Hence, the joint distribution is
written

$$g(x,y;p,m) = \frac{e^{-m} m^x}{x!} \binom{x}{y} p^y q^{x-y} \tag{4.54}$$

where y = 0, 1, ..., x; x = 0, 1, ... A rearrangement of terms
show that (4.54) can be written

$$g(x,y;p,m) = \frac{e^{-m}(mp)^y}{y!} \frac{(mq)^{x-y}}{(x-y)!}$$

The domain of X for fixed y is easily seen to be x = y, y + 1, ...
Hence, the marginal model for Y is

$$g_1(y;p,m) = \frac{e^{-m}(mp)^y}{y!} \sum_{x-y=0}^{\infty} \frac{(mq)^{x-y}}{(x-y)!}$$

$$= \frac{e^{-mp}(mp)^y}{y!} \tag{4.55}$$

where y = 0, 1, 2, ... Equation (4.55) shows that the probability
model for the number of persons within a family that is cured of

the disease, Y, is Poisson with parameter (or mean) mp.

Another compound probability model can be generated by considering X as the number of bacterial colonies on a slide. (See
Section 4.3.) At a given or fixed time, it was seen that under
fairly general conditions, X obeys the Poisson model (4.13). However, it is possible that m, the mean number of colonies per slide,
itself varies over time and among slides. In fact, suppose that
M is a random variable that obeys the gamma model

$$g(m;\lambda,r) = \frac{\lambda^r m^{r-1} e^{-\lambda m}}{(r-1)!} \qquad (4.56)$$

where $m \geq 0$; $\lambda \geq 0$, $r \geq 1$, where r is a positive integer. In this
case, one may think of M, the mean number of bacterial colonies,
as having a positively skewed distribution since $M \geq 0$ and very
large values of M are not as likely as small to moderate values.

The joint density function of X and M is then

$$h(x,m;\lambda,r) = \frac{\lambda^r}{x!(r-1)!} m^{r+x-1} e^{-(\lambda+1)m}$$

Thus the marginal model for x is

$$h_1(x;\lambda,r) = \frac{\lambda^r}{x!(r-1)!} \int_0^\infty m^{r+x-1} e^{-(\lambda+1)m} \, dm$$

Use of the properties of the gamma function and the fact that r is
an integer shows that

$$\int_0^\infty m^{r+x-1} e^{-(\lambda+1)m} \, dm = \frac{(x+r-1)!}{(\lambda+1)^{r+x}}$$

Hence,

$$h_1(x;\pi,r) = \binom{x+r-1}{x} \pi^r (1-\pi)^x \qquad (4.57)$$

where $x = 0, 1, 2, \ldots,$ and $\pi = \lambda/(\lambda+1)$. If one compares (4.57)
with (4.37), it is not difficult to see they are the same model.

Thus, $h_1(x;\pi,r)$ is the *negative binomial* model. That is, if the
Poisson model is compounded with the gamma model, the resulting
model is the negative binomial model.

The compound Poisson process leads to another class of compound models. An example of such a process is now discussed. It
was shown in Section 4.3 that if telephone calls arrive randomly
and uniformly throughout a finite time period $(0,t)$, then the
number of calls that arrive obeys the Poisson model with mean λt,
where λ is the number of calls per unit time. Suppose Y is the
length of a call which is assumed to be independent of X, the
number of calls made during $(0,t)$. If Y has continuous distribution $F(y)$ with mean $E(Y)$ and finite moment-generating function
$m_\theta(Y)$, the problem is to find the model that $\sum_{i=1}^{X} Y_i$ obeys. That
is, what is the model for the total length of calls made during
$(0,t)$? To answer this question, note that

$$m_\theta(\sum_{i=1}^{X} Y_i | X = x) = \{m_\theta(Y)\}^X$$

If $U = \sum_{i=1}^{X} Y_i$, it then follows in a manner similar to (4.48) that

$$m_\theta(U) = E_x[m_\theta(U) | X = x] = E_x\{m_\theta(Y)\}^X$$

$$= e^{-\lambda t} \sum_{x=0}^{\infty} \frac{\{m_\theta(Y)\lambda t\}^X}{x!}$$

$$= \exp\{-\lambda t + \lambda t m_\theta(Y)\} \qquad (4.58)$$

Thus, (4.58) is the moment-generating function of the model that
describes the total length of all calls made during $(0,t)$. The
mean length of all calls made and its variance which are easily
obtained from (4.58) are, respectively,

$$E(U) = \lambda t E(Y) \qquad (4.59)$$

and

$$\text{var } U = \lambda t E(Y^2) \qquad (4.60)$$

Example 4.7.3 Suppose at a switchboard calls arrive at a mean rate
of three per minute and that the mean and variance of the length of
a call are 2.2 and 0.8 minutes, respectively. What is the probability that the total length of calls will not exceed 225 minutes
during a given 30-minute period of call arrivals? If U is assumed to
to be approximately normally distributed, then since E(U) = 198 min.
and $\sqrt{\text{var } U}$ = 22.53 min., it follows that pr$\{U \leq 225\}$ = pr$\{Z \leq 1.2\}$ =
0.885, where Z obeys the standard normal model.

Problems

1. If one uses the fact $\int_0^1 p^{\alpha-1}(1-p)^{\beta-1} \, dp = \Gamma(\alpha)\Gamma(\beta)/\Gamma(\alpha+\beta)$,
    show that for x and n − x integers $\int_0^1 p^x(1-p)^{n-x} \, dp =$
    $(n+1)^{-1} \binom{n}{x}^{-1}$.

2. Prove the results shown in equations (4.48) and (4.49).

3. Show that the mean and variance of the beta model are $\alpha/(\alpha+\beta)$
    and $\alpha\beta/[(\alpha+\beta)^2(\alpha+\beta+1)]$, respectively. Hence, or other-
    wise, verify equations (4.48a) and (4.49a) for the binomial-beta
    model.

4. Verify the probabilities for the binomial-beta model in Table
    4.1.

5. Suppose in a given neighborhood influenza attacks individuals
    according to the Poisson model with mean 20. Given an indivi-
    dual in the neighborhood has the disease the probability of
    recovery is 0.995, if there are 300 people in the neighborhood
    at risk, what is the probability of at least one death due to
    influenza?

6. In the previous problem if there are 500 people in the neighbor-
    hood at risk, does the probability of at least one death due to
    influenza change? Explain.

7. Calculate the mean and variance of X for the beta-binomial
    model (i) directly; and (ii) by applying (4.48) and (4.49).

8. Under what condition does the compound Poisson model become a
    simple Poisson model?

9. Suppose calls arrive at a switchboard according to the Poisson
    model at a mean rate of six per minute. It is also assumed
    that the length of calls is distributed according to the
    exponential model with mean length of 1.5 min. What is the
    probability the total length of calls will not exceed 300 min.
    during the next ten hour period?

10. The number of female insects in a given region follows a Poisson
    distribution with mean m_1. The number of eggs laid by each
    insect follows a Poisson distribution with mean m_2. Find the
    moment-generating function of the probability model of the
    number of eggs in the region and hence find its mean and vari-
    ance. (Diploma in Statistics, Oxford University, England,
    1951.)

4.8   POWER SERIES MODELS

Some of the models that are discussed in this chapter belong to a
more general class of discrete models. This class is called the
power series models.

The general formulation of a power series model is the following. Let X be the number of events that occur (at a specific time);
e.g., number of phone calls, number of bacterial colonies observed
on a slide, or number of defects in an inspection of a passenger
airliner. Then if

$$\text{pr}\{X = x\} = \alpha \, a_x \lambda^x \qquad\qquad (4.61)$$

where $\alpha = (\Sigma_{x=0}^{\infty} a_x \lambda^x)^{-1}$; $x = 0, 1, \ldots$, the random variable X is
said to obey the power series model. Clearly, the series $\Sigma_{x=0}^{\infty} a_x \lambda^x$
must be convergent and the coefficients a_x (which are independent
of λ) are nonnegative for all x.

To demonstrate that some of the earlier models discussed in
this chapter are contained in the power series class, consider the
following examples:

Example 4.8.1 If $a_x = (x!)^{-1}$, then

$$pr\{X = x\} = \frac{\alpha \lambda^x}{x!}$$

where $x = 0, 1, \ldots$. Now $\alpha = \left(\Sigma_{x=0}^{\infty} \lambda^x/x! \right)^{-1}$. However, this series converges to $e^{-\lambda}$. Thus,

$$pr\{X = x\} = \frac{e^{-\lambda} \lambda^x}{x!}$$

where $x = 0, 1, \ldots$, which is the Poisson model [see (4.13)].

Example 4.8.2 Suppose $a_0 = 0$ and $a_x = 1$ for all integers $x > 0$. Then under the assumption that $0 < \lambda < 1$, (4.61) is the geometric model.

Example 4.8.3 Let $a_x = \binom{x - 1}{n - 1}$ for $x > n$, and $a_x = 0$ for $x < n$, where x is a nonnegative integer. If $0 < \lambda < 1$, (4.61) is the negative binomial model.

Example 4.8.4 Suppose $a_x = \binom{n}{x}$, $x = 0, 1, \ldots, n$, and $a_x = 0$ for all integers $x > n$. Then if $\lambda = p/q$, where $0 < p < 1$, (4.61) is the binomial model (4.1).

Some general properties of the power series model are now put forth. First of all, define $g(\lambda) \equiv \alpha^{-1} \equiv \Sigma_{x=0}^{\infty} a_x \lambda^x$. If $g(\lambda)$ is used in obtaining the moment-generating function of X, then

$$m_\theta(X) = \frac{\Sigma_{x=0}^{\infty} e^{\theta x} a_x \lambda^x}{\Sigma_{x=0}^{\infty} a_x \lambda^x}$$

$$= \frac{\Sigma_{x=0}^{\infty} a_x (\lambda e^\theta)^x}{\Sigma_{x=0}^{\infty} a_x \lambda^x} = \frac{g(\lambda e^\theta)}{g(\lambda)}$$

$$(4.62)$$

It then follows that

$$E(X) = \frac{\Sigma_{x=0}^{\infty} \, xa_x \lambda^x}{\Sigma_{x=0}^{\infty} \, a_x \lambda^x} \tag{4.63}$$

and

$$\text{var } X = \frac{\left(\Sigma_{x=0}^{\infty} \, x^2 a_x \lambda^x\right)\left(\Sigma_{x=0}^{\infty} \, a_x \lambda^x\right) - \left(\Sigma_{x=0}^{\infty} \, xa_x \lambda^x\right)^2}{\left(\Sigma_{x=0}^{\infty} \, a_x \lambda^x\right)^2} \tag{4.64}$$

The mode $\hat{x}$ of the power series model is that value of x such that

$$\frac{a\hat{x} - 1}{a\hat{x}} < \lambda < \frac{a\hat{x}}{a\hat{x} + 1}$$

when conditions permit this inequality.

In many characterizations of the power series model it is possible to obtain the maximum likelihood estimator of λ.   The likelihood function $L(x_1, \ldots, x_k; \lambda)$ given a random sample $(x_1, \ldots, x_k)$ is

$$L(x_1, \ldots, x_k; \lambda) = \sum_{i=1}^{k} \frac{a_{x_i} \lambda^{x_i}}{g(\lambda)}$$

$$\ln L = \sum_{i=1}^{k} \ln a_{x_i} + \sum_{i=1}^{k} x_i \ln \lambda - k \ln g(\lambda)$$

Thus,

$$\frac{d \ln L}{d\lambda} = \sum_{i=1}^{k} \frac{x_i}{\lambda} - k \frac{g'(\lambda)}{g(\lambda)}$$

The value $\hat{\lambda}$ is that value of $\hat{\lambda}$ that is the solution to the equations

$$\frac{\bar{x}}{\hat{\lambda}} - \frac{g'(\hat{\lambda})}{g(\hat{\lambda})} = 0$$

and that maximizes L with respect to λ.

Suppose $a_0 = 0$, and $a_x = x^{-1}$, where $x > 0$ is an integer.   In
this case (4.61) becomes the logarithmic series model provided
$0 < \lambda < 1$.   The reason for this name is that the constant of proportionality $g(\lambda) = [\ln(1 - \lambda)]$; hence the individual probabilities
are proportional to the terms in the logarithmic series.   The
moment-generating function, mean, and variance of the logarithmic
series model are not difficult to obtain, either directly or by
substituting into (4.62), (4.63), and (4.64), respectively.   The
derivation is left as an exercise to the reader.   The formulas are,
respectively

$$m_\theta(X) = \frac{\ln(1 - \lambda e^\theta)}{\ln(1 - \lambda)}$$

where θ is chosen such that $\lambda e^\theta < 1$,

$$E(X) = \frac{\lambda}{g(\lambda)(1 - \lambda)}$$

and

$$\text{var } X = \frac{\lambda}{(1 - \lambda)^2 g(\lambda)}\left[1 - \frac{\lambda}{g(\lambda)}\right]$$

Fisher et al. (1943) point out an application of the logarithmic series model in the collection of species of butterflies.   They
assume that the distribution of the j-th species has a Poisson
model with parameter m_j.   In turn, the distribution of m_j is
assumed to be gamma with scale and shape parameters λ and r,
respectively.   If one uses the results in the previous section
(4.57), the unconditional distribution of the number of butterflies
of each species is represented by the negative binomial where
$\lambda = q$ and $(1 - \lambda) = p$.   However, $x = 0$ would be truncated since
zero butterflies represent no species.   Thus, if $h(x;\pi,r)$ is the
truncated model, then

$$h(x;\pi,r) = \frac{1}{(1 - \pi^r)}\binom{x + r - 1}{x}\pi^r(1 - \pi)^x \tag{4.65}$$

where $x = 1, 2, \ldots$. It is not difficult to show, applying l'
Hôspital's rule, that the limiting distribution of $h(x;\pi,r)$, as
$r \to 0$ is the logarithm series model; i.e., if $f(x;\pi) =$
$\lim_{r \to 0} h(x;\pi,r)$, then

$$f(x;\pi) = \frac{-(\ln \pi)^{-1}(1 - \pi)^x}{x} \tag{4.66}$$

where $x = 1, 2, \ldots$. As Johnson and Kotz (1969) point out, letting
r approach zero is increasing the variability in the parameters m_j.
Hence, (4.66) implies that there is considerable variability among
species.

Problems

In Problems 1-4 describe the distribution (name it if possible),
obtain the moment-generating function, the mean and variance for
the values of a_x given, where

$$\text{pr}\{X = x\} = \alpha a_x \lambda^x \qquad \alpha = \left(\sum_{x=0}^{\infty} a_x \lambda^x \right)^{-1}$$

is the power series model.

1. $a_0 = 0$, $a_x = a > 0$ for integers $x > 0$, and $0 < \lambda < 1$.
2. $a_x = a > 0$ for all integers $x \geq 0$, and $0 < \lambda < 1$.
3. $a_0 = 0$, $a_x = \{x(x + 1)\}^{-1}$ for integers $x > 0$.
4. $a_x = x!/(x + r)!$ for all integers $x \geq 0$, and $r \geq 1$ a fixed
   integer.
5. Suppose in the power series model $\lambda = 1$, $a_0 = 0$, $a_x = x^{-2}$ for
   all integer values $x > 1$. Can an appropriate power series model
   be obtained? If so, can one find its mean variance and
   moment-generating function?
6. Obtain formulas for the moment-generating function, mean, and
   variance of the general power series model; i.e., derive (4.62),
   (4.63), and (4.64), respectively.
7. Obtain the formula for $\hat{\lambda}$, the maximum likelihood estimator of
   λ, the parameter of the general power series model.

8. Obtain the moment-generating function, the mean, and variance of the logarithmic series model.

9. Show that if $a_x = \binom{n}{x}$, $x = 0, 1, \ldots, n$, $a_x = 0$, $x > n$, and if $\lambda = p/q$, $0 < p < 1$, $q = 1 - p$, the power series model coincides with the binomial model.

10. If $h(x;\pi,r)$ is given by (4.65) show that $\lim_{r \to 0} h(x;\pi,r) = f(x;\pi)$ which is given by (4.66).

REFERENCES

Bishop, Y. M. M., Feinberg, S. E., and Holland, P. W. (1975). *Discrete Multivariate Analysis: Theory and Practice*, MIT Press, Canbridge, Mass.

Dahiya, R. C., and Gross, A. J. (1973). Estimating the zero class from a truncated Poisson sample, *J. Amer. Stat. Assoc. 68*, No. 343, pp. 731-733.

Dixon, W. J., and Massey, F. J. (1969). *Introduction to Statistical Analysis* (3rd ed.), McGraw-Hill Book Company, New York.

Feller, W. (1957). *An Introduction to Probability Theory and Its Applications*, Vol. I (2nd ed.), John Wiley and Sons, Inc., New York.

Fisher, R. A., Corbet, A. S., and Williams, C. B. (1943). The relationship between the number of species and the number of individuals in a random sample of an animal population, *J. Animal Ecology 12*, pp. 42-57.

Fox, B. L. (1966). *A Bayesian Approach to Reliability Assessment,* RM-5084-NASA, RAND Corporation, Santa Monica, Calif.

Goodman, L. A. (1965). On simultaneous confidence intervals for multinomial proportions, *Technometrics 7*, No. 2, pp. 247-254.

Gross, A. J. (1971). The application of exponential smoothing to reliability assessment, *Technometrics 13*, No. 4, pp. 877-883.

Gross, A. J., and Hosmer, D. W., Jr. (1978). Approximating tail areas of probability distributions, *Ann. Stat. 6*, pp. 1352-1359.

Hahn, G., and Shapiro, S. S. (1967). *Statistical Models in Engineering*, John Wiley & Sons, Inc., New York.

Haight, F. (1967). *Handbook of the Poisson Distribution*, John Wiley and Sons, Inc., New York.

Harter, H. L. (1964). *New Tables of the Incomplete Gamma-Function Ratio and Percentage Points of the Chi-square and Beta Distribution*, U.S. Government Printing Office, Washington, D.C.

Hogg, R. V., and Craig, A. T. (1970). *Introduction to Mathematical Statistics*, MacMillan Publishing Company, Inc., New York.

Johnson, N. L., and Kotz, A. (1969). *Distributions in Statistics: Discrete Distributions*, Houghton Mifflin Company, Boston.

Lieberman, G. J., and Owen, D. B. (1961). *Tables of the Hypergeometric Probability Distribution*, Stanford University Press, Stanford, Calif.

McKendrick, A. G. (1926). Application of mathematics to medical problems, *Proc. Edinburgh Math. Soc. 44*, pp.98-103.

Molina, E. C. (1942). *Poisson's Exponential Binomial Limit*, D. Van Nostrand Company, Inc., Princeton, N.J.

Patil, G. P. (1961). Asymptotic bias and variance of ratio estimates in generalized power series distributions and certain applications, *Sankhyā, Ser. 23*, pp. 269-280.

Patil, G. P. (1962). Certain properties of the generalized power series distribution, *Ann. Inst. Stat. Math. 14*, pp. 179-182.

Patil, G. P. (1963). On the multivariate generalized power series distribution and its application to the multinomial and negative multinomial, *Proc. Int. Symp. Discrete Distribution*, Montreal, pp. 183-194.

Patil, G. P. (1964a). Estimation for the generalized power series distribution with two parameters and its application to the binomial distribution, *Contributions to Statistics on the 70th Birthday of P.C. Mahalanobis*, Pergamon Press, Oxford, and Statistical Publishing Society, Calcutta, pp. 335-344.

Patil, G. P. (1964b). On certain compound Poisson and Binomial distributions, *Sankhyā, Ser. A 26*, pp. 293-294.

Pearson, E. S., and Harley, H. O. (1966). *Biometrika Tables for Statisticians, Vol. 1* (3rd ed.), Cambridge University Press, Cambridge.

Quesenberry, C. P., and Hurst, D. C. (1964). Large sample simultaneous confidence intervals for multinomial proportions, *Technometrics 6*, No. 2, pp. 191-195.

Raiffa, H., and Schlaifer, R. (1961). *Applied Statistical Decision Theory*, Harvard University Press, Cambridge, Mass.

Teicher, H. (1955). An inequality on Poisson probabilities, *Ann. Math. Stat. 26*, No. 1, pp. 147-149.

5.1 INTRODUCTION

In previous chapters the concern has been describing the basic
characteristics of statistical models so that an analyst could use
his insight of the mechanism under study in order to choose a
specific model. However, in some instances, due to the complexity
of the situation, lack of understanding of the model, or simply
the desire to find a "best" approximation to a set of data, a
statistical model is selected for its ability to fit a given set
of data rather than from its relationship to the mechanism under
study. The term "empirical models" has been used to describe
families of distributions which can serve this purpose. In most
cases such empirical models give no insight into the mechanism
behind the data; they simply serve as a means of summarizing the
output and can be used for interpolation and extrapolation, i.e.,
to obtain probability values corresponding to points not contained
in the original set of data.

Example 5.1.1 The Johnson S_B distribution (to be described subsequently) was used to approximate the data obtained from a system
under test. The resulting density function

$$f(x) = \frac{3}{\sqrt{2\pi}} \; \frac{1}{x(1-x)} \; \exp\left\{-\frac{1}{2}\left[0.5 + 3 \; \ln\left(\frac{x}{1-x}\right)\right]^2\right\}$$

$$0 \leq x \leq 1$$

can be used to summarize the output data. Thus, the entire data
set can be replaced by the above expression. The density function
can be used to find, for example, the probability that x < 1/2 by
finding F(1/2), i.e., $\int_0^{1/2}$ f(x) dx.

Probabilities for values outside the range of the set of data
can be obtained; however, great care must be exercised in using
such extrapolation results since the model is selected based on
its ability to fit the results in hand and not from a basic understanding of the mechanism giving rise to the output. In such cases
extrapolation can yield results that may be misleading.

For example, an empirical distribution that has been fitted to
lifetime data over a time interval (0,1000) hr. may give poor
predictions of lifetimes of 2000 hr. or more, if, for example,
the mode of failure changed at 1500 hr. It may happen that after
1500 hr. of operation the system, due to the length of operation,
may show a higher likelihood of failure than earlier. Empirical
models are used in conjunction with tolerance or propagation
studies (see Chapter 7). In such analyses the moments of the output distribution are generated from a knowledge of the mathematical
relationship between the output and the design variables and a
knowledge of the tolerance of the design variables. The results
of such a study provide estimates of the first four moments of the
output distribution. An empirical distribution can then be fitted
to these estimates and be used to model the system output. This
can be done without having any data on system output; in fact, it
is often done prior to the building of the system and is used to
evaluate the proposed specifications. Empirical distributions
are also used in a similar manner to summarize the data generated
via a Monte Carlo analysis (see Chapter 7).

An empirical distribution is in reality a family of distributions which can assume a wide variety of shapes that are determined by the values of its parameters. All the families discussed
in this chapter will have four parameters that can be adjusted to

match the first four moments of the data or be adjusted to match the
data at two to four selected percentiles. The following are the
most important criteria in the selection of a specific family:

1. The family includes shapes that vary as widely as possible.
2. It is easy to select a member of the family.
3. It is easy to approximate the data set.
4. It is easy to determine desired percentiles.

    If the model is to be used in a Monte Carlo study, ease of
generating random deviates is an additional criterion. Three
families are described in this chapter.

5.2 SYSTEMS FOR CLASSIFICATION OF DISTRIBUTIONS

A distribution is uniquely defined by its moments, i.e., $E(X^k)$. If
one knows all the moments of a distribution, it is equivalent to
knowing the distribution itself. The same is true also for the
moments about the mean $E[(X - \mu)^k]$ since there is a direct relationship to the moments about zero, $E(X^k)$, e.g., $E(X - \mu)^2 = E(X^2) - \mu^2$.
Therefore, a procedure for classifying distributions can be based
on the moments about the mean. In most studies sufficient accuracy
in approximating a distribution can be achieved by considering only
the first four moments. In fact, since the first two moments contain information about location and spread but not shape, we can
restrict the classification system to the third and fourth moments.
Thus, once a shape is selected the location and scaling can be
adjusted to fit the specific problem. The natural quantities to
use in such a classification system are the standardized third and
fourth moments.

Definition 5.2.1 The third standardized moment of a distribution
is defined as

$$\sqrt{\beta_1} = \frac{E[(X - \mu)^3]}{\{E[(X - \mu)^2]\}^{3/2}} = \frac{\mu_3}{\mu_2^{3/2}} \tag{5.1}$$

and the fourth standardized moment is given as

$$\beta_2 = \frac{E[(X - \mu)^4]}{\{E[(X - \mu)^2]\}^2} = \frac{\mu_4}{\mu_2^2} \tag{5.2}$$

A convenient method of describing families of distributions is by means of a plot of $\sqrt{\beta_1}$ vs. β_2. Such a plot is shown in Figure 5.2, where β_1 is used instead of $\sqrt{\beta_1}$ to avoid problems with the sign. The value of β_2 must be greater than or equal to $\beta_1 + 1$ as shown by the following theorem.

Theorem 5.2.1[†] β_2 must be greater than or equal to $\beta_1 + 1$.

Since β_1 and β_2 are origin-invariant, we assume without loss of generality that $\mu = 0$.

Proof: Let $f(x)$ be a continuous probability distribution with mean zero where X is defined over a domain D. The set of all polynomials $ax^2 + bx + c$ with real number coefficients forms a three-dimensional vector space V under the usual addition and scalar multiplication of polynomials. Define the dot product

$$P \cdot Q = \int_D P(x)Q(x)f(x) \, dx \qquad \text{and} \qquad |P| = \sqrt{P \cdot P}$$

Consider

$$A(x) = x$$
$$B(x) = x^2$$
$$I(x) = 1$$

Then

$$\mu = A \cdot I = 0$$
$$\mu_2 = |A|^2 = B \cdot I$$
$$\mu_3 = B \cdot A$$
$$\mu_4 = |B|^2$$

[†]Proof due to J. Slifker.

We wish to show that

$$\beta_2 \geq 1 + \beta_1$$

Hence

$$\frac{\mu_4}{\mu_2^2} \geq 1 + \frac{\mu_3^2}{\mu_2^3} \qquad \text{or} \qquad \mu_2\mu_4 \geq \mu_2^3 + \mu_3^2$$

In terms of vectors defined above,

$$|A|^2 \, |B|^2 \geq |A|^2(B \cdot I)^2 + (B \cdot A)^2$$

Division by $|A|^2|B|^2$ yields

$$1 \geq (B_1 \cdot I)^2 + (B_1 \cdot A_1)^2 \tag{5.3}$$

where

$$B_1 = \frac{B}{|B|} \qquad \text{and} \qquad A_1 = \frac{A}{|A|}$$

are unit vectors. This is just a form of the Pythagorean theorem. This is true since I, A_1, B_1 are unit vectors with I and A_1 orthogonal and $B_1 \cdot I$ and $B_1 \cdot A_1$ are just the projections of B_1 onto I and A_1, respectively. Considering the right triangle R whose legs are the projections of B_1 onto A_1 and I, then the hypotenuse of R has the same length as the projection of B_1 onto the plane spanned by I and A_1. Since B_1 is a unit vector, this projection has length less than or equal to 1. Hence, (5.3) holds by the Pythagorean theorem. This equality holds if and only if B lies in the plane spanned by A and I. But this means that there are real numbers a and b so that

$$x^2 = ax + b$$

for all x. This is impossible for continuous distributions; hence, $\beta_2 > 1 + \beta_1$. The equality can hold for the discrete case whose proof follows the above procedure.

The (β_1, β_2) plot can be used to locate some of the standard distributions previously studied. For example, the normal distribution has $\beta_1 = 0$, $\beta_2 = 3$ and is represented by a single point in the plane. All distributions with no shape parameters are clearly represented by a single point. Since neither β_1 nor β_2 are functions of the parameters they are represented by single numbers. Distributions with a single shape parameter will lie on a line in the (β_1, β_2) plane.

Example 5.2.1 Find the point in the (β_1, β_2) plane corresponding to the exponential model. By use of (3.7) this point is (4,9).

Example 5.2.2 Determine the line on which all gamma distributions lie in a (β_1, β_2) plot. The standardized third and fourth moments of the gamma distribution are, using (3.36), (5.1), and (5.2),

$$\sqrt{\beta_1} = \frac{2}{\sqrt{\gamma}}$$

and

$$\beta_2 = \frac{3(\gamma + 2)}{\gamma}$$

To express the relationship between β_1 and β_2 one substitutes $\gamma = 4/\beta_1$ in the expression of β_2 to obtain

$$\beta_2 = 3\left(1 + \frac{\beta_1}{2}\right)$$

which is the equation of a straight line and describes all gamma distributions. Note the limits of β_1 and β_2 as $\gamma \to \infty$ are 0 and 3, respectively. Thus, the line for the gamma distribution goes through the point corresponding to the normal distribution. One can also show that in the limit as $\gamma \to \infty$ the gamma goes to the normal distribution.

The procedures of Example 5.2.2 can be used to obtain the equation for other one-shape parameter distributions. In the problem section, the reader is asked to show that the parametric

equations for the curve in the (β_1, β_2) plane for lognormal distributions are

$$\beta_1 = (w - 1)(w + 2)^2$$

and

$$\beta_2 = w^4 + 2w^3 + 3w^2 - 3 \qquad\qquad (5.4)$$

where

$$w = e^{\sigma^2} \qquad \sigma > 0$$

In order to plot the curve defined by equation (5.4) one selects
values for w satisfying $w > 1$ (since $\sigma^2 > 0$) and solves for β_1 and
β_2. For example, one point on the curve for the lognormal distribution can be obtained by setting $w = 1.1$. Equation (5.4)
yields $\beta_1 = 0.961$ and $\beta_2 = 4.756$. The curve can be generated
over a range of (β_1, β_2) values by choosing other values of w and
fitting a smooth curve to the points.

Distributions that have two shape parameters cover regions in
the (β_1, β_2) plane. For example, members of the beta family (3.67),

$$f(x; \alpha, \beta) = \frac{\Gamma(\alpha + \beta)}{\Gamma(\alpha)\Gamma(\beta)} x^{\alpha-1}(1 - x)^{\beta-1} \qquad 0 < x < 1$$

all lie in the region between the lines

$$\beta_2 = 3\left(1 + \frac{\beta_1}{2}\right)$$

and

$$\beta_2 = 1 + \beta_1$$

Given values of β_1 and β_2 the plot can be used to identify a
specific distribution; however, since β_1 and β_2 do not uniquely
determine a distribution, there can be several distributions that
satisfy the desired values.

The (β_1, β_2) plot is of little value in identifying distributions when only the sample estimates $\sqrt{b_1}$ and b_2 (see below) are available, since:

1. The probability that the random pair $(\sqrt{b_1}, b_2)$ falls exactly on a specific point or line is negligible. Thus, most of the common distributions studied in prior chapters would never be identified.

2. The variance of high-order moments such as $\sqrt{b_1}$ and b_2 are large even for moderately large sample sizes; hence, these estimators are generally not used as a criteria for choosing specific distributions. The estimates are also highly biased for small samples. Johnson and Lowe (1979) have shown that $b_2 \leq n$ and $|\sqrt{b_1}| \leq \sqrt{n - 1}$, where n is the number of observations. Thus, the maximum value of b_2 for a sample of 50 from a lognormal distribution ($\mu = 0$, $\sigma^2 = 1$) is 50 while the β_2 value for the distribution is 113.9.

Other techniques for choosing models based on a set of data are described in Chapter 6. In some cases $\sqrt{b_1}$ and b_2 are used to select specific members of an empirical distribution. Thus for example, the percentage points of the Pearson family (Section 5.5) are tabulated according to their third and fourth standardized moments.

Two common procedures for fitting an empirical model to a set of data once the family or family subset is selected are matching of moments and matching of percentiles.

Definition 5.2.2   The kth sample moment is defined as

$$m_k' = \frac{\sum_{i=1}^{n} x_i^k}{n}$$

The kth sample moment about the sample mean is defined as

$$m_k = \sum_{i=1}^{n} \frac{(x_i - \bar{x})^k}{n}$$

The sample measures of skewness and kurtosis are

$$\sqrt{b_1} = \frac{m_3}{m_2^{3/2}}$$

and

$$b_2 = \frac{m_4}{m_2^2}$$

The method of matching moments requires that we equate one sample moment to the corresponding moment of the empirical distribution for each unknown parameter. Since the moments of the empirical distribution are expressions containing the parameters, the resulting system of equations is solved to obtain estimates of the parameters. Generally for moments of order 2 or higher we use the moments about the mean. Thus, we match

$$\mu = m_1' = \bar{x}$$

and letting

$$\mu_k = \text{kth population moment about mean} = E(X - \mu)^k$$

we set, if there are r unknown parameters,

$$m_k = \mu_k \qquad k = 2, 3, 4, \ldots, r$$

If there are four unknown parameters this generates four equations in four unknowns, namely,

$$\sum_{i=1}^{n} \frac{x_i}{n} = \bar{x} = \mu_1 \qquad \sum_{i=1}^{n} \frac{(x_i - \bar{x})^2}{n} = \mu_2 \qquad \sum_{i=1}^{n} \frac{(x_i - \bar{x})^3}{n} = \mu_3$$

and

$$\sum_{i=1}^{n} \frac{(x_i - \bar{x})^4}{n} = \mu_4$$

Since there are an infinite number of data moments and there can be an infinite number of population moments, there are many possible parameter estimators. However, due to the stability (smaller variances) of the lower moments and the desire for the mean, variance, skewness, and kurtosis of the sample to match that of the data, one always uses the rth lowest moments to estimate r parameters.

Example 5.2.3 Generate the equations for the estimation of the three parameters of a gamma distribution using the method of matching moments. The three-parameter gamma distribution is defined by

$$f(x;\varepsilon,\lambda,\gamma) = \frac{\lambda^\gamma}{\Gamma(\gamma)} (x - \varepsilon)^{\gamma-1} e^{-\lambda(x-\varepsilon)} \qquad x > \varepsilon$$

The moments of this distribution are, from (3.36),

$$\mu = \frac{\gamma}{\lambda} + \varepsilon$$

$$\mu_2 = \frac{\gamma}{\lambda^2}$$

$$\mu_3 = \frac{2\gamma}{\lambda^3}$$

If we match these moments with the corresponding sample moments we have

$$\bar{x} = \frac{\gamma^*}{\lambda^*} + \varepsilon^*$$

$$m_2 = \sum_{i=1}^{n} \frac{(x_i - \bar{x})^2}{n} = \frac{\gamma^*}{\lambda^{*2}}$$

$$m_3 = \sum_{i=1}^{n} \frac{(x_1 - \bar{x})^3}{n} = \frac{2\gamma^*}{\lambda^{*3}}$$

Solution of these equations for the three parameters yields

$$\lambda^* = \frac{2m_2}{m_3} \qquad \gamma^* = \lambda^{*2} m_2 \qquad \varepsilon^* = \bar{x} - \frac{\gamma^*}{\lambda^*}$$

where * indicates the requisite parameter estimator.
Assuming that $\bar{x} = 10.0$, $\Sigma(x_i - \bar{x})^2/n = 4.0$, and $\Sigma(x_i - \bar{x})^3 = 9.0$,
then the estimators of the parameters are

$$\lambda^* = \frac{2(4.0)}{9.0} = 0.89 \qquad \gamma^* = \frac{64}{81}(4.0) = 3.16$$

$$\varepsilon^* = 10 - \frac{3.16}{0.89} = 6.45$$

The technique of matching of moments sometimes leads to a set
of nonlinear simultaneous equations that may be difficult to solve
and must be done numerically.

The procedure is limited to distributions which have moments;
for example, it cannot be used for the Cauchy distribution (see
Problem 5b, Section 2.5). One criticism of the technique is that
the higher moments m_3 and m_4 have large variances and are very sensitive to outliers. Hence, the resulting estimates may be poor
and lead to poor fits to the data. It is important to check the
resulting approximation throughout the range of the data to
ascertain whether or not the fit is reasonable.

An alternate method is the matching of percentiles.

Definition 5.2.3 The pth percentile of a distribution is defined
as $x_p = F^{-1}(p)$ and is the value x_p that satisfies the equation

$$p = \int_{-\infty}^{x_p} f(u) \, du$$

Definition 5.2.4 The pth percentile in the sample is the ith ranked
observation (ranking from smallest to largest) satisfying

$p = i/n \times 100\%$. Thus, if there are 100 observations the 90th
largest is the 90th percentile. In most cases linear interpolation
is used to determine a value for the percentile. For example, if

$$n = 50 \quad \tilde{} \quad \text{and} \quad p = 0.95$$

then

$$0.95 = \frac{i}{50} \quad \text{and} \quad i = 47.5$$

Thus, 1/2 the distance between the 47th and 48th observation is used
as the 95th data percentile. If the 47th largest observation is
1.073 and the 48th largest is 1.081, then

$$Y_{0.95} = 1.073 + \frac{1}{2} (1.081 - 1.073) = 1.077$$

The method of matching percentiles matches distribution percentiles which are in terms of the parameters with those of the sample
percentiles. One percentile is needed for each unknown parameter.
The resulting set of equations is solved for the parameter values.
It is general practice to choose the percentiles symmetrically
from the distribution and to spread them out so as to attempt to
obtain a uniform fit over the entire range. Since the empirical
distribution fits the data exactly at the matching points, it is
good practice to choose as percentiles for matching areas where inferences are drawn. Therefore, if tail areas are of interest, choose
tail percentiles. However, the sample size will limit how extreme
a percentile should be estimated from the data. A rule of thumb is
never estimate a percentile where interpolation is required between
the two highest or two lowest values. Therefore, if one has 100
observations, the most extreme percentiles to use would be 0.01
and 0.99; however, if there are 1000 observations, one could use
0.001 and 0.999. General practice is to use for moderate sample
sizes $p = 0.05$ and $p = 0.95$ while for somewhat larger sample size
one uses $p = 0.01$ and $p = 0.99$. If there are two unknown parameters, then one set of the above percentiles would be used.

If there are three unknown parameters, these points would be augmented by the median p = 0.50. It there were four unknown parameters, the two extreme points may be augmented by the quartiles p = 0.25 and p = 0.75.

One advantage of matching percentiles is that the procedure is far less sensitive to an outlier than the moment-matching procedure. However, it does not guarantee that the mean and variance of the approximating distribution correspond to the sample values. One disadvantage of this procedure is that if x_p cannot be expressed in a simple form (for example, the normal and gamma distributions), the technique cannot be used. It is also possible that the resulting equations are nonlinear in a form that cannot be solved analytically, hence limiting the usefulness of the procedure.

Example 5.2.4 Fit a two-parameter Weibull distribution to a set of data using the matching of percentiles procedure. The Weibull distribution (3.23) is defined by

$$f(x;\lambda,\gamma) = \lambda\gamma x^{\gamma-1} e^{-\lambda x^{\gamma}} \qquad x > 0$$

$$p = F(x_p) = 1 - e^{-\lambda x_p^{\gamma}}$$

and

$$x_p = \left[-\frac{1}{\lambda} \ln(1 - p) \right]^{1/\gamma}$$

Assume a sample of size 200 and an interest in tail areas. The values p = 0.05 and p = 0.95 are chosen for matching. The data values corresponding to i = 10 and i = 190 are matched. Assume the 10th ordered observation is 1.3 and the 190th is 15.3, i.e.,

$$Y_{0.95} = 15.3 \qquad \text{and} \qquad Y_{0.05} = 1.3$$

The matching equations are

$$Y_{0.05} = \left[-\frac{1}{\lambda^*} \ln(0.95) \right]^{1/\gamma^*} = (0.0513)^{1/\gamma^*} \left(\frac{1}{\lambda^*}\right)^{1/\gamma^*}$$

and

$$Y_{0.95} = \left[-\frac{1}{\lambda^*} \ln(0.05)\right]^{1/\gamma^*} = (2.9957)^{1/\gamma^*}\left(\frac{1}{\lambda^*}\right)^{1/\gamma^*}$$

Elimination of $1/\lambda^*$ yields

$$(58.3957)^{1/\gamma^*} = \frac{Y_{0.95}}{Y_{0.05}} \qquad \text{or} \qquad \gamma^* = \frac{\ln(58.3957)}{\ln\left(Y_{0.95}/Y_{0.05}\right)}$$

and

$$\lambda^* = \frac{2.9957}{\left(Y_{0.95}\right)^{\gamma^*}}$$

The substitution of the values for $Y_{0.95}$ and $Y_{0.05}$ yields

$$\gamma^* = 1.65$$

$$\lambda^* = 0.033$$

where * indicates these values are estimates of the parameters.

Problems

1. Determine the coordinates in the (β_1, β_2) plane for
   (a) The uniform distribution.
   (b) The half-normal distribution. Obtain the half-normal model from the transformation $Y = |X|$, where X has the standard normal distribution.

2. Determine the curve in the (β_1, β_2) plane for
   (a) Beta distributions for $\alpha = 2$ and $\beta \geq 1$.
   (b) Lognormal distribution, equation (5.4).

3. Using equation (5.4) and the information that $\beta_1 = 4.62$, find β_2 for a lognormal distribution and determine the value of the parameter σ^2.

4. Ten metal specimens are tested for corrosion resistance. The times (in hours) until breakage in an acid bath are recorded. The resulting data yield the following summary statistics:

$$\sum_{i=1}^{10} t_i = 10.0 \qquad \sum_{i=1}^{10} t_i^2 = 40.0$$

$$\sum_{i=1}^{10} t_i^3 = 173.0 \qquad \sum_{i=1}^{10} t_i^4 = 1116.0$$

Choose an approximating distribution from the above exercises to use for the data with $\sqrt{b_1}$ and b_2 as the selection criteria. Determine the probability that breakage will occur prior to 0.6 hr.

5. Obtain the equations for estimating the parameters α and β of the beta distribution defined on the unit interval using the moment-matching technique.

6. A work study analysis was performed to determine the proportion of an 8-hour shift (p) lathe operators were busy with tasks other than tending their machines. The results of observing 100 workers were as follows:

$$\sum_{i=1}^{100} p_i = 25.0 \qquad \sum_{i=1}^{100} p_i^2 = 10.0$$

Use a beta distribution and the equations generated in the previous problem to estimate the parameters of the model.

7. A company is doing a study of its inventory control procedures. The number of days to order and receive a shipment from a vendor for a major part was recorded as indicated below. A gamma distribution is chosen as a model of the reorder time. Since it is impossible to order and receive a part in zero days, a three-parameter model must be used where the location parameter represents the minimum time to order and receive a part. Use the data to estimate the three parameters.

Time (in days) to Order and Receive Part xyz

4.7	6.0	11.1	5.8	14.4
11.8	7.3	5.6	20.3	9.9
7.4	6.3	6.7	8.5	5.3
14.8	13.2	6.4	11.6	11.1
9.3	7.8	6.9	8.8	8.5

Use the method of matching moments to obtain formulas for estimating the parameters of the extreme-value distribution for largest elements.

9. One test of potency of an antibacterial agent is the amount of time required to kill a standard quantity of bacteria in a culture. Since the time recorded is the maximum survival time of the bacteria in the culture, an extreme-value distribution is chosen to model this phenomenon. Estimate the parameters of the model from the results (in hours) of 10 tests shown below.

$$\sum_{i=1}^{10} t_i = 15.3 \qquad \sum_{i=1}^{10} t_i^2 = 24.95$$

Find the probability that the time to kill all the bacteria in a culture will exceed 2.5 hr.

10. Use the method of matching percentiles to derive estimates for the extreme-value distribution for largest elements.

11. A two-parameter Weibull distribution has been used in the past to model the time-to-failure of an electronic switching device. The failure times of 275 units are given below. Use the method of matching percentiles to estimate the parameters of the distribution. Plot the histogram of the data and sketch in the approximating model to determine whether the fit is reasonable.

Time-to-Failure (Hours): Electronic Switching Device

Time-to-failure	Frequency	Time-to-failure	Frequency
0- 1.99	16	36.00-40.99	10
2.00- 3.99	30	41.00-45.99	13
4.00- 5.99	25	46.00-50.99	12
6.00-10.99	21	51.00-55.99	9
11.00-15.99	27	56.00-60.99	7
16.00-20.99	26	61.00-65.99	10
21.00-25.99	25	66.00-70.99	5
26.00-30.99	21	71.00-75.99	2
31.00-35.99	15	76.00-80.99	1

12. In the design of the cabin of a large jumbo jet an important
     safety factor is the time it takes to evacuate the passengers
     from a loaded plane. The critical value is the time it takes
     to get the last passenger off the plane. A total of 70 tests
     was run. The times it took the last passenger to exit the
     plane are shown below. Fit an extreme-value distribution to

Time (in Minutes) to Evacuate Aircraft

1.81	1.11	1.29	2.05	1.36	1.49	1.62
2.29	1.45	1.54	1.30	1.49	1.32	1.44
1.07	2.33	1.44	1.70	1.71	2.14	1.53
0.96	1.16	1.66	1.67	1.64	1.04	1.49
2.53	1.21	0.93	1.68	2.76	1.17	1.18
1.69	1.23	3.08	1.42	1.52	1.08	1.71
1.97	1.27	1.99	1.77	1.08	1.76	2.11
1.00	1.59	2.40	1.46	0.97	1.12	2.84
2.14	1.22	2.50	0.42	0.98	1.93	1.56
1.21	0.67	1.47	2.38	1.40	1.41	2.43

the data, first using the method of moments and then the method
of matching percentiles. The design criterion states that the
plane must be capable of being evacuated in less than 3 min.
Determine the probability that the evacuation time will be
less than 3 min. using both sets of estimates.

5.3   THE GENERALIZED LAMBDA FAMILY

This section introduces a family of distributions which is not only
flexible, i.e., takes on a wide variety of shapes, but is useful
in Monte Carlo simulation studies due to the ease of generating
random variates from it.   The family was originally suggested in a
paper by Hastings et al. (1947), was used by J. Tukey in some
unpublished work, and was generalized and developed by Dudewicz
et al. (1974).   The distribution was developed in an attempt to
obtain a simple function which could be used as an approximation
to other distributions.   It may be used to compute percentiles or
generate random deviates in place of using the more complex standard distributions such as the normal or the gamma.   It has a
major advantage in that it is not necessary to use $\sqrt{b_1}$ or b_2 in
selecting subgroups of families and hence, eliminates any problems
in the use of these quantities.   The density function for this
family cannot be expressed in closed form.   However, it can be
expressed in terms of its percentile function.   Its major disadvantage is that either the higher moments must be used in estimating the parameters or simultaneous nonlinear equations using
percentiles.

Definition 5.3.1   The percentile function R(p) of a random variable
X is the inverse of the distribution function

$$x_p \equiv R(p) = F^{-1}(p) \qquad 0 \le p \le 1 \qquad\qquad (5.5)$$

The percentile function yields the value of the random variable
such that the area under the curve to the left of x_p is p.

Example 5.3.1   The percentile function for the exponential distribution

$$f(x;\lambda) = \lambda e^{-\lambda x} \qquad x \geq 0$$

is obtained as follows:

$$p = F(x_p) = \int_0^{x_p} \lambda e^{-\lambda t} \, dt = 1 - e^{-\lambda x_p}$$

$$x_p = -\frac{1}{\lambda} \ln(1 - p) = R(p) \tag{5.5a}$$

There is a direct relationship between R(p) and f(x).

Lemma   Suppose y = f(x) is a 1-1 onto transformation which transforms x to y.   Then dx/dy = 1/(dy/dx).   (a well-known calculus result).

Theorem 5.3.1   Suppose F(x) = p and $x = F^{-1}(p)$, defines a 1-1 onto transformation from x to p.   Then f(x) = 1/R´(p), where $R(p) \equiv F^{-1}(p)$; see (5.5).

Proof: For some x and p,

$$x = R(p)$$

Therefore,

$$\frac{dx}{dp} = R´(p)$$

But by the lemma,

$$\frac{dx}{dp} = \frac{1}{dp/dx}$$

Now dp/dx = F´(x) = f(x), the density.   Therefore,

$$f(x) = \frac{1}{R´(p)} \qquad \text{or} \qquad R´(p) = \frac{1}{f(x)} \tag{5.6}$$

Example 5.3.2 To illustrate the theorem the exponential distribution is considered:

$$R(p) = -\frac{1}{\lambda} \ln(1 - p) \qquad \text{and} \qquad p = 1 - e^{-\lambda x}$$

$$R'(p) = \frac{1}{\lambda(1 - p)} = \frac{1}{\lambda e^{-\lambda x}} = \frac{1}{f(x)}$$

where x is the pth percentile. The moments of f(x) can also be obtained directly from R(p).

Theorem 5.3.2

$$E(X^k) = \int_0^1 [R(p)]^k \, dp$$

First of all, by definition,

$$E(X^k) = \int_{-\infty}^{\infty} x^k f(x) \, dx$$

Let

$$x = R(p)$$

$$dx = R'(p) \, dp$$

$$E(X^k) = \int_0^1 [R(p)]^k f(x) R'(p) \, dp$$

but from Theorem 5.3.1, $R'(p) = 1/f(x)$; hence,

$$E(X^k) = \int_0^1 [R(p)]^k \, dp$$

Example 5.3.3 Find $E(X^2)$ for the exponential distribution using the percentile function

$$E(X^2) = \int_0^1 [R(p)]^2 \, dp = \int_0^1 \frac{1}{\lambda^2}[\ln(1 - p)]^2 \, dp = \frac{2}{\lambda^2}$$

The generalized lambda distribution (GLD) [Dudewicz et al.

(1974)] is defined in terms of its percentile function

$$x_p = R(p) = \lambda_1 + \frac{p^{\lambda_3} - (1 - p)^{\lambda_4}}{\lambda_2} \qquad 0 \le p \le 1 \qquad (5.7)$$

The density of x_p can be obtained by use of Theorem 5.3.1 as

$$f(x) = \frac{1}{R'(p)}$$

$$= \frac{\lambda_2}{\lambda_3 p^{\lambda_3 - 1} + \lambda_4 p^{\lambda_4 - 1}} \qquad 0 \le p \le 1 \qquad (5.7a)$$

Restrictions must be placed on λ_2, λ_3, and λ_4 so that $f(x) \ge 0$, which is a requirement for any density function. The domain of the random variable X will also depend on the choice of the λ_i's, i = 2, 3, 4. For example, to determine the bounds of X when λ_2, λ_3, and λ_4 are all negative, we investigate what occurs when p = 0 and p = 1:

$$x_p = \lambda_1 + \frac{1/p^{-\lambda_3} - 1/(1-p)^{-\lambda_4}}{\lambda_2}$$

For p = 0, $x_0 = -\infty$

For p = 1, $x_1 = \infty$

Thus, the lower and upper bounds are $(-\infty, \infty)$

If $\lambda_2 < 0$, $\lambda_3 = 0$, and $\lambda_4 < 0$, and then $x_p = \lambda_1 + \dfrac{1 - 1/(1-p)^{-\lambda_4}}{\lambda_2}$

For p = 0, $x_0 = \lambda_1$, and for p = 1, $x_1 = \infty$

Thus, the domain of X is $[\lambda_1, \infty]$

A table of lower and upper bounds was computed by Ramberg et al. (1974) and is shown in Table 5.1. The reader should refer to the above article for a more complete discussion.

TABLE 5.1   Lower and Upper Bounds of the Generalized Lambda
Distribution

λ_2	λ_3	λ_4	Lower bound	Upper bound
<0	<-1	>1	$-\infty$	$\lambda_1 + 1/\lambda_2$
<0	>1	<-1	$\lambda_1 - 1/\lambda_2$	∞
>0	>0	>0	$\lambda_1 - 1/\lambda_2$	$\lambda_1 + 1/\lambda_2$
>0	=0	>0	λ_1	$\lambda_1 + 1/\lambda_2$
>0	>0	=0	$\lambda_1 - 1/\lambda_2$	λ_1
<0	<0	<0	$-\infty$	∞
<0	=0	<0	λ_1	∞
<0	<0	=0	$-\infty$	λ_1

The density f(x) and the random variable X have been defined
parametrically in terms of the distribution function F(x).   In
order to make a plot of the density function it is therefore
necessary to choose a range of p values on the unit interval [0,1],
compute x_p and $f(x_p)$ for each value, and plot the resulting pairs
of points.

The parameter λ_1 of the GLD is a location parameter.   This
can be seen from an examination of the moments of the distribution.
From Theorem 5.3.2 we have

$$E(X^k) = \int_0^1 \left[\lambda_1 + \frac{p^{\lambda_3} - (1-p)^{\lambda_4}}{\lambda_2} \right]^k dp$$

Hence,

$$E(X) = \int_0^1 \left[\lambda_1 + \frac{p^{\lambda_3} - (1-p)^{\lambda_4}}{\lambda_2} \right] dp$$

$$= \lambda_1 + \frac{1}{\lambda_2} \left[\frac{p^{\lambda_3+1}}{\lambda_3+1} + \frac{(1-p)^{\lambda_4+1}}{\lambda_4+1} \right]_0^1$$

$$= \lambda_1 + \frac{1}{\lambda_2} \left(\frac{1}{\lambda_3+1} - \frac{1}{\lambda_4+1} \right)$$

Since $E(X - \lambda_1)$ does not contain λ_1, ones sees that λ_1 is a location parameter which shifts the distribution along the x axis.

In calculating the second, third, and fourth moments about the mean, one sets $\lambda_1 = 0$ since it is a location parameter and does not affect the values of $E(X - \mu)^k$:

$$E(X^k) = \frac{1}{\lambda_2^k} \int_0^1 \left[p^{\lambda_3} - (1 - p)^{\lambda_4} \right]^k dp$$

Using the binomial theorem, expanding and interchanging summation and integration operations, one obtains

$$E(X^k) = \lambda_2^{-k} \sum_{i=1}^k \binom{k}{i} (-1)^i \int_0^1 p^{\lambda_3(k-i)} (1 - p)^{\lambda_4 i} dp \qquad (5.8)$$

Since $\lambda_2^k E(X^k)$ does not contain λ_2, it follows that $1/\lambda_2$ is a scale parameter. The integrand is a beta density, and using results from Section 4.7 one has

$$E(X^k) = \lambda_2^{-k} \sum_{i=0}^k \binom{k}{i} (-1)^i \frac{\Gamma[\lambda_3(k - i) + 1]\Gamma(\lambda_4 i + 1)}{\Gamma[\lambda_3(k - i) + \lambda_4 i + 2]} \qquad (5.9)$$

Letting $k = 2$, one finds

$$E(X^2) = \lambda_2^{-2} \left[\frac{\Gamma(2\lambda_3 + 1)}{\Gamma(2\lambda_3 + 1)} - \frac{2\Gamma(\lambda_3 + 1)\Gamma(\lambda_4 + 1)}{\Gamma(\lambda_3 + \lambda_4 + 2)} + \frac{\Gamma(2\lambda_4 + 1)}{\Gamma(2\lambda_4 + 2)} \right]^\dagger$$

$$= \lambda_2^{-2} \left[\frac{1}{2\lambda_3 + 1} - 2\frac{\lambda_3! \, \lambda_4!}{(\lambda_3 + \lambda_4 + 1)!} + \frac{1}{2\lambda_4 + 1} \right]$$

Thus,

$$\text{var}(X) = \frac{1}{\lambda_2^2} \left[\frac{1}{2\lambda_3 + 1} - 2\frac{\lambda_3! \, \lambda_4!}{(\lambda_3 + \lambda_4 + 1)!} + \frac{1}{2\lambda_4 + 1} \right.$$

$$\left. - \left(\frac{1}{\lambda_3 + 1} - \frac{1}{\lambda_4 + 1} \right)^2 \right]$$

---

†In what follows $\Gamma(x)$ is replaced by $(x - 1)!$ for notational simplification.

Use of (5.9), shows the third and fourth moments about the mean are:

$$\mu_3 = \frac{1}{\lambda_2^3}\left\{\frac{1}{3\lambda_3+1} - 3\frac{(2\lambda_3)!\lambda_4!}{(2\lambda_3+\lambda_4+1)!} + \frac{3\lambda_3!(2\lambda_4)!}{(\lambda_3+2\lambda_4+1)!}\right.$$

$$- \frac{1}{3\lambda_4+1} - 3\left[\frac{1}{2\lambda_3+1} - 2\frac{\lambda_3!\lambda_4!}{(\lambda_3+\lambda_4+1)!} + \frac{1}{2\lambda_4+1}\right]$$

$$\left. \times \left[\frac{1}{\lambda_3+1} - \frac{1}{\lambda_4+1}\right] + 2\left[\frac{1}{\lambda_3+1} - \frac{1}{\lambda_4+1}\right]^3\right\} \qquad (5.10)$$

$$\mu_4 = \frac{1}{\lambda_2^4}\left\{\frac{1}{4\lambda_3+1} - 4\frac{(3\lambda_3)!\lambda_4!}{(3\lambda_3+\lambda_4+1)!} + 6\frac{(2\lambda_3)!(2\lambda_4)!}{(2\lambda_3+2\lambda_4+1)!}\right.$$

$$- 4\frac{\lambda_3!(3\lambda_4)!}{(\lambda_3+3\lambda_4+1)!} + \frac{1}{4\lambda_4+1} - 4\left[\frac{1}{3\lambda_3+1} - 3\frac{(2\lambda_3)!\lambda_4!}{(2\lambda_3+\lambda_4+1)!}\right.$$

$$\left. + 3\frac{\lambda_3!(2\lambda_4)!}{(\lambda_3+2\lambda_4+1)!} - \frac{1}{3\lambda_4+1}\right]\left[\frac{1}{\lambda_3+1} - \frac{1}{\lambda_4+1}\right]$$

$$+ 6\left[\frac{1}{2\lambda_3+1} - \frac{2\lambda_3!\,\lambda_4!}{(\lambda_3+\lambda_4+1)!} + \frac{1}{2\lambda_4+1}\right]\left[\frac{1}{\lambda_3+1} - \frac{1}{\lambda_4+1}\right]^2$$

$$\left. - 3\left[\frac{1}{\lambda_3+1} - \frac{1}{\lambda_4+1}\right]^4\right\}$$

The existence of these moments again depends on the values of λ_2, λ_3, and λ_4. Ramberg et al. (1974) showed that all positive moments (k > 0) exist in region 3 (see Table 5.1) and no positive moments exist in regions 1 and 2. There is no point in region 4 where all positive moments exist, although many moments do exist near the point $\lambda_2 = \lambda_3 = \lambda_4 = 0$. In such cases the integral in (5.8) must be checked for specific values of k. The kth moment does not exist when the argument of the beta function is negative.

The formulas for the moments can be used to compute the standardized third and fourth moments $\sqrt{\beta_1}$ and β_2. Note that these two measures are only functions of λ_3 and λ_4, since λ_2 is a scaling parameter. The values of $\sqrt{\beta_1}$ and β_2 can be used to classify the members of the GLD family and cover a wide region of the (β_1,β_2) plane. The members of the family can be symmetric or asymmetric. If $\lambda_3 = \lambda_4 = \lambda$, then

$$\mu_3 = \frac{1}{\lambda_2^3}\left\{\frac{1}{3\lambda + 1} - \frac{3(2\lambda)!\lambda!}{(3\lambda + 1)!} + \frac{3\lambda!(2\lambda)!}{(3\lambda + 1)!} - \frac{1}{3\lambda + 1}\right.$$

$$- 3\left[\frac{1}{2\lambda + 1} - \frac{2\lambda!\lambda!}{(2\lambda + 1)!} + \frac{1}{2\lambda + 1}\right]\left[\frac{1}{\lambda + 1} - \frac{1}{\lambda + 1}\right]$$

$$\left. + 2\left[\frac{1}{\lambda + 1} - \frac{1}{\lambda + 1}\right]^3\right\}$$

or

$$\mu_3 = 0$$

and GLD is symmetric.

The two methods of fitting empirical distributions to a set of data, matching percentiles or moments, can be used with the GLD. The percentile matching method is particularly convenient since the GLD is defined in terms of its percentiles; however, it requires the solution of a pair of nonlinear simultaneous equations. The procedure requires matching four percentiles and solving for λ_1, λ_2, λ_3, and λ_4. For example, the four equations using data percentiles, y_p corresponding to $p = 0.95$, 0.75, 0.25, and 0.05, would be

$$y_{0.95} = \lambda_1^* + \frac{0.95^{\lambda_3^*} - 0.05^{\lambda_4^*}}{\lambda_2^*}$$

$$y_{0.75} = \lambda_1^* + \frac{0.75^{\lambda_3^*} - 0.25^{\lambda_4^*}}{\lambda_2^*} \qquad (5.11)$$

$$y_{0.25} = \lambda_1^* + \frac{0.25^{\lambda_3^*} - 0.75^{\lambda_4^*}}{\lambda_2^*}$$

$$y_{0.05} = \lambda_1^* + \frac{0.05^{\lambda_3^*} - 0.95^{\lambda_4^*}}{\lambda_2^*}$$

These equations are solved for λ_1^*, λ_2^*, λ_3^*, and λ_4^*, the percentile estimators of λ_1, λ_2, λ_3, and λ_4, respectively. These equations can be reduced to two equations in two unknowns by subtraction and division to yield

$$w_1 = \frac{y_{0.95} - y_{0.05}}{y_{0.95} - y_{0.25}} = \frac{0.95^{\lambda_3^*} + 0.95^{\lambda_4^*} - 0.05^{\lambda_3^*} - 0.05^{\lambda_4^*}}{0.95^{\lambda_3^*} + 0.75^{\lambda_4^*} - 0.25^{\lambda_3^*} - 0.05^{\lambda_4^*}}$$

(5.12)

$$w_2 = \frac{y_{0.75} - y_{0.25}}{y_{0.95} - y_{0.25}} = \frac{0.75^{\lambda_3^*} + 0.75^{\lambda_4^*} - 0.25^{\lambda_3^*} - 0.25^{\lambda_4^*}}{0.95^{\lambda_3^*} + 0.75^{\lambda_4^*} - 0.25^{\lambda_3^*} - 0.05^{\lambda_4^*}}$$

These two equations must be solved numerically to obtain λ_3^* and λ_4^* based on w_1 and w_2 and p = 0.95, 0.75, 0.25, and 0.05. Thus, to fit a GLD to a set of data it is first necessary to calculate w_1 and w_2 from equation (5.12). Next, λ_2^* is calculated from

$$w_3 = y_{0.95} - y_{0.05}$$

and

$$\lambda_2^* = \frac{0.95^{\lambda_3^*} + 0.95^{\lambda_4^*} - 0.05^{\lambda_3^*} - 0.05^{\lambda_4^*}}{w_3}$$

(5.13)

Finally, λ_1^* is obtained from

$$\lambda_1^* = y_{0.95} - \frac{0.95^{\lambda_3^*} - 0.05^{\lambda_4^*}}{\lambda_2^*}$$

(5.14)

Example 5.3.4 Fit a GLD to a set of data where $y_{0.95} = 10.062$, $y_{0.75} = 8.618$, $y_{0.25} = 6.668$, $y_{0.05} = 5.776$. First calculate, from (5.12),

$$w_1 = \frac{10.062 - 5.776}{10.062 - 6.668} = \frac{4.286}{3.394} = 1.263$$

$$w_2 = \frac{8.618 - 6.668}{10.062 - 6.668} = \frac{1.950}{3.394} = 0.575$$

Solution of (5.12) by the Newton-Raphson method yields

$$\lambda_3^* = 0.062$$

$$\lambda_4^* = 0.415$$

Next,

$$w_3 = 10.062 - 5.776 = 4.286$$

and, using (5.13),

$$\lambda_2^* = \frac{0.95^{0.062} + 0.95^{0.415} - 0.05^{0.062} - 0.05^{0.415}}{4.286} = \frac{0.8567}{4.286}$$

$$= 0.200$$

Finally, one obtains, using (5.14),

$$\lambda_1^* = 10.062 - \frac{0.95^{0.062} - 0.05^{0.415}}{0.200} = 6.521$$

A plot of the density function is obtained as follows. Choose values of p of 0.0 (0.1) 1.0 and solve for x_p and $f(x_p)$ using (5.7) and (5.7a). The following values are obtained:

p	x_p	$f(x_p)$	p	x_p	$f(x_p)$
0.0	1.521	0.000	0.6	7.947	0.303
0.1	6.070	0.094	0.7	8.378	0.335
0.2	6.488	0.149	0.8	8.889	0.364
0.3	6.849	0.194	0.9	9.566	0.392
0.4	7.200	0.234	1.0	11.521	0.419
0.5	7.560	0.270			

A plot of x_p versus $f(x_p)$, the density function, is shown in Figure 5.1.

The procedure for fitting a GLD by matching of moments can also be done in a similar manner. Ramberg et al. (1979) have suggested the following procedure and give tables to solve the resulting equations. First calculate the sample moments

$$\bar{x} = \frac{1}{n} \sum_{i=1}^{n} x_i$$

$$m_2 = \sum_{i=1}^{n} \frac{(x_i - \bar{x})^2}{n}$$

$$m_3 = \sum_{i=1}^{n} \frac{(x_i - \bar{x})^3}{n} \tag{5.15}$$

$$m_4 = \sum_{i=1}^{n} \frac{(x_i - \bar{x})^4}{n}$$

Calculate $\sqrt{b_1}$ and b_2 from

$$\sqrt{b_1} = \frac{m_3}{m_2^{3/2}}$$

$$b_2 = \frac{m_4}{m_2^2}$$

See Table A.3 (Appendix A) and select values of $\tilde{\lambda}_3$ and $\tilde{\lambda}_4$ having values closest to the computed values of $\sqrt{b_1}$ and b_2. If $\sqrt{b_1}$ is negative, look up the absolute value of $\sqrt{b_1}$ to determine $\tilde{\lambda}_3$ and $\tilde{\lambda}_4$ and then interchange these values and change the sign of $\tilde{\lambda}_1$. That is, if $\sqrt{b_1} < 0$, look up $|\sqrt{b_1}|$ and b_2; select from the table λ_3^* and λ_4^* and then use $\tilde{\lambda}_3 = \lambda_4^*$ and $\tilde{\lambda}_4 = \lambda_3^*$, where * indicates the value found in the table. The table also gives values for λ_1^* and λ_2^*. The desired values are then obtained by letting

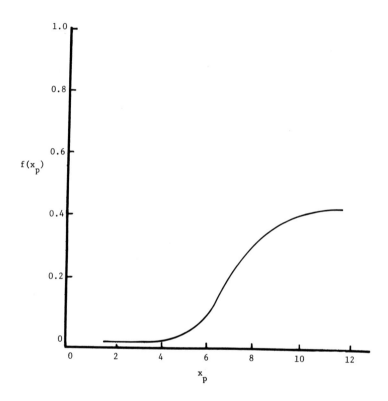

Figure 5.1 Plot of density function from Example 5.3.4.

$$\tilde{\lambda}_2 = \frac{\lambda_2^*}{\sqrt{m_2}}$$

and   (5.16)

$$\tilde{\lambda}_1 = \lambda_1^*\sqrt{m_2} + \bar{x}$$

noting that if $\sqrt{b_1}$ is negative, $-\lambda_1^*$ is used.

Example 5.3.5 Fit a GLD to a set of data where $\bar{x}$ = 20.1, m_2 = 5.2, m_3 = -4.31, and m_4 = 86.1. The first step is to calculate $\sqrt{b_1}$ and b_2.

$$\sqrt{b_1} = \frac{-4.31}{(5.2)^{3/2}} = \frac{-4.31}{11.86} = -0.363$$

$$b_2 = \frac{86.1}{(5.2)^2} = 3.18$$

We find, referring to Table A.3 for $\sqrt{b_1}$ = 0.35 and b_2 = 3.2, that

$$\lambda_3^* = 0.0758$$

$$\lambda_4^* = 0.1462$$

$$\lambda_1^* = -0.3430$$

$$\lambda_2^* = 0.1665$$

Since $\sqrt{b_1}$ is negative, we have

$$\tilde{\lambda}_3 = 0.1462$$

$$\tilde{\lambda}_4 = 0.0758$$

and from (5.16), after changing the sign of $\tilde{\lambda}_1$,

$$\tilde{\lambda}_2 = \frac{0.1665}{\sqrt{5.2}} = 0.073$$

$$\tilde{\lambda}_1 = 0.343\sqrt{5.2} + 20.1 = 20.88$$

Once the parameters are fitted to the data, then one can use (5.7a) as a summary of the observations. The percentiles of the distribution can be obtained from the defining relationship

$$x_p = \lambda_1 + \frac{p^{\lambda_3} - (1 - p)^{\lambda_4}}{\lambda_2} \qquad 0 \leq p \leq 1$$

Example 5.3.6 Find the 80th and 99th percentiles of the GLD derived in Example 5.3.5.

$$x_{0.80} = 20.88 + \frac{0.80^{0.1462} - 0.2^{0.0758}}{0.073}$$

$$= 22.01$$

$$x_{0.99} = 20.88 + \frac{0.99^{0.1462} - 0.01^{0.0758}}{0.073}$$

$$= 24.90$$

One disadvantage of the GLD distribution is that if $F(x)$ is wanted, it is necessary to solve for p in the percentile function, that is,

$$p^{\lambda_3} - (1 - p)^{\lambda_4} = \lambda_2(x_p - \lambda_1) \tag{5.17}$$

This must be done on a trial-and-error basis or solved numerically on a computer.

Example 5.3.7 For Example 5.3.5 find the probability that X is less than 18.00.

Use of (5.17) with the values of λ_1, λ_2, λ_3, and λ_4 from Example 5.3.5 requires the solution of the following equation for p:

$$p^{0.1462} - (1 - p)^{0.0758} = 0.073(18.00 - 20.88) = -0.2102$$

As a starting value, we set p = 0.30 and determine the value of left-hand side to be -0.1347. This value is too high; thus, a lower p value is needed. The choice p = 0.25 yields a value of

-0.1619, which is still too high. Next, the choice p = 0.17 yields
a value of -0.2142, which for most purposes is close enough; however,
if greater accuracy is desired, this process can continue. As a
check one calculates the actual value for x_p for the chosen p value
of 0.17. In this case,

$$x_{0.17} = 20.88 + \frac{0.17^{0.1462} - 0.83^{0.0758}}{0.073}$$

$$= 20.88 - \frac{0.2142}{0.073} = 17.95$$

which is close to the value of 18 stated in the problem.

A major benefit of using the GLD in a Monte Carlo study is that
generation of random variates from this distribution can be done
directly and easily from uniform random variates by means of the
percentile function. Using the fact that $p = F(x_p)$ is uniformly
distributed on (0,1) (see Theorem 6.3.1), a random deviate from the
GLD for given values of λ_1, λ_2, λ_3, and λ_4 is obtained by generating
a uniform variate and inserting it for p in the function

$$x_p = \lambda_1 + \frac{p^{\lambda_3} - (1 - p)^{\lambda_4}}{\lambda_2}$$

Problems

1. Determine $E(X^k)$ for a Weibull distribution using Theorem 5.3.2.
2. Determine the values for $\sqrt{\beta_1}$ and β_2 for the GLD for parameters
    $\lambda_2 = 1$, $\lambda_3 = 1$, $\lambda_4 = 2$.
3. Use the method of matching percentiles to determine the values
    λ_1^*, λ_2^*, λ_3^*, and λ_4^* which should be used to approximate a stand-
    ard normal distribution by the GLD.
4. Use the method of moments to determine values of $\tilde{\lambda}_1$, $\tilde{\lambda}_2$, $\tilde{\lambda}_3$,
    and $\tilde{\lambda}_4$ which should be used to approximate an exponential dis-
    tribution with a minimum value of 2.0 and a scale parameter of
    0.5.
5. The number of customers entering a store between the hours of
    3:00 p.m. and 5:00 p.m. was recorded for 30 days. Use a GLD

to approximate the distribution of the number of customers entering and find the probability that there are less than 88 persons who enter during the selected hours.

Number of Persons Entering Store
Between 3:00 p.m. and 5:00 p.m.

78	55	52	66	60
78	68	55	58	65
61	62	37	47	69
66	65	63	72	43
69	73	54	51	75

6. A sociologist obtained the number of years a resident of a specific area lived at his current address. Use a GLD to model the data and find the number of years of residence which was exceeded by only 10% of the population. The statistics from the data were

$$\bar{x} = 7.0 \text{ yr} \qquad \sqrt{b_1} = 0.25$$

$$m_2 = 2.5 \qquad b_2 = 3.20$$

7. Tenants in an office building were complaining about the poor elevator service resulting in an excessive time elapsing between entering the building and arrival at their office. A study was made on the elapsed time between entering the building to arrival at an office for 20 people randomly selected over a period of a week in the "rush hour" of 8:00 to 9:00 a.m. Approximate the elapsed time distribution with a GLD and find the probability that it will take more than 10 min. from arrival at building to entering one's office.

Elapsed Time (Minutes)

8.3	7.7	4.9	8.1
13.5	12.2	7.4	7.6
7.0	5.9	8.2	10.4
10.4	6.3	9.6	5.6
8.0	7.1	9.2	5.8

8. The yields of 20 batches of a chemical process are shown below.
    Use the GLD as an approximating model and determine the prob-
    ability of getting a yield below 40%.

    Percentage Yield

95.1	93.5	97.2	91.7
96.4	80.5	80.7	77.9
80.2	87.6	88.6	93.6
90.3	93.6	90.8	85.2
72.1	55.6	83.5	97.6

9. A retail firm has a large number of outlets throughout the
    country. They are considering adding a new product to their
    line. They randomly select 15 stores to test-market the new
    product for a 2-week period. The results are given below. In
    order to be profitable an outlet must sell at least $10,000
    worth of the product in a 2-week period. Use the GLD to
    approximate the distribution of total dollar sales in a 2-week
    period and estimate the proportion of stores which will not
    surpass this minimum.

    Total Sales (Thousands of Dollars) -- 2-Week Period

15.3	17.6	17.7
12.7	16.9	21.0
13.1	13.3	19.1
12.9	18.0	11.1
10.3	17.3	14.2

5.4 THE JOHNSON SYSTEM

In developing families to be used as empirical distributions one
attempts to make them as easy as possible to use and yet have a
wide flexibility so that they can cover most of, if not all of, the
shapes represented in the (β_1, β_2) plane. The GLD family of distributions discussed in Section 5.3 has the advantage that its

percentiles were easily obtainable and one form included all the
members of the family covering most of the (β_1, β_2) plane. Another
family or actually a system containing three families was introduced
by N. L. Johnson (1949). This system has the advantage that there
is a direct relationship between its members and the standard normal
distribution and it covers the entire allowable range of the (β_1, β_2)
plane. The first property makes it easy to obtain both percentiles
and cumulative probabilities for the distribution by use of a normal
table. The one disadvantage of the system is that it contains
three separate subfamilies and methods for fitting the distribution
to a set of data varying depending on which of the three forms is
appropriate.

The Johnson system has the advantage that neither special
tables nor solution of nonlinear equations are required to estimate
the parameters as in the case of the GLD.

Johnson introduced the idea that a general system of distributions could be generated by the transformation

$$z = \gamma + \eta k_i(x;\varepsilon,\lambda) \qquad i = 1, 2, 3 \qquad\qquad (5.18)$$

where

Z is a standard normal variate

γ, η, ε, and λ are parameters

X is the random variable

$k_i(x;\varepsilon,\lambda)$ is a functional form defining the three groups of
subfamilies in the system

The functions $k_i(x;\varepsilon,\lambda)$ were chosen so that one subfamily group
would lie on a line in the (β_1, β_2) plane and it would serve as the
dividing line between the other two subfamilies in the system
which cover the area above and below this line. The function which
serves as the dividing line is

$$k_1(x;\varepsilon,\lambda) = \ln\left(\frac{x - \varepsilon}{\lambda}\right)$$

Substitution of this function in (5.18) yields the transformation

$$z = \gamma + \eta \ \ln\left(\frac{x - \epsilon}{\lambda}\right) \tag{5.19}$$

In order to determine the density function of X, it is necessary to transform Z, a standard normal variable using (5.19).

Theorem 5.4.1 If the probability density of a continuous random variable Z is f(z) and the distribution of X = h(Z) is desired where h(z) is differentiable and either increasing or decreasing for all values within the domain of z for which $f(z) \neq 0$, then the density of X is given by

$$g(x) = f(z)\left|\frac{dz}{dx}\right| \qquad \text{provided that } \left|\frac{dx}{dz}\right| \neq 0$$

Proof of this theorem appears in most standard introductory mathematical statistics books such as Freund (1971) and Mood et al. (1974).

First of all, since Z has the standard normal, the density f(z) is

$$f(z) = \frac{1}{\sqrt{2\pi}} \ e^{-(1/2)z^2} \qquad -\infty < z < \infty$$

An application of Theorem 5.4.1 yields

$$\left|\frac{dz}{dx}\right| = \frac{d}{dx}\left[\gamma + \eta \ \ln\frac{x - \epsilon}{\lambda}\right] = \frac{\eta}{x - \epsilon}$$

and

$$g_1(x) = \frac{\eta e^{-(1/2)\{\gamma + \eta \ \ln[(x-\epsilon)/\lambda]\}^2}}{\sqrt{2\pi}(x - \epsilon)} \tag{5.20}$$

The domain of x is determined directly from z.

When $z = -\infty$, $x = \epsilon$ [Note: $\ln(0) = -\infty$]

When $z = \infty$, $x = \infty$

Thus, the domain of x is (ϵ,∞). Equation (5.20) can be simplified by rewriting the exponent in the form

$$\gamma + \eta \ln(x - \varepsilon) - \eta \ln \lambda$$

and letting

$$\delta = \gamma - \eta \ln \lambda$$

Thus,

$$g_1(x) = \frac{\eta e^{-(1/2)[\delta+\eta \ln(x-\varepsilon)]^2}}{\sqrt{2\pi}(x - \varepsilon)} \qquad (5.21)$$

$$\varepsilon < x < \infty \qquad \eta > 0 \qquad -\infty < \delta < \infty \qquad -\infty < \varepsilon < \infty$$

which is a three-parameter version of the lognormal distribution discussed in Section 3.5 with the distribution origin being at ε. This form is called the Johnson S_L distribution and its members lie on a line in the (β_1, β_2) plane defined by equation (5.4).

The function for the distribution lying above the lognormal line in Figure 5.2 is given by

$$k_2(x;\varepsilon,\lambda) = \ln \frac{x - \varepsilon}{\lambda + \varepsilon - x}$$

The density function can be derived using Theorem (5.4.1), where

$$\frac{dz}{dx} = \frac{d}{dx}\{\eta[\ln(x - \varepsilon) - \ln(\lambda + \varepsilon - x)]\}$$

$$= \eta\left(\frac{1}{x - \varepsilon} + \frac{1}{\lambda + \varepsilon - x}\right) = \frac{\eta\lambda}{(x - \varepsilon)(\lambda + \varepsilon - x)}$$

and

$$g_2(x) = \frac{\eta\lambda e^{-(1/2)\{\gamma+\eta \ln[(x-\varepsilon)/(\lambda+\varepsilon-x)]\}^2}}{\sqrt{2\pi}(x - \varepsilon)(\lambda + \varepsilon - x)}$$

$$\qquad (5.22)$$

$$\eta > 0 \qquad -\infty < \gamma < \infty \qquad -\infty < \varepsilon < \infty \qquad \lambda > 0$$

The domain of x is obtained by examining the domain of z. When $z = -\infty$, $\ln[(x - \varepsilon)/(\lambda + \varepsilon - x)] = -\infty$ and this occurs when $x = \varepsilon$. When $z = \infty$, $\ln[(x - \varepsilon)/(\lambda + \varepsilon - x)] = \infty$ and this occurs when $x = \lambda + \varepsilon$. Thus, the domain of x is $(\varepsilon, \lambda + \varepsilon)$ and this bounded subfamily is called the Johnson S_B distribution.

The third subfamily of distributions which lie between the lognormal line in Figure 5.2 is obtained from the function

$$k_3(x;\varepsilon,\lambda) = \sinh^{-1}\left(\frac{x-\varepsilon}{\lambda}\right) = \ln\left\{\frac{x-\varepsilon}{\lambda} + \left[\left(\frac{x-\varepsilon}{\lambda}\right)^2 + 1\right]^{1/2}\right\}$$

Use of Theorem 5.4.1 and the fact that

$$\frac{d}{dx}\,\eta\,\sinh^{-1}(u) = \frac{\eta(du/dx)}{\sqrt{1+\mu^2}} = \frac{\eta}{\sqrt{\lambda^2 + (x-\varepsilon)^2}}$$

leads to

$$g_3(x) = \frac{\eta\,e^{-(1/2)\{\gamma+\eta\,\sinh^{-1}[(x-\varepsilon)/\lambda]\}}}{\{2\pi[\lambda^2 + (x-\varepsilon)^2]\}^{1/2}} \qquad\qquad (5.23)$$

$$\eta > 0 \qquad -\infty < \gamma < \infty \qquad -\infty < \varepsilon < \infty \qquad \lambda > 0$$

When $z = -\infty$, $x = -\infty$ since

$$\lim_{u\to-\infty}\ln(u + \sqrt{u^2 + 1}) = -\infty$$

When $z = \infty$, $x = \infty$. Thus x is unbounded and this subfamily is called the Johnson S_U distribution.

The methods for fitting each of the three families differ. Therefore, it is necessary to decide from the data which of the three families to select. The following procedure was developed by Slifker and Shapiro (1980) for the selection of a member of the Johnson system and can also be used in the estimation of the parameters.

1. Choose a value of z > 0 where z is a standard normal variate. The choice is motivated by two considerations, the number of observations and (for subsequent use) the points at which the percentiles will be matched. In the estimation procedure the matching occurs at the percentiles corresponding to 3z, z, -z, and -3z. Therefore, z should be chosen so that the fit covers areas of interest. Generally for moderate-size data sets, n < 1000, a value of z < 1 should be selected. A choice of z > 1 would make it difficult to estimate percentage points

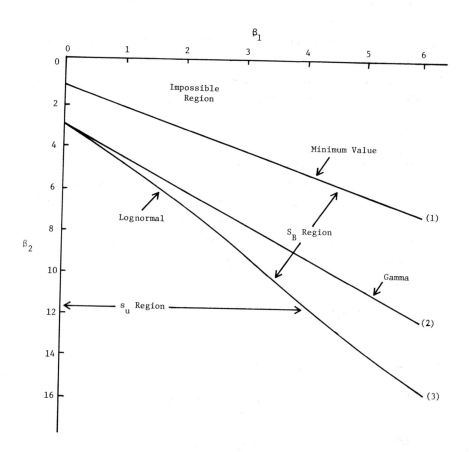

Figure 5.2 Standardized moment plot β_1 versus β_2.

from the data corresponding to $\pm 3z$ points, i.e., points beyond the 0.0014 and 0.9986 percentiles. Values of z close to 0.5 are recommended for moderate-size samples. For example, the choice of $z = 0.524$, which yields $3z = 1.572$, means the use of the 5.8%, 30%, 70%, and 94% points.

2. Once z is chosen, determine from Table A.1 $F(3z)$, $F(z)$, $F(-z)$, and $F(-3z)$.

3. Obtain from the data the x percentiles corresponding to $\pm z$ and $\pm 3z$ by solving for i using the relationship

$$\frac{i}{n} = F(k)$$

where $k = \pm 3z$ and $\pm z$ and n is the sample size. Thus, x_k equals the ith ordered observation corresponding to the value of k. If i is not an integer, interpolation is used (see Definition 5.2.4). Let x_{3z}, x_z, x_{-z}, and x_{-3z} be the percentiles corresponding to $F(3z)$, $F(z)$, $F(-z)$, and $F(-3z)$, respectively.

4. Compute the following quantities:

$$m = x_{3z} - x_z$$

$$n = x_{-z} - x_{-3z}$$

$$p = x_z - x_{-z}$$

5. Form the discriminant ratio mn/p^2 and if

$$\frac{mn}{p^2} = 1 \qquad \text{use a Johnson } S_L \text{ distribution}$$

$$\frac{mn}{p^2} > 1 \qquad \text{use a Johnson } S_U \text{ distribution}$$

$$\frac{mn}{p^2} < 1 \qquad \text{use a Johnson } S_B \text{ distribution}$$

It should be noted that since m, n, and p are random variables, the probability that $mn/p^2 = 1$ is nil. However, the S_L distribution can be used for values of the discriminant ratio close to 1. How much of a tolerance should be allowed depends

on the sample size, the application, and the value of reducing
the parameters to three instead of four. One should always
check the resulting fit. If one restricts his choice to the
S_U and S_B, the problem of how big a tolerance can be avoided.
If the Johnson S_L distribution is selected, the next step is to
estimate the parameters ε, η, and δ. Estimates can be obtained
by either matching of moments or percentiles (or by the methods
of Section 3.5). If ε is known, the relationship between the
normal and lognormal can be used to estimate η and δ. The
moments of the Johnson S_L distribution can be obtained from
direct evaluation of the integrals.

Theorem 5.4.2   If Y is normally distributed with mean μ and variance σ^2, then

$$E(e^{tY}) = e^{\mu t + (1/2)\sigma^2 t^2}$$

Proof.   See (3.49).

Now

$$E(X) = \int_\varepsilon^\infty \frac{\eta x e^{-1/2[\delta + \eta \, \ln(x-\varepsilon)]^2}}{\sqrt{2\pi}(x - \varepsilon)} \, dx$$

If one lets $y = \ln(x - \varepsilon)$, uses Theorem 5.4.1, notes that $y = -\infty$
when $x = \varepsilon$, and factors out η from the exponent, the expression for
$E(X)$ is

$$E(X) = \int_{-\infty}^\infty \frac{\eta e^y e^{-1/2\eta^2[(\delta/\eta)+y]^2}}{\sqrt{2\pi}} \, dy + \varepsilon \int_{-\infty}^\infty \frac{\eta e^{-1/2\eta^2[(\delta/\eta)+y]^2}}{\sqrt{2\pi}} \, dy$$

The first integral is just $E(e^Y)$ for a normal density and the
second integral is unity. Since $E(e^Y)$ is obtained from Theorem
5.4.2 with $t = 1$, then noting that $\mu = -\delta/\eta$ and $\sigma = 1/\eta$,

$$E(X) = e^{-\delta/\eta + (1/2)\eta^{-2}} + \varepsilon \tag{5.24}$$

Use of the same procedure provides

$$E(X^2) = \int_{-\infty}^{\infty} \frac{\eta}{\sqrt{2\pi}} (e^y - \epsilon)^2 e^{-(1/2)\eta^2[(\delta/\eta)+y]^2} dy$$

which can be written as

$$E(X^2) = E(e^{2Y}) - 2\epsilon E(e^Y) + \epsilon^2$$

Theorem 5.4.2 with t = 2 and t = 1 yields

$$E(X^2) = e^{-(2\delta/\eta)+(2/\eta^2)} + 2\epsilon e^{-(\delta/\eta)+(1/2)\eta^{-2}} + \epsilon^2$$

Then,

$$\mu_2(X) = var(X) = e^{-(2\delta/\eta)+(1/\eta^2)}(e^{1/\eta^2} - 1) \qquad (5.25)$$

In a similar manner

$$\mu_3(X) = e^{-(3\delta/\eta)+(3/2\eta^2)}(e^{1/\eta^2} - 1)(e^{1/\eta^2} + 2)$$

and

$$\sqrt{\beta_1} = (e^{1/\eta^2} - 1)^{1/2}(e^{1/\eta^2} + 2) \qquad (5.26)$$

An estimate of η is obtained by using the sample moments

$$m_2 = \sum_{i=1}^{n} \frac{(x_i - \bar{x})^2}{n}$$

and

$$m_3 = \sum_{i=1}^{n} \frac{(x_i - \bar{x})^3}{n}$$

and matching $\sqrt{\beta_1}$ to

$$\sqrt{b_1} = \frac{m_3}{m_2^{3/2}}$$

Thus, the estimate of η is obtained by solving the equation

$$\sqrt{b_1} = (e^{1/\eta^2} - 1)^{1/2}(e^{1/\eta^2} + 2) \tag{5.27}$$

for the positive root η^*. This can be done iteratively. Once η^* is obtained it is substituted into (5.25) and is matched with m_2 to solve for δ^*:

$$\delta^* = \frac{-\eta^*}{2}\left\{ \ln\left[\frac{m_2}{e^{\eta^{*-2}} - 1} \right] - \eta^{*-2} \right\} \tag{5.28}$$

and then using η^* and δ^* we solve for ε^* matching $\bar{x}$ with (5.24). Thus,

$$\varepsilon^* = \bar{x} - e^{-\delta^*/\eta^* + 1/(2\eta^{*2})} \tag{5.29}$$

If ε is known, then the relationship between the lognormal and normal distributions can be utilized to estimate δ and η. When X is lognormally distributed, then $Y = \ln(X - \varepsilon)$ is normally distributed. We set the parameters of the normal distribution to be

$$\mu = \frac{-\delta}{\eta} \qquad \text{and} \qquad \sigma^2 = \frac{1}{\eta^2}$$

The maximum likelihood estimators of μ and σ^2 are $\hat{\mu} = 1/n \sum_{i=1}^{n} \ln(x_i - \varepsilon)$ (see 3.52), and

$$\hat{\sigma}^2 = \sum_{i=1}^{n} \frac{(y_i - \bar{y})^2}{n} = \sum_{i=1}^{n} \frac{[\ln(x_i - \varepsilon)]^2}{n}$$

Then

$$\hat{\eta} = \frac{1}{\hat{\sigma}} \tag{5.30}$$

and

$$\hat{\delta} = -\hat{\mu}\hat{\eta}$$

See Section 3.5 for further discussion of maximum likelihood estimators.

The method of matching percentiles is very simple to use
since the percentiles of the Johnson S_L are related directly to
the standard normal percentiles by equation (5.19). If all three
parameters are unknown, then three matches are made. The solution
of the equations is facilitated if the median and two symmetrical
points corresponding to x_p and x_{1-p} are used. From equation
(5.19) the equations are

$$z_{1-p} = \delta + \eta \ln(x_{1-p} - \varepsilon)$$
$$0 = \delta + \eta \ln(x_{0.50} - \varepsilon) \tag{5.31}$$
$$z_p = \delta + \eta \ln(x_p - \varepsilon)$$

where x_p is the pth percentage point of the data as given in
Definition 5.2.4, and z_p is the pth percentile of the standard
normal distribution. Note that $z_{1-p} = -z_p$. Solution of these
equations yields the following estimates:

$$\varepsilon^* = \frac{x_p x_{1-p} - x_{0.50}^2}{x_p + x_{1-p} - 2x_{0.50}}$$

$$\eta^* = \frac{z_{1-p}}{\ln[(x_{1-p} - \varepsilon^*)/(x_{0.50} - \varepsilon^*)]} \tag{5.32}$$

$$\delta^* = -\eta^* \ln(x_{0.50} - \varepsilon^*)$$

Example 5.4.1 Use the method of moments to fit a Johnson S_L distribution to a set of data where $\bar{x} = 27.5$, $m_2 = 3.1$, and $m_3 = 25.2$.
Use of (5.27) gives

$$\sqrt{b_1} = \frac{25.2}{(3.1)^{3/2}} = 4.617 = (e^{1/\eta^2} - 1)^{1/2}(e^{1/\eta^2} + 2)$$

and yields by iteration

$$\eta^* = 1.124$$

Now use of (5.28) shows that

$$\delta^* = \frac{-1.124}{2}\left\{\ln\left[\frac{3.1}{\exp(1/(1.124)^2) - 1}\right] - \frac{1}{(1.124)^2}\right\} = -0.0857$$

and from (5.29),

$$\varepsilon^* = 27.5 - \exp\left[\frac{-0.0857}{1.124} + \frac{1}{2(1.124)^2}\right] = 26.123$$

Example 5.4.2 Use the method of matching percentiles to fit a Johnson S_L distribution to a data set where $x_{0.05} = 15.43$, $x_{0.50} = 18.75$, and $x_{0.95} = 26.10$. The corresponding values of z are -1.645, 0, and 1.645. Use of (5.32) yields

$$\varepsilon^* = \frac{25.1(15.43) - (18.75)^2}{26.1 + 15.43 - 2(18.75)} = 12.695$$

$$\eta^* = \frac{1.645}{\ln[(26.1 - 12.695)/(18.75 - 12.695)]} = 2.070$$

$$\delta^* = 2.70 \ln(18.75 - 12.695) = 3.728$$

The cumulative distribution function for a given value of X can be obtained directly from Table A.1 by computing

$$Z = \delta + \eta \ln(X - \varepsilon) \tag{5.33}$$

which has a standard normal distribution. Likewise to obtain x_p, the percentile for a given value of p one merely finds the corresponding z_p from a standard normal table and solves for x_p using (5.33):

$$x_p = \exp\left[\frac{z_p - \varepsilon}{\eta}\right] + \varepsilon \tag{5.34}$$

Example 5.4.3 Find the probability that X will be less than 26.20 in Example 5.4.1. First calculate the normal deviate,

$$Z = -0.0857 + 1.124 \ln(26.20 - 26.123) = -2.97$$

then, using Table A.1,

$$\text{pr}\{X < 26.20\} = \text{pr}\{Z < -2.97\} = 0.0015$$

To determine the 70th percentile of the distribution for Example
5.4.1, equation (5.34) is used with the normal percentile $z_{0.70} = 0.52$:

$$x_{0.70} = \exp\left[\frac{0.52 + 0.0857}{1.124}\right] + 26.123$$

$$= 27.84$$

Random variables from the Johnson S_L are generated directly from
standard normal deviates using equation (5.34) by generating the
random normal variate Z and solving for X.

The curves falling above the lognormal line in Figure 5.2
belong to the Johnson S_B family. These curves are all bounded on
$(\varepsilon, \lambda + \varepsilon)$. The density function can be unimodal or bimodal
depending on whether there are one or three intersections of curve

$$U = \eta^2 \ln\left[\frac{1 + Y'}{1 - Y'}\right]$$

and the line

$$U = Y' - \gamma\eta$$

where $Y' = 2Y - 1$, and Y is the standardized variate on the unit
interval, i.e., $\varepsilon = 0$, $\lambda = 1$.

Computation of the moments can be obtained from the recursion
relationship given by Johnson (1949):

$$\mu'_{r+1} = \mu'_r + \frac{\eta}{r}\frac{\partial \mu'_r}{\partial \gamma}$$

The form of these expressions is quite complex and not required in
subsequent derivations. The interested reader can find these
expressions in Johnson's article. The method of moments is not
particularly useful in fitting the S_B distribution to a set of
data due to its complex form. Instead the method of matching of
percentiles is described. The procedure depends on whether none,
one, or both endpoints are known.

If the range of variation is known then ε is the minimum value
of the variate and $\varepsilon + \lambda$ is the maximum. The values of γ and η
are obtained by matching two percentiles; usually symmetrical
points are chosen from each tail. Use of the relationship

$$z = \gamma + \eta \ln\left(\frac{x - \varepsilon}{\varepsilon + \lambda - x}\right)$$

allows one to match, using symmetrical points in each tail,

$$-z_{1-p} = \gamma + \eta \ln\left(\frac{x_p - \varepsilon}{\varepsilon + \lambda - x_p}\right)$$

$$z_{1-p} = \gamma + \eta \ln\left(\frac{x_{1-p} - \varepsilon}{\varepsilon + \lambda - x_{1-p}}\right)$$

This yields

$$\eta^* = \frac{2z_{1-p}}{\ln[(x_{1-p} - \varepsilon)(\varepsilon + \lambda - x_p)/(x_p - \varepsilon)(\varepsilon + \lambda - x_{1-p})]} \qquad (5.35)$$

and

$$\gamma^* = z_{1-p} - \eta^* \ln\left(\frac{x_{1-p} - \varepsilon}{\varepsilon + \lambda - x_{1-p}}\right) \qquad (5.35a)$$

If only one terminus of the distribution is known, for example,
the minimum value ε, then three percentiles must be matched.
Use of the median ($z_{0.50} = 0$) and symmetric percentiles in each
tail results in the estimator

$$\lambda^* = a\left[\frac{a(b + c) - 2bc}{a^2 - bc}\right] \qquad (5.36)$$

where

$$a = x_{0.5} - \varepsilon$$

$$b = x_p - \varepsilon$$

$$c = x_{1-p} - \varepsilon$$

If λ is known, then ε is estimated from the larger root (satisfying $x_i \geq \varepsilon$ for all i) of

$$\varepsilon^* = -b \pm \sqrt{b^2 - a} \qquad (5.36a)$$

where

$$b = \left(\frac{x_{0.50}^2 - x_p x_{1-p}}{x_p + x_{1-p} - 2x_{0.50}} \right) \frac{1}{x_{0.50}}$$

$$a = \lambda b + \frac{2x_p x_{1-p} - x_{0.50}(x_p + x_{1-p})}{x_p + x_{1-p} - x_{0.50}}$$

Then equations (5.35) and (5.35a) are used to obtain estimators of η and γ by using either λ^* or ε^*. In the case where none of the parameters is known, four percentiles are matched. The resulting equations can be solved analytically if symmetric percentiles in each tail are chosen and the percentiles are selected so that the z value for the point in the extreme tail is 3 times the value of the point close to the center of the distribution as was suggested in the selection process above. Hence, matching on the percentiles corresponding to

$$3z, z, -z, \text{ and } -3z$$

enables one to solve the equations analytically.

Use of the notation described in the selection procedure shows the estimators of the parameters are

$$\eta^* = \frac{z}{\cosh^{-1}\{(1/2)[(1 + p/n)(1 + p/m)]^{1/2}\}}$$

$$\gamma^* = \eta^* \sinh^{-1}\left\{ \frac{(p/n - p/m)[(1 + p/n)(1 + p/m) - 4]^{1/2}}{2(p^2/mn - 1)} \right\}$$

$$(\gamma^* > 0 \text{ if } m > n \text{ and } \gamma^* < 0 \text{ if } m < n) \qquad (5.37)$$

$$\lambda^* = \frac{p\{[(1 + p/n)(1 + p/m) - 2]^2 - 4\}^{1/2}}{p^2/mn - 1}$$

$$\varepsilon^* = \frac{x_z + x_{-z}}{2} - \frac{\lambda^*}{2} + \frac{p(p/m - p/n)}{2(p^2/mn - 1)}$$

Example 5.4.4 A sample of 200 observations is obtained of the current passing through a resistor (see Table 5.2). The minimum value of the current is zero and due to a fuse in the circuit the maximum value is 5 A. It is desired to fit a Johnson S_B distribution to this set of data. Since there are but 200 observations it is not feasible to determine percentage points that are extreme as 1%; reasonable choices for the symmetrical points are the 6.7 and the 93.3 percentiles corresponding to $z = \pm 1.5$. The percentiles from the data are obtained by solving for i, the order number, in the relationship

$$p = \frac{i}{n}$$

where

$p = 0.933$ and 0.067

$n = 200$

Table 5.2 Current in Resistors (A). (Partial Listing)

Order #	Reading	Order #	Reading
12	2.186	137	3.398
13	2.187	138	3.400
14	2.193	139	3.400
15	2.195	140	3.401
.	.	.	.
.	.	.	.
60	2.800	185	3.881
61	2.803	186	3.887
62	2.815	187	3.892
63	2.817	188	3.904
.	.	.	
.	.	.	
99	3.150	.	
100	3.150		
101	3.151		

and then obtaining x_p corresponding to the order number. Since i
in most cases will be fractional, it is usually necessary to
interpolate to obtain x_p. In this example i_1 = 186.6 and i_2 = 13.4.
Thus, $x_{0.933}$ is obtained via interpolation as the 186th and 187th
ordered value. Likewise $x_{0.067}$ is obtained via interpolation from
the 13th and 14th ordered values. For this data set $x_{0.933}$ = 3.89
and $x_{0.67}$ = 2.19. Since 0 and 5 are the minimum and maximum values,

$$\varepsilon = 0 \qquad \text{and} \qquad \lambda = 5$$

Use of equation (5.35) yields

$$\eta^* = \frac{2(1.5)}{\ln[(3.89)(5.0 - 2.19)/(2.19)(5.0 - 3.89)]} = 1.996$$

and equation (5.35a) yields

$$\gamma^* = 1.5 - 1.996 \ln\left(\frac{3.89}{5 - 3.89}\right) = 1.003$$

If only the minimum value were known (which would occur if there
were no fuse in the device), then another equation matching the
50th percentile would be used to obtain an estimate of λ. Given
that $x_{0.50}$ = 3.15, then an estimate of λ is obtained from equation
(5.36):

$$\lambda^* = 3.15 \frac{3.15(3.89 + 2.19) - 2(3.89)(2.19)}{(3.15)^2 - (3.89)(2.19)} = 4.74$$

This estimate of λ is used in the preceeding equations to get estimates of η and γ, i.e.,

$$\eta^* = \frac{2(1.5)}{\ln[(3.89)(4.74 - 2.19)/(2.19)(4.74 - 3.89)]} = 1.793$$

$$\gamma^* = 1.5 - 1.793 \ln\left(\frac{3.89}{4.74 - 3.89}\right) = -1.227$$

If none of the parameters is known, it is necessary to match four
percentiles. If z = 0.5, then 3z = 1.5 allows for the direct parameter solution. From the previous parts of this problem, x_{3z} =
3.89 and x_{-3z} = 2.19. The percentage points corresponding to
z = ±0.5 are 0.692 and 0.308 and thus the appropriate values for

x_z and x_{-z} from Table 5.2 are $x_2 = 3.40$ and $x_3 = 2.81$.   Now

$$m = 3.89 - 3.40 = 0.49$$
$$n = 2.81 - 2.19 = 0.62$$
$$p = 3.40 - 2.81 = 0.59$$

and

$$\frac{mn}{p^2} = 0.873$$

Therefore, an S_B distribution is appropriate since the discriminant ratio is less than 1.   Use of the equations (5.37) yields

$$\eta^* = \frac{0.5}{\cosh^{-1}\{(1/2)(1.952)(2.205)^{1/2}\}} = 1.836$$

$$\gamma^* = 1.836 \sinh^{-1} \frac{(0.952 - 1.205)[(1.952)(2.205) - 4]^{1/2}}{2(1.146 - 1)}$$

$$= -0.847$$

Note, since $m < n$, $\gamma^* < 0$.

$$\lambda^* = \frac{0.59\{[(1.952)(2.205) - 2]^2 - 4\}^{1/2}}{1.146 - 1} = 4.624$$

$$\epsilon^* = \frac{3.40 + 2.81}{2} - \frac{4.625}{2} + \frac{0.59(0.952 - 1.205)}{2(1.146 - 1)} = 0.282$$

The value of the cumulative distribution function for a given X can be obtained directly from Table A.1 by transforming back to Z using

$$z = \gamma + \eta \ln\left(\frac{x - \epsilon}{\lambda + \epsilon - x}\right) \tag{5.38}$$

Likewise, to obtain x_p corresponding to a desired percentage point p one determines the value z_p from a standard normal table corresponding to p and solves for x in (5.38).   Thus,

$$x_p = \frac{\epsilon + (\epsilon + \lambda)e^w}{1 + e^w} \tag{5.39}$$

where

$$w = \frac{z_p - \gamma}{\eta}$$

Example 5.4.5 Find the probability that X will be less than 3.00
in Example 5.4.4 for the case where none of the parameters is known.
Using (5.38) and transforming to the standard normal deviate, one
obtains

$$z = -0.847 + 1.836 \, \ln\left(\frac{3.0 - 0.282}{4.906 - 3.0}\right)$$

$$= -0.195$$

Thus,

$$\text{pr}\{X < 3.0\} = \text{pr}\{Z < -0.195\} = 0.422$$

To determine the 70th percentile of the distribution of X for this
same example we use equation (5.39) with

$$z_{0.70} = 0.52$$

$$x_{0.70} = \frac{0.282 + (4.906) \, \exp[(0.52 + 0.847)/1.836]}{1 + \exp[(0.52 + 0.847)/1.836]} = 3.42$$

Finally, it should be noted that since normal variables are available on a computer, one can use (5.39) to transform these to
Johnson S_B variates.

The curves falling below the lognormal line in Figure 5.2 belong to the Johnson S_U family. These curves are unimodal with a
domain of $(-\infty,\infty)$. The moments of the distribution can be obtained
directly from the normal distribution by solving for x, where

$$z = \gamma + \eta \, \ln\left\{\left(\frac{x - \epsilon}{\lambda}\right) + \left[\left(\frac{x - \epsilon}{\lambda}\right)^2 + 1\right]^{1/2}\right\} \tag{5.40}$$

Let

$$w = (z - \gamma)/\eta. \quad \text{Then}$$

$$x = \frac{\lambda[e^{2w} - 1]}{2e^w} + \epsilon \tag{5.41}$$

$$E(X^r) = E\left[\frac{\lambda(e^{2w} - 1)}{2e^w} + \varepsilon\right]^r$$

Since ε is a location parameter and λ a scale parameter, one need consider only the moments of the standardized variate with $\varepsilon = 0$ and $\lambda = 1$. Let

$$Y = \frac{X - \varepsilon}{\lambda}$$

$$E(Y^r) = \frac{1}{\sqrt{2\pi}} \int_{-\infty}^{\infty} \left[\frac{e^{2(y-\gamma)/\eta} - 1}{2e^{(y-\gamma)/\eta}}\right]^r e^{-(1/2)y^2} dy$$

$$= \frac{1}{2^{[(2r+1)/2]}\sqrt{\pi}} \int_{-\infty}^{\infty} \left[e^{(y-\gamma)/\eta} - e^{-(y-\gamma)/\eta}\right]^r e^{-(1/2)y^2} dy$$

If one sets $r = 1$,

$$E(Y) = -a \sinh b$$

where

$$a = e^{(1/2)\eta^2}$$

and

$$b = \frac{\gamma}{\eta}$$

Thus,

$$E(X) = -\lambda a \sinh b + \varepsilon \qquad\qquad (5.42)$$

Likewise,

$$\text{var}(Y) = \frac{1}{2}(a^2 - 1)(a^2 \cosh 2b - 1)$$

$$\text{var}(X) = \lambda^2 \text{var}(Y) \qquad\qquad (5.42a)$$

$$\mu_3(Y) = -\frac{1}{4} a^2(a^2 - 1)^2 [a^2(a^2 + 2) \sinh 3b + 3 \sinh b] \qquad (5.42b)$$

Note that when sinh 3b is positive, the skewness is negative.
Since $b = \gamma/\eta$ and $\eta > 0$ the direction of skewness is determined by
the sign of γ. If γ is positive, then sinh b is positive and μ_3
is negative. If γ is negative, the skewness is positive.

$$\mu_4(Y) = \frac{1}{8}(a^2 - 1)^2 [a^4(a^8 + 2a^6 - 3a^4 - 3)\cosh 4b$$

$$+ 4a^4(a^2 + 2)\cosh 2b + 3(2a^2 + 1)]$$

The method of percentiles is the easiest way to estimate the parameters of the distribution.

Again, using the notation which was developed in the selection
process, the formulas for the matching of four percentiles are as
follows:

$$\eta^* = \frac{2z}{\cosh^{-1}[(1/2)(m/p + n/p)]}$$

$$\gamma^* = \eta^* \sinh^{-1}\left[\frac{n/p + m/p}{2(mn/p^2 - 1)^{1/2}}\right]$$

(If $m < n$, γ is positive; otherwise it is negative.)

$$\lambda^* = \frac{2p(mn/p^2 - 1)^{1/2}}{(m/p + n/p - 2)(m/p + n/p + 2)^{1/2}} \tag{5.43}$$

and

$$\varepsilon^* = \frac{x_z + x_{-z}}{2} + \frac{p(n/p - m/p)}{2(m/p + n/p - 2)}$$

The method of moments can also be used but requires a set of
tables which have been constructed by Johnson (1965) that simplify
the necessary calculations. Estimates of η and γ can be obtained
directly from these tables based on the values of $\sqrt{b_1}$ and b_2 calculated from the data (see Table A.4). The sign of γ is taken as
opposite of the sign of $\sqrt{b_1}$, since $-\gamma$ implies positive skewness
and conversely. Next (5.42) and (5.42a) are matched to the sample

mean and variance respectively to obtain λ^* and ε^*:

$$\lambda^* = \frac{m_2^{1/2}}{[(1/2)(a^2 - 1)(a^2 \cosh 2b - 1)]^{1/2}} \tag{5.44}$$

and

$$\varepsilon^* = \bar{x} + \lambda^* a \sinh b \tag{5.45}$$

where γ^* and η^* are substituted for γ and η, respectively.

Example 5.4.6   The following data were abstracted from Johnson (1949) concerning the measurements of the lengths of 9440 beans. Due to the large number of observations, it is reasonable to use a value of $z = 1$.   Then

$$x_{3z} = x_{0.9986} = 16.68$$
$$x_z = x_{0.8413} = 15.24$$
$$x_{-z} = x_{0.1587} = 13.58$$
$$x_{-3z} = x_{0.0014} = 10.16$$

and

$$m = 1.447$$
$$n = 3.172$$
$$p = 1.661$$

The discriminant ratio

$$\frac{mn}{p^2} = 1.664$$

and hence the use of an S_U distribution is appropriate.   Use of equations (5.43) yields

$$\eta^* = \frac{2(1)}{\cosh^{-1}[(1/2)(0.87 + 1.910)]} = 2.333$$

$$\gamma^* = 2.333 \sinh^{-1}\left[\frac{1.910 - 0.871}{2[(1.910)(0.871) - 1]^{1/2}}\right] = 1.402$$

γ^* is positive since m < n,

$$\lambda* = \frac{2(1.66)(1.664 - 1)^{1/2}}{(0.871 + 1.910 - 2)(0.871 + 1.910 + 2)^{1/2}} = 1.585$$

and

$$\epsilon* = \frac{15.242 + 13.581}{2} + \frac{1.661(1.910 - 0.871)}{2(1.910 + 0.871 - 2)} = 15.516$$

Example 5.4.7 The following values were calculated from a set of data:

$$\sqrt{b_1} = 1.3 \qquad \bar{x} = 23.2$$

$$b_2 = 8.1 \qquad m_2 = 4.1$$

Select the appropriate Johnson family of distributions and estimate the parameters using the method of moments. Since $\sqrt{b_1}$ = 1.3 and b_2 = 8.1 the Johnson S_U distribution is selected, since the point (1.3,8.1) lies below the lognormal line in the (β_1,β_2) plane. One finds using Table A.4,

$\eta* = 1.672$

$\gamma* = -0.971$

The sign for $\gamma*$ is negative since $\sqrt{b_1}$ is positive. From (5.44) the estimate of λ is obtained:

a = 1.430

b = -0.581

and hence

$$\lambda* = \frac{(4.1)^{1/2}}{\{(1/2)[(1.430)^2 - 1][(1.430)^2 \cosh(-1.161) - 1]\}^{1/2}} = 3.556$$

and the estimate of ϵ is from (5.45) is

$\epsilon* = 23.2 + (3.556)(1.430) \sinh(-0.581) = 20.077$

The value of the cumulative distribution function for a given value of X can be obtained directly from a table of areas of the

normal distribution by transforming back to Z using (5.40). Likewise, to obtain x_p corresponding to a desired percentage point p,
equation (5.41) can be used setting $x = x_p$ and $z = z_p$, where z_p
is the standard normal percentile corresponding to p. The relationship (5.41) can also be used to generate random deviates from the
Johnson S_U distribution from normal deviates.

Problems

1. Find the appropriate subfamily (S_L, S_U, or S_B) of the Johnson
    system for the following values of $\sqrt{\beta_1}$ and β_2:

    (a) $\sqrt{\beta_1} = 0.89$, $\beta_2 = 3.6$

    (b) $\sqrt{\beta_1} = 1.10$, $\beta_2 = 7.3$

    (c) $\sqrt{\beta_1} = -3.87$, $\beta_2 = 34.0$

    (d) $\sqrt{\beta_1} = 7.01$, $\beta_2 = 159$

2. Use Theorem 5.4.2 to derive $\mu_3(X)$, equation (5.26), for the
    Johnson S_L distribution.

3. A new clinical drug is being investigated. One of its charac-
    teristics is the time to reduce a patient's temperature to
    normal. The drug was administered to 15 sick patients and the
    time required to reduce a person's temperature to normal were
    recorded as given below. Use a Johnson S_L distribution to find
    the probability that a patient's fever could last for over 47
    hrs. after given the drug. Use the method of matching moments.

Time to Reduce Fever to Normal (hr)

4.1	12.6	8.4
44.0	3.4	4.2
31.3	10.1	4.3
19.2	16.5	0.8
11.9	28.0	3.8

4. One of the inputs into setting up a machine maintenance schedule
    is the number of hours it takes to overhaul a machine. The
    overhaul time for a group of 24 lathes was recorded. Fit a
    Johnson S_L distribution to the data given below and estimate

the 90th percentile, the repair time which will be exceeded
only 10% of the time.

Time to Overhaul Machine (hr.)

16.50	16.41	16.91	16.94	16.65
16.19	15.67	16.23	15.70	15.69
16.72	18.90	16.19	19.37	17.26
23.01	16.63	15.80	18.20	16.48
19.68	16.49	16.03	16.87	

5. Redo Problem 4 under the assumption that the minimum time to
    overhaul a machine is 15 hrs. Use the maximum likelihood
    estimators of the parameters.

6. A company is engaged in offshore mining of nodules. They wish
    to estimate the possible yield of a given area of the ocean.
    They randomly sample 50 nodules off the ocean floor in a 1/4-
    square-mile area and determine the weight of each nodule. Fit
    a Johnson distribution to the data and estimate the probability
    of recovering a nodule weighing in excess of 200 lb. Use the
    method of matching percentiles with z = 0.5.

Weight of Nodules (lb)

20.8	25.9	24.3	82.2	60.1	39.3	103.1	44.1	122.5	29.4
22.3	17.9	17.6	19.8	27.1	118.5	38.5	193.1	21.5	23.3
18.0	28.1	26.1	27.9	111.8	22.1	23.2	70.9	22.0	29.7
21.6	38.7	117.5	32.2	32.6	22.6	105.6	26.2	21.1	35.5
20.7	30.6	22.2	30.8	36.1	22.6	43.1	171.4	149.0	35.9

7. A company has developed a new cord which it plans to market.
    A random sample of 30 1-ft sections are selected from an initial
    production lot for use in a strength-of-materials test. A
    section of cord is placed in a machine and a load of 25 lb is
    placed on it. The load is gradually increased until the cord
    breaks. The breaking loads are given below. Fit a Johnson
    distribution to the data and determine what proportion of the
    population will be able to withstand a load of 62 lb. Estimate

all of the parameters.

Breaking Strength of Cord (lb)

37	46	38	29	40	35
43	31	51	45	47	34
44	40	48	41	61	39
46	39	50	42	35	41
48	53	59	47	50	40

8. The efficiency of a motor is a key measurement. The efficiencies of a random sample of 25 motors were measured; the results are shown below. If the company must certify that the motor has a minimum efficiency of 0.50, what percentage of the motors will fail to meet this criterion? Use a Johnson distribution to answer this question.

Efficiency of Motor

0.59	0.81	0.59	0.52	0.72
0.46	0.62	0.65	0.61	0.56
0.69	0.73	0.57	0.58	0.46
0.76	0.75	0.64	0.62	0.63
0.52	0.63	0.49	0.66	0.76

9. A CPA took a random sample of size 200 of a brokerage firm's small-order accounts. Select a Johnson distribution to approximate the distribution of account sizes and estimate the median account size. The 93rd, 69th, 31st, and 7th percentiles of the sample were (in thousands of dollars) $4.20, $3.42, $2.72, and $2.50.

10. Derive the $E(X^2)$ for the lognormal distribution and then determine the equation (5.25).

11. Derive the equation for estimating ε for the Johnson S_B distribution given that λ is known. Use the matching of percentile technique.

12. Derive the $E(X)$ for the Johnson S_U distribution. Note that $\sinh x = (e^x - e^{-x})/2$.

13. A simulation model was being constructed of a complex steam
     generation plant. One of the important elements was the time
     it took for a safety valve to close. A series of 20 tests
     was performed; the closure times are shown below. Due to a
     readily available source of normal deviates, it was decided
     to use a Johnson distribution to model the time it took the
     valve to close. Estimate the parameters of the model and
     determine the probability that the closure time will exceed
     2.60 sec.

     Time to Close (sec.)

2.23	2.21	2.41	2.27
2.28	2.35	2.59	2.34
2.22	2.38	2.37	2.43
2.36	2.39	2.52	2.22
2.22	2.31	2.45	2.58

14. A psychologist who developed a new testing procedure found that
     the Johnson S_U distribution gave a good approximation to his
     test results. If the approximating distribution had parameters
     $\epsilon = 70$, $\lambda = 3$, $\gamma = -1$, $\eta = 2$, find the mean and variance of the
     test scores. Also find the 95th percentile of the distribution.

5.5 THE PEARSON FAMILY

Another system that is used to approximate a set of data is the
Pearson family. There are 12 distinct members of the family, all
of which can be derived as solutions of the differential equation

$$\frac{df(x)}{dx} = \frac{(x - a)f(x)}{b + cx + dx^2} \tag{5.46}$$

The solutions depend on the values of the four parameters a, b, c,
and d. Some of the models of Chapter 3 such as the normal, gamma,
and beta distributions are included in the system.

     The major advantage of the use of this system is the existence
of tables of its percentiles, indexed as a function of β_1 and β_2.

Therefore, if the user is only interested in the percentiles corresponding to the tabulated percentage points, the system provides
a simple method of obtaining these subject to the problems of
using estimates of β_1 and β_2. However, if other information is
desired, such as the actual density functions, parameter estimates,
or other percentiles that are not tabulated, the family is not as
easy to use as those described previously. This is due to the fact
that there are 12 members in the family that cover the (β_1, β_2)
plane (see Figure 5.2) and it is necessary first to select one of
the 12, then use the parameter estimating equations for the specific member, and finally obtain a set of tables or integrate, where
possible, the density function for that member. One also has the
problem that some of the members are only points or lines in the
(β_1, β_2) plane and if these are to be given a chance of being
selected from sample estimates of β_1 and β_2 it is necessary to put
a tolerance interval about each. Thus, the problem of selection
is more complex than in the three-member Johnson system.

If only information on the tabulated percentiles is desired
then the following procedure can be used.[†]

First, calculate the first four data moments $\bar{x}$, m_2, m_3, and
m_4. Obtain $\sqrt{b_1}$ and b_2. Next, using Table A.5 obtain the standardized percentile z_p (location parameter assumed to be zero and
scale parameter assumed to be 1). If $\sqrt{b_1}$ is greater than zero,
then the deviates in the lower tail ($p < 0.50$) are negative. The
desired percentiles are given by

$$x_p = \sqrt{m_2}\, z_p + \bar{x} \qquad\qquad (5.47)$$

Table A.5 gives the percentiles corresponding to $p = 0.005$, 0.01,
0.025, 0.05, 0.95, 0.975, 0.99, and 0.995. A more comprehensive
table is given by Amos and Daniel (1971).

[†]Techniques for selecting a Pearson curve based on three moments and
an endpoint are discussed in Hoadly (1968) and Muller and Vahl
(1976).

Example 5.5.1 Find the 5th and 95th percentiles of the approximating Pearson distribution for which the following sample estimates were obtained; $\bar{x} = 16.3$, $m_2 = 3.1$, $\sqrt{b_1} = 0.316$, and $b_2 = 4.2$. Referring to Table A.5 one finds for $\beta_1 = 0.10$ and $\beta_2 = 4.2$,

$$z_{0.05} = -1.55 \qquad \text{and} \qquad z_{0.95} = 1.68$$

Use of (5.47) shows the desired percentiles:

$$x_{0.05} = -1.55\sqrt{3.1} + 16.3 = 13.57$$
$$x_{0.95} = 1.68\sqrt{3.1} + 16.3 = 19.26$$

The members of the family are designated type I, type II, etc. Types I and II are the beta distributions, type III is a gamma distribution, and type VII is the normal distribution. More details of these distribution are found in Kendall and Stuart (1958).

Example 5.5.2 In equation (5.46) let $-b$ replace b, and $c = d = 0$, and solve for $f(x)$:

$$\frac{df(x)}{dx} = \frac{-(x - a)f(x)}{b} \qquad \text{or} \qquad \frac{df(x)}{f(x)} = \frac{-(x - a)}{b} dx$$

Integrating both sides yields

$$\log f(x) = \frac{-(x - a)^2}{2b} + K$$

$$f(x) = Ke^{-(x-a)^2/2b}$$

This yields the normal density function when $b > 0$ and K is chosen so that the area under $f(x)$ is unity, i.e., $K = 1/\sqrt{2\pi b}$ (see equation 2.26). Other members can be derived by an appropriate choice of the parameters. See the problems section.

Problems

1. Solve equation (5.46) under the following conditions and identify the resulting distributions:

    (a) Let $a = b = d = 0$, $c < 0$.

    (b) Let $b = d = 0$, $c < 0$, and $a < c$.

2. Given that $\bar{x} = 37.5$, $m_2 = 4.2$, $\sqrt{b_1} = 0.63$ and $b_2 = 4.25$, fit a Pearson distribution using these summary statistics and find the 95th percentile of the approximating distribution.

3. In a study of pain and discomfort a psychologist records the time at which 30 subjects push a panic button in order to be released from an unpleasant experimental condition. He wants to determine the time by which 99% of the population will opt for termination. Using a Pearson distribution approximation estimate this value based on the following data:

$$\bar{x} = 23.2, \quad m_2 = 4.1, \quad m_3 = 7.4, \text{ and } m_4 = 67.2$$

4. Use the Pearson approximation with the data in Problem 6 in Section 5.3 to determine the 5th percentile of the distribution.

REFERENCES

Amos, D. E., and Daniel, S. L. (1971). Tables of percentage points of standardized Pearson distributions, Research Report, Sandia Laboratories, Albuquerque, New Mexico.

Dudewicz, E. J., Ramberg, J. S., and Tadikamalla, P. R. (1974). A distribution for data fitting and simulation, *An. Tech. Conf. Trans. Amer. Soc. Qual. Control 28*, pp. 407-418.

Freund, J. E. (1971). *Mathematical Statistics*, Prentice-Hall, Inc., Englewood Cliffs, N.J., p. 120.

Hastings, C., Mosteller, F., Tukey, J., and Winsor, C. (1947). Low moments for small samples: A comparative study of order statistics, *Ann. Math. Stat. 18*, pp. 413-426.

Hoadley, A. B. (1968). Use of the Pearson densities for approximating a skew density whose left terminal and first three moments are known, *Biometrika 55*, pp. 559-563.

Johnson, M. E., and Lowe, V. W. (1979). Bounds on the sample skewness and kurtosis, *Technometrics 21*, pp. 377-378.

Johnson, N. L. (1965). Tables to facilitate fitting S_U frequency curves, *Biometrika 52*, pp. 547-558.

Johnson, N. L. (1949). Systems of frequency curves generated by methods of translation, *Biometrika 36*, pp. 274-282.

Kendall, M. G., and Stuart, A. (1958). *The Advanced Theory of Statistics*, Charles Griffin & Company Limited, London.

Mood, A. M., Graybill, F. A., and Boes, D. C. (1974). *Introduction to the Theory of Statistics*, McGraw-Hill Book Company, New York, p. 200.

Muller, P. H., and Vahl, H. (1976). Pearsons system of frequency curves whose left boundary and first three moments are known, *Biometrika 63*, pp. 191-194.

Ramberg, J. S., and Schmeiser, B. W. (1974). An approximate method for generating asymmetric random variables, *Comm. ACM 17*, pp. 78-82.

Ramberg, J. S., Tadikamalla, P. R., Dudewicz, E. J., and Mykytha, E. F. (1979). A probability distribution and its uses in fitting data, *Technometrics 21*, pp. 201-214.

Slifker, J. F., and Shapiro, S. S. (1980). The Johnson system: selection and parameter estimation, *Technometrics 22*, pp. 239-246.

Chapter 6

TESTING MODEL ASSUMPTIONS

6.1 INTRODUCTION

Previous chapters have dealt with the selection of statistical
models based on knowledge of the underlying mechanism or on common
usage. This chapter describes techniques that are used to evaluate
the reasonableness of such a selection when the analyst has available measurements. These techniques can also be used to aid in the
model selection process by determining, in light of the data, which
models are reasonable to select with regard to a system under study.
Thus, if in a study of failure times of a particular mechanism, it
was determined that the lognormal distribution is a likely candidate to model a process in a simulation study, the techniques presented here may be used to evaluate this choice by obtaining a
sample of observations from the process and testing whether such an
assumption is reasonable. In another situation suppose there is
a set of life test data for which there is no information, a priori,
to select a specific distributional form. The techniques presented
here are useful in ascertaining which models (from a list of possibilities) are reasonable candidates to use in further analysis of
the data. Most of these procedures follow the classical test of
hypothesis procedures. Namely, one chooses a test level α and
assumes the model of interest to be correct (null hypothesis), calculates a test statistic based on this assumption, determines
whether the computed statistic falls within a critical region of
size α (type I error, see Section 2.7) and rejects the model if it

does or accepts the model if it does not.

In testing models the following points should be noted:

(a) Statistical models are idealizations of real-life phenomena and don't really exist. Thus, the purpose of a distributional test is to assess whether the use of a selected model yields reasonable inferences. The procedure is not to test whether the output is actually distributed as specified. Remember, no real-world data are exactly normally distributed.

(b) It is quite likely that one may find several possible models for which a null hypothesis is not rejected. That is, one might not reject the hypothesis of a lognormal, Weibull, or gamma distribution for the repair time distribution of a given assembly. All three may appear to be "reasonable" based on the criteria of a statistical test. In such cases the selection should be based on analysis of the mechanism, on what has been used in previous studies, and/or on the principle of simplicity, i.e., use the model which has the simplest and easiest applicable form.

(c) In setting the type I error for a distributional test it is often reasonable to use α values larger than the traditional 0.05 because consequences of falsely rejecting a model are usually small, a search for another model or transformation of the data is made. On the other hand, the use of an erroneous model can be costly especially if predictions concerning tail areas are to be made. Therefore, α values of 0.10 to 0.20 are more reasonable to use.

(d) Testing models without shape parameters (or with known shape parameters) is conceptually quite different than testing distributions with unknown shape parameters. For example, testing whether the hypothesis of an exponential distribution is supported by the data limits the shape of the null distribution to a single form. On the other hand, a test for the Weibull distribution is equivalent to asking whether one of the infinitely many shapes of the Weibull distribution fit the given

data. These shapes, in fact, include the exponential distribution. Thus, the test of models with no unspecified shape
parameters is a more stringent procedure than testing an entire
family of distributions that can have a wide range of shapes.

(e) In testing a distributional null hypothesis against an unspecified alternative (for example, in testing the null hypothesis
or normality vs. nonnormality) there is no uniformly most
powerful nor optimal test. Due to this, large numbers of test
procedures exist and, depending on the null hypothesis, alternate distribution, and sample size, one procedure can be shown
to be more powerful in one situation and less powerful in
another. Only a small number of the available procedures is
described in this chapter. These procedures have been selected
for their informative value (probability plotting) or generally
high power (see Definition 2.7.4) against a wide range of
alternatives as compared to others available.

Tests for distributional assumptions have been developed based
on a variety of properties of distributions. Two properties that
underlie many test procedures are:

1. If the ordered observations are plotted or regressed against
the expected values of the order statistics from the population
from which the sample was drawn, the resulting relationship is
linear.

2. If the data are transformed by means of the probability integral
transformation, the resulting data are uniformly distributed
(see Theorem 6.3.1) under the assumption that the correct model
has been used.

These two principles are described subsequently along with
some of the tests which are based on them. A discussion of the
time-honored chi-square procedure is also included.

6.2 REGRESSION TESTS

A number of techniques used for testing model assumptions is based on the concept of regression with order statistics. A thorough discussion of regression and order statistics is not included here as it requires more specialized background than needed for this text. Readers who are interested should consult Kendall and Stuart (1961), Aitken (1935), and David (1970). A very brief summary of these concepts is given in the following paragraphs.

Definition 6.2.1   Consider a set of paired values $\{(x_i y_i); i = 1, 2, \ldots, n\}$, where the x_i's are known values and the y_i's are measurements that are linearly related to the x_i via the model

$$y_i = \alpha + \beta x_i + e_i$$

where α and β are constants and e_i is a random variable with the following properties:

$$E(e_i) = 0 \qquad \text{for all i}$$
$$\text{var}(e_i) = \sigma^2 \qquad \text{for all i}$$
$$\text{cov}(e_i e_j) = 0 \qquad \text{for all } i \neq j$$

It follows that the conditional mean $E(Y_i | x_i)$ is the simple linear regression of Y on x, that is, $E(Y_i | x_i) = \alpha + \beta x_i$ for all i.

This is a restricted definition since it applies only to regression relationships that are linear in x.

The regression line is the expected or average value of Y given known values of x. Least-squares procedures are used to estimate the parameters α, β, and σ^2. The resulting formulas are found in almost any elementary statistical text.

Sometimes the e_i's are not uncorrelated, nor are they homoscedastic. That is, $\text{cov}(e_i, e_j) \neq 0$ and $\text{var}(e_i) \neq \text{var}(e_j)$. When this is the case the estimates of α, β, and σ^2 become more complex requiring a technique known as generalized least squares. An important application of the regression technique is when the x_i's are the expected value of the order statistics from a given distribution.

Definition 6.2.2 Given a random sample of size n $\{x_i, i = 1, 2,$ ..., n\} from a population $f(x;\mu,\sigma)$, the order statistics $x_{(i)}$ are defined by arrangement of the observations such that $x_{(1)} \leq x_{(2)} \leq$ ... $\leq x_{(n)}$. Here, it is assumed μ and σ are respectively the location and scale parameters of the distribution.

The order statistics are constrained in their sample space by the previous cited inequalities and hence are not independent. The joint distribution of the order statistics can be specified [see Freund (1971)], and the marginal distributions for any subset can be obtained. The mean of the marginal distribution of $X_{(i)}$ is denoted m_i, which is the expected value of the ith order statistic.

Now consider the regression of the ordered observations i.e., the order statistics, on their expected values. Let $X_{(i)}$ be the ith order statistic from a standardized distribution $f(x;0,1)$ and denote $E(X_{(i)}) = m_i$. Then any measurement from $f(y;\mu,\sigma)$, a distribution with arbitrary location and scale parameters, is written as

$$Y_{(i)} = \mu + \sigma X_{(i)}$$

and

$$E(Y_{(i)}) = \mu + \sigma m_i \qquad (6.1)$$

Thus, one can consider the model $y_{(i)} = \mu + \sigma m_i + e_i$ as the linear regression of the ordered observations $\{y_{(i)}\}$ on the expected value of the order statistics $\{m_i\}$ from the standardized distribution. Since the $y_{(i)}$'s are ordered, the e_i's are not independent; and therefore, generalized least squares must be used to estimate μ and σ.

The regression-type tests use the concept that the regression of $y_{(i)}$ on m_i is linear if the m_i are the expected values of the order statistics of the standardized distribution from which the $y_{(i)}$'s are a sample.

The technique of probability plotting does not follow the formalism of a test of hypothesis described previously. It is,

however, one of the most informative yet simplest techniques for
evaluating distributional assumptions that has become more popular
in its usage in recent years.   The technique is based on a
graphical assessment of the linearity that is expected by equation
(6.1).

If the ordered sample values $\{y_{(i)}, i = 1, 2, \ldots, n\}$ are
plotted against the expected values of the order statistics of the
appropriate standardized distribution $\{m_i, i = 1, 2, \ldots, n\}$, then
the resulting plot should be linear.   If the data do not come from
the distribution represented by the m_i's then the points will not
fall in a linear pattern.   The assessment of linearity is done by
eye and a subjective judgment is made as to whether to "accept"
the hypothesized model.   The procedure does not follow the formal,
objective framework of tests of hypothesis and has no specified,
quantifiable type I error.   The test procedure may be carried out
without knowing μ and σ; all that is necessary are the m_i's.
These expected values can be determined only for distributions
that do not have any unknown shape parameters, which include the
normal, exponential, and extreme-value models.   It can be also
used with the Weibull and lognormal distributions by using a logarithmic transformation that converts the Weibull to the extreme
value and the lognormal to the normal.

Values of the m_i's are tabulated for several distributions;
however, the plotting procedure is simplified by using the following approximation suggested by Blom (1958).   Blom showed that an
approximation of the expected value of an order statistic may be
obtained from

$$m_i \approx F^{-1}(\pi_i)$$

where

$$\pi_i = \frac{i - \alpha_i}{n - \alpha_i - \beta_i + 1}$$

and $F^{-1}(\pi_i)$ is the inverse of the distribution function, i.e., the

$100\pi_i$ percentile of the distribution, that value of the random variable X for which $P\{X < F^{-1}(\pi_i)\} = \pi_i$. The values of α_i and β_i vary for different distributions, but a useful compromise for many distributions is to set $\alpha_i = \beta_i = 1/2$. Thus,

$$m_i = F^{-1}\left(\frac{i - 1/2}{n}\right) = F^{-1}(\gamma)$$

Since $F^{-1}(\gamma)$ is a function of only i and n for a distribution with no unknown shape parameters, it is possible to scale a sheet of graph paper so that $\gamma = (i - 1/2)/n$ can be plotted directly on the x-axis and the scaling converts it to $F^{-1}(\gamma)$.

Example 6.2.1 Construct a sheet of probability paper for use in testing the hypothesis that an exponential model fits a given set of data.

Solution: The cdf of the standard exponential is $F(x) = 1 - e^{-x}$; $x > 0$. Let $x_\gamma = F^{-1}(\gamma)$ with $\gamma = (i - 1/2)/n$. Then

$$\gamma = F(x_\gamma) = 1 - e^{-x_\gamma} \qquad \text{or} \qquad x_\gamma = -\ln(1 - \gamma) \qquad (6.2)$$

Therfore, by using a logarithmic spacing on the x-axis the scaling transforms γ to $F^{-1}(\gamma)$. The spacing on the y-axis is arithmetic. A sheet of exponential paper can be constructed from the semi-logarithmic graph paper. A sheet of three-cycle semilog paper can be converted into exponential probability paper as follows. First invert the paper so the cycles run backward $[-\ln(1 - \gamma)$ is to be plotted]. (See Figure 6.1.) The start of the first cycle is labeled 0.01 (instead of 10), the next division is labeled 0.10 instead of 9, the next 0.20 instead of 8, etc., until the end of the first cycle the line is labeled 0.90 instead of 1. The second cycle starts with 0.90 and each major division is incremented by 0.01; hence the major line in the middle of the cycle corresponding to 5 is labeled 0.95. The increments for the next cycle are 0.001. Hence the line corresponding to 5 in the third cycle is labeled 0.995 and the end of the cycle is 0.999.

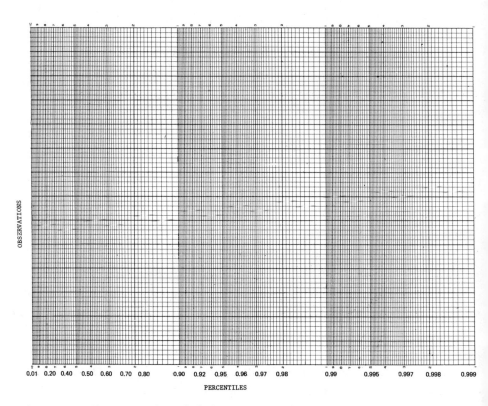

Figure 6.1 Exponential probability paper.

Example 6.2.2 Prepare a sheet of probability paper for the normal distribution.

Solution: The same procedure is followed as for the exponential case except the values of $F^{-1}(\gamma)$ must be obtained from a table of areas of the normal distribution. Using values of 0.001(0.001) 0.01(0.01)0.20(0.02)0.80(0.01)0.98(0.001)0.99, the value of x_γ is obtained and used for scaling the x-axis. For example, $x_{0.05} = -1.645$, $x_{0.50} = 0$, etc. Therefore, the ordinate labeled 0.05 would be 1.645 units below the ordinate labeled 0.50. By symmetry of the normal the ordinate labeled 0.95 would be 1.645 units above the ordinate labeled 0.50. The size of a unit would depend on the desired size of the paper. The y-axis would have an arithmetic scaling. (See Figure 6.2.)

Example 6.2.3 Prepare a sheet of probability paper for the Weibull distribution (3.23). The Weibull distribution contains a shape parameter, hence, probability paper is not directly available. However, if one uses a logarithmic transformation, the Weibull is converted to the extreme-value distribution, which has no shape parameter. Thus, if Y is Weibull with shape parameter η and scale parameter λ so that

$$f(y;\eta,\lambda) = \lambda\eta y^{\eta-1} e^{-\lambda y^\eta} \qquad y > 0$$

then $Z = \ln Y$ has the distribution

$$g(z;\eta,\lambda) = \eta \exp\left\{\eta\left(z + \frac{\ln \lambda}{\eta}\right) - \exp\left[\eta\left(z + \frac{\ln \lambda}{\eta}\right)\right]\right\} \qquad (6.3)$$

where $1/\eta$ is now the scale parameter and $-\ln \lambda/\eta$ is the location parameter. This is the extreme-value distribution (3.56). The standardized extreme-value distribution is given as

$$F(x) = 1 - \exp(-e^x)$$

$$x_\gamma = F^{-1}(\gamma) = \ln[-\ln (1 - \gamma)] \qquad (6.4)$$

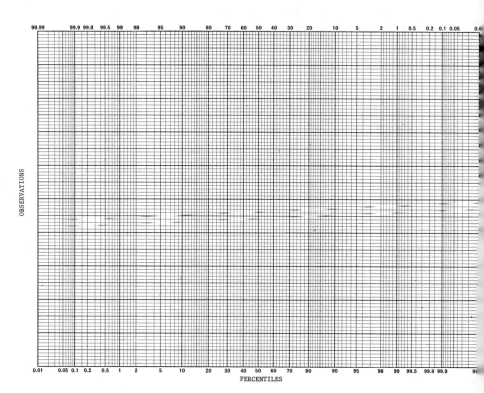

Figure 6.2   Normal probability paper.

Thus, the spacing on the x-axis is scaled by x_γ in (6.4) and the y-axis has a logarithmic scale (see Figure 6.3).

Fortunately it is not necessary to construct probability paper; paper for most common distributions is commercially available, e.g., it is available for the normal, exponential, lognormal, Weibull, extreme-value, chi-square (for specified degrees of freedom), and logistic distributions.

Once the paper is obtained, the actual procedure for performing the test is straightforward:

(a)   First order the observations from smallest to largest. Let i be the rank of the observation, then $y_{(1)} \leq y_{(2)} \leq \cdots \leq y_{(n)}$ are the ranked ordered observations.

(b)   Plot the ordered observations $\{y_{(i)}\}$ vs. $\gamma_i = (i - 1/2)/n$. The $y_{(i)}$'s are plotted on the arithmetic scale and γ_i's are plotted on the probability scale. As indicated previously for some distributions (lognormal and Weibull), the $y_{(i)}$'s are plotted on a log scale. For some prescaled paper the axes are interchanged, i.e., the probability scale is placed on the ordinate axis and the observations are plotted on the abscissa.

(c)   Evaluate the linearity of the plot. The test is carried out "by eye." Drawing a best-fitting line by eye is often helpful in making the evaluation. The following should be kept in mind when evaluating the plot:

(1)   The data are outcomes of random variables and will not fall perfectly on a straight line due to random fluctuations. The larger the sample size, the greater the tendency toward linearity given that the appropriate model was chosen.

(ii)   The data points are not independent because of the ordering and therefore even if the hypothesized model is correct one would not necessarily find the plotted values randomly scattered about the line. Runs of points above and below the line often occur even when the proper model is chosen.

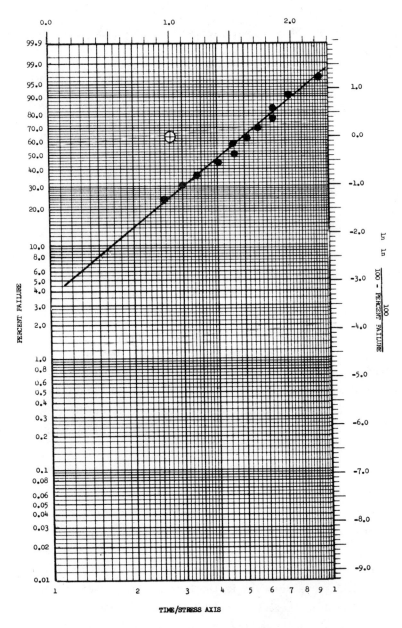

Figure 6.3 Weibull probability paper (Example 6.2.6).

(iii) Systematic departures from linearity indicate that a wrong model was selected.

(iv) The variances of the plotted points are not equal. Points in the center of the distribution or near a terminal point will have smaller variances than points in an unrestricted tail. Therefore, in evaluating a plot allow a greater departure from linearity for those points where the variance is higher. For example, for the normal distribution a greater "tolerance" (divergence from linearity) should be allowed for points in either tail, while for the exponential distribution only points in the upper tail should by allotted this extra "tolerance."

(d) If the plot is reasonably linear, then the hypothesis is not rejected. If there are systematic departures from linearity, the hypothesis is rejected. While this procedure is subjective, with some practice and experience meaningful judgments can be made.

(e) If the model is not rejected, then estimates of the location parameter μ and the scale parameter σ can be made. This procedure can be aided by drawing a best-fitting line "by eye" on the probability paper. An estimate of σ can be obtained from the slope of the line and an estimate of μ from the y-intercept (see equation 6.1). In obtaining these estimates the actual values of $F^{-1}(\gamma)$ must be used for the x values and not the values of γ which were used to make the plot. Likewise, if the observations were transformed as in the case of the Weibull distribution, then the values of ln y should be used and not the values of y which were used in the plotting.

Example 6.2.4 Estimate the scale parameter of the exponential distribution from a probability plot.

The exponential distribution is defined by

$$f(y;\theta) = \frac{1}{\theta} e^{-y/\theta} \qquad y > 0$$

where $\theta = \lambda^{-1}$ is the scale parameter (see 3.4). Its estimate is obtained from the slope of the line which can be calculated from the coordinates of two points on the plotted line (the point-slope formula for a straight line):

$$\tilde{\theta} = \frac{y_{\gamma_1} - y_{\gamma_2}}{x_{\gamma_1} - x_{\gamma_2}}$$

Generally values of γ_1 and γ_2 are chosen from opposite tails of the distribution. If $\gamma_1 = 0.95$ and $\gamma_2 = 0.05$ are used then

$$\tilde{\theta} = \frac{y_{0.95} - y_{0.05}}{x_{0.95} - x_{0.05}}$$

From (6.2) we have that

$$x_{0.95} = -\ln(0.05) = 2.9957$$

and

$$x_{0.05} = -\ln(0.95) = 0.0513$$

Hence,

$$\tilde{\theta} = \frac{y_{0.95} - y_{0.05}}{2.9940}$$

or approximately

$$\tilde{\theta} = \frac{1}{3}\left(y_{0.95} - y_{0.05}\right) \tag{6.5}$$

The values of $y_{0.95}$ and $y_{0.05}$ are read directly from the plotted line; they are the y coordinates corresponding to the chosen values of x.

Example 6.2.5 Estimate the parameters of the normal distribution from a probability plot. Since μ is the location parameter, it can be estimated from the y intercept, the value of y where x_γ is equal to zero. Since $F^{-1}(0.50) = 0$, then $y_{0.50}$ gives an estimate of μ. This is read off the plotted line. The estimate of the scale parameter σ is obtained from the slope of the line, and if γ_1 = 0.90 and γ_2 = 0.10 are used, then

$$\tilde{\sigma} = \frac{y_{0.90} - y_{0.10}}{2.56} \approx \frac{2}{5}\left(y_{0.90} - y_{0.10}\right) \tag{6.6}$$

Since $x_{0.90} = 1.28$ and $x_{0.10} = -1.28$.

Example 6.2.6 Estimate the parameters of the Weibull distribution by means of a probability plot. As discussed in Example 6.2.3, the Weibull probability paper is prepared using the scale ln y and $\ln[-\ln(1 - \gamma)]$. The estimate of the scale parameter $1/\eta$ is obtained from the slope of the line, and hence the estimate of γ is obtained from the reciprocal of the slope

$$\tilde{\eta} = \frac{x_{\gamma_1} - x_{\gamma_2}}{z_{\gamma_1} - z_{\gamma_2}} = \frac{x_{\gamma_1} - x_{\gamma_2}}{\ln y_{\gamma_1} - \ln y_{\gamma_2}}$$

A choice of $\gamma_1 = 0.90$ and $\gamma_2 = 0.10$ yields, using the fact that $x_\gamma = \ln[-\ln(1 - \gamma)]$,

$$\tilde{\eta} = \frac{3.084}{\ln y_{0.90} - \ln y_{0.10}} \approx \frac{3}{\ln y_{0.90} - \ln y_{0.10}} \tag{6.7}$$

The location parameter[†] $-\ln \lambda/\eta$ can be estimated by the y intercept, that is, the value of γ such that $F^{-1}(\gamma) = 0$. This occurs at $\gamma = 0.6322$ since $F^{-1}(\gamma) = \ln[-\ln(1 - \gamma)]$. Hence, $z_{0.6322} = \ln y_{0.6322}$ is used to estimate the location parameter. An estimate of the parameter λ is obtained from

$$\tilde{\lambda} = e^{-\tilde{\eta}z_{0.6322}} = (y_{0.6322})^{-\tilde{\eta}} \tag{6.8}$$

(f) If the model is rejected the next step is either to search for a new model or look for a transformation so the data can be

---

[†]Note that $(-\ln \lambda)/\eta$ and $1/\eta$ are the location and scale parameters of the extreme-value distribution, while $(\lambda^{-1})^{\eta-1}$ and η are the scale and shape parameters of the Weibull distribution.

converted to the desired model. Techniques for transformations
are discussed by Fraser (1967) and Andrews (1971) and are
beyond the scope of this text. The actual plot can be used
to suggest possible alternate models. For example, if the
plotted points deviate from linearity in both tails by curving
upward in the upper tail and downward in the lower tail, then
a distribution with longer tails than the specified model
should be investigated. If the deviation in the upper tail is
as described previously but the points curve upward from the
line in the lower tail, look for an alternative with a longer
upper tail and a shorter lower tail. If the null distribution
were symmetric, this would indicate that a skewed distribution
should be selected. Thus, by examing the plot one can obtain
suggestions for possible alternatives. Probability paper can
be used to make a test of a distributional assumption even if
all the observations are not known, such as in the case of a
censored sample where the r smallest or largest observations
are missing. All that is needed to construct a plot and
estimate the parameters is a knowledge of the total sample
size and the rank order number of each of the known observations. Therefore, if a test of n units is run and terminated
at the rth failure time (r < n), the plotting procedure can
still be used to test the reasonableness of the model and
estimate the parameters. Another similar technique, hazard
plotting, was suggested by Nelson (1972).

Example 6.2.7 Fifteen samples of water were tested for a certain
impurity. The test device cannot measure impurity levels below a
given threshhold value of 20 ppm. There were 3 samples for which
it was not possible to get an actual measurement. Use the data to
see if a Weibull distribution can be used to model the system and
if the model is acceptable estimate the parameters. The data after
ordering are

No reading   34.9   56.3

No reading   40.5   61.5

No reading   45.0   62.8

  25.1   45.2   70.8

  30.9   50.1   92.1

The data have been plotted on Figure 6.3, where the first known
value, 25.1, has been plotted vs. $(4-1/2)/15 = 0.23$.   Note the x-axis
is the observational scale and the y is the probability scale.   The
plotted points are quite close to the line drawn by eye and there
is no evidence that the model does not fit the data.   Use of (6.7),
shows the estimate of η is

$$\tilde{\eta} = \frac{3.084}{\ln(76.0) - \ln(18.5)} = \frac{3.084}{1.413} = 2.183$$

From (6.8) the estimate of λ is $\tilde{\lambda} = (51)^{-2.183} = 18.72 \times 10^{-5}$.

Probability plots are very useful and informative.   However,
sometimes in borderline cases where it is difficult to decide
whether to reject the model, it is useful to augment them with an
objective test.   This section discusses techniques that specify
a type I error and make an objective decision whether to reject
the hypothesized model.   The so-called regression tests are based
on the relationship specified in (6.1).   These tests evaluate the
linearity of the regression line by estimating the scale parameter
from (6.1) using generalized least squares and comparing it to
another estimate that is not based on the regression line.   The
first statistic gives an appropriate estimate of the scaling parameter only if the relationship is linear, that is, only if the
hypothesized distribution is correct.   (The relationship is linear
only when the m_i's were selected from the hypothesized distribution.)   The second statistic yields a proper estimate of the scale
parameter regardless of whether the model chosen is correct.   The
test procedure consists of comparing these two estimates.

Suppose $\hat{b}$ is the estimate of the scaling parameter σ based on the regression of the ordered observations on the expected values of the order statistics. If generalized least squares is applied, this estimator is given by

$$\hat{b} = \frac{1'V^{-1}(1m' - m1')V^{-1}y}{1'V^{-1}1m'V^{-1}m - (1'V^{-1}m)^2} \qquad (6.9)$$

where

    1 is a 1 × n vector of ones

    m is a 1 × n vector of the expected values of the order statistics

    V is an n × n matrix of the covariances between the order statistics

    i.e., $\text{cov}(e_i e_j) = \sigma_{ij}$, and $V = (\sigma_{ij})$

    y is the 1 × n vector of the ordered observations

The second estimate used is the total sum of squares about the mean,

$$s^2 = \sum_{i=1}^{n} (y_i - \bar{y})^2 = \sum_{i=1}^{n} y_i^2 - \frac{\left(\sum_{i=1}^{n} y_i\right)^2}{n} \qquad (6.10)$$

which gives an estimate of $(n - 1)\sigma^2$. These two estimators, after scaling $\hat{b}$ and squaring, are compared by forming the ratio W described below. The distribution of W will depend only on the sample size and the form of the null distribution. It is independent of any location and scale parameters. Tables of m, V^{-1}, and the percentiles of the null distribution of W are needed for each distributional form to be tested.

The development of the W test for the normal distribution proceeds as follows.

For any symmetrical distribution $m_i = m_{n-i+1}$ from which it can be shown that $1'V^{-1}m = 0$, substitution of this into (6.9) yields

$$\hat{b} = \frac{m'V^{-1}y}{m'V^{-1}m} = \sum_{i=1}^{n} c_i y_i$$

In the development that follows it is convenient to scale $\hat{b}$ so that the coefficients $\{c_i, \; i = 1, 2, \ldots, n\}$ have the property that their sum of squares is unity. Thus the scaled estimator of σ is defined to be

$$b = \frac{m'V^{-1}y}{(m'V^{-1}V^{-1}m)^{1/2}} = \sum_{i=1}^{n} a_i y_i$$

where $\{a_i, \; i = 1, 2, \ldots, n\}$ are the required replacement values for $\{c_i, \; i = 1, 2, \ldots, n\}$. Furthermore, since the normal is symmetric, $a_i = -a_{n-i+1}$. Hence,

$$b = \sum_{i=1}^{k} a_{n-i+1}(y_{n-i+1} - y_i) \tag{6.11}$$

where

$$k = \begin{cases} \dfrac{n}{2} & \text{if } n \text{ is even} \\[2ex] \dfrac{n-1}{2} & \text{if } n \text{ is odd} \end{cases}$$

The statistic for the test is defined as

$$W = \frac{b^2}{s^2} \tag{6.12}$$

The values of a_{n-i+1} and the percentage points of W are given in Tables A.6 and A.7.

The W test is lower-tailed, that is, nonnormality is indicated by too-small values of W. Hence, if W is less than W_{α}, then the hypothesis of normality is rejected on an $(1 - \alpha)100\%$ test. [See Shapiro and Wilk (1965) for more details.]

Example 6.2.8   Test the following 20 readings for normality using
the W test:

1.6925	1.6985	1.7015	1.7055
1.6952	1.6988	1.7016	1.7060
1.6970	1.6992	1.7026	1.7083
1.6974	1.7001	1.7030	1.7085
1.6981	1.7012	1.7050	1.7107

Equation (6.10) yields

$$s^2 = 57.90485 - \frac{(34.0307)^2}{20} = 0.000423$$

Use of (6.11) and the values of a_i from Table A.6 shows that

$$b = 0.4734(1.7107 - 1.6925) + 0.3211(1.7085 - 1.6952) + \cdots$$
$$+ 0.0422(1.7016 - 1.7001) + 0.0141(1.7015 - 1.7012) = 0.0204$$

and from (6.12),

$$W = \frac{0.000416}{0.000423} = 0.983$$

A reference to Table A.7 for $n = 20$ shows when $\alpha = 0.05$ that $W_{0.05}$
is 0.905. Since the computed value of W is larger than this value,
the hypothesis of normality is not rejected based on a 5% level test.

This same technique applies for developing a test for any
distribution with only location and scale parameters. Thus, if
one wants to develop a test for the exponential distribution, he
would proceed as follows: Let

$$f(x_i;\mu,\theta) = \frac{1}{\theta} e^{-[(x-\mu)/\theta]} \qquad x > \mu$$

The order statistics from the exponential have a simple form
and have the following properties [see Shapiro and Wilk (1972) or
Kendall and Stuart (1961) for more details]:

(a) $m' = 1'V$ $1' = m'V^{-1}$

(b) $1'm = n$

(c) $1'V^{-1} = (n^2, 0, 0, \ldots, 0)$

(d) $m'V^{-1}m = n$

(e) $1'V^{-1}m = n$

(f) $1'V^{-1}1 = n^2$

(g) $1 V^{-1}y = n^2 y_{(1)}$

Substitution of these relationships into (6.9) yields

$$\hat{b} = \frac{n}{n - 1}[\bar{y} - y_{(1)}]$$

where $\bar{y} = (1/n) \sum_{i=1}^{n} y_i$ and $y_{(1)}$ is the smallest observation. If

$$b = \sqrt{\frac{n}{n - 1}}[\bar{y} - y_{(1)}] \tag{6.13}$$

then

$$W = \left(\frac{n}{n - 1}\right)\frac{[\bar{y} - y_{(1)}]^2}{s^2} \tag{6.14}$$

In this case no special coefficients are needed for the calculation of b. The W test for the exponential is two-sided, that
is, nonexponentiality is indicated by too-high or too-low values
of W. The percentage points of W for the exponential case are
given in Table A.8. [For further details see Shapiro and Wilk
(1972).]

Often in testing the exponential model the value of μ, say μ_0,
is known. The following procedure that was developed by Stephens
(1978) modifies the W test. Include the known minimum value of μ_0
in the sample, yielding a sample size of n + 1.

$$b_0 = \sqrt{\frac{n + 1}{n}}(\bar{y} - \mu_0) \tag{6.15}$$

and

$$W = \frac{b_0^2}{s^2} \tag{6.16}$$

where s^2 and $\bar{y}$ are calculated based on n + 1 values.

The same table of percentages points is used as before except that one considers the sample to be of size n + 1 instead of n.

Example 2.6.8 Test the following 10 readings for exponentiality. Assume the location parameter μ is unknown.

25	169
122	178
147	191
150	205
163	255

Use of equations (6.13) and (6.14) yields

$$W = \frac{(10/9)\,[(160.5 - 25)^2]}{289963 - (1605)^2/10} = 0.6304$$

By use of Table A.8 for n - 10, it is seen that the computed value of W exceeds the upper tail 0.975 percentile of W, which is 0.2897, and hence the null hypothesis is rejected on a two-tail 5% level test. If it is assumed that the minimum value is known and is equal to 10, this value is included with the above observations and W is recomputed:

$$W = \frac{(146.818 - 10)^2}{290063 - (1615)^2/11} = 0.3889$$

Referring to Table A.8 one notes that this value exceeds the 0.975 percentile, $W_{0.975} = 0.2619$, for n = 11; hence, the exponential model is rejected based on a two-tail 5% level test.

Procedures for other distributions can be developed from the same general format. In Problem 12 the reader is asked to develop the W statistic for the uniform distribution.

The concept of the regression tests can be even further extended. One can fit higher-order models to the plot of the

ordered observations regressed on the expected value of the order
statistics and compare the higher-order fits to the linear fit.
LaBrecque (1977) has developed this extension. Others using this
principle have included Filliben (1975) and Shapiro and Francia
(1972). The power properties of all these procedures are very
similar, since all derive from the same principle; however, some
are easier to use than others and apply to sample sizes for which
tables for the W test have not as yet been developed.

Problems

1. Construct a sheet of probability paper for the normal distribution so that it is scaled to cover most of an 8 × 11 inch
   sheet of paper.

2. The pulse rates of a random sample of 20 high school students
   were measured and are shown below. Test whether a normal distribution is a reasonable model for these data using probability paper and, if so, estimate the two parameters of the
   distribution from the graph.

Pulse Rate			
72	66	83	83
85	69	76	70
81	53	66	63
63	75	71	67
69	54	74	66

3. In the study of a machine shop the number of minutes a repair
   mechanic waits for his first service call of the day is recorded for a period of 25 days. Determine, by means of probability
   paper, whether an exponential distribution is a reasonable
   choice for a model. Assume the origin is zero. If the model
   is appropriate, estimate λ.

Waiting Time for First Repair (min)

5	15	25	2	8
12	46	14	16	12
7	19	13	5	16
22	8	7	4	23
11	18	28	10	7

4. Twenty-five TV sets are put on test. The testing is stopped after the 20th failure. The results are given below. Use the probability-plotting technique to determine if it is reasonable to use an exponential model, and if so, estimate the mean time to failure of the set.

Failure Time of TV Sets (hr)

Unit	Failure time	Unit	Failure time	Unit	Failure time
1	1819	9	2400	17	441
2	2887	10	2330	18	Did not fail
3	Did not fail	11	721	19	3613
4	2160	12	3035	20	Did not fail
5	1205	13	810	21	650
6	1982	14	535	22	1662
7	125	15	Did not fail	23	4610
8	Did not fail	16	3388	24	840
				25	4915

5. The lognormal distribution is often used in connection with studies of income. A random sample of 14 incomes was selected from a wealthy community. The results are shown below. Prepare a probability plot to test whether the lognormal is a reasonable model. If it is, estimate the parameters of the distribution.

Income (000 omitted)

30	14	167	159
120	28	29	74
192	45	99	
384	81	69	

6. In a boring operation the specifications calls for a hole to
    be drilled in the center of a disk. A quality-control inspector
    took a random sample of 25 units and measured the distance be-
    tween the center of the disk and the center of the drilled
    hole. Since it is reasonable to assume for the boring machine
    that the vertical and horizontal errors are independent normal
    variates, then a plausible model to use for the measurements
    is the Rayleigh distribution. Use the data recorded below and
    the appropriate probability paper to test the assumption that
    the data came from a Rayleigh distribution and if so estimate
    its parameter. (Hint: Note the relationship of the Rayleigh
    to one of the distributions for which commercial probability
    paper is available.)

Distance Off-center (μ)

2.24	3.63	2.02	1.24	1.54
0.41	1.38	1.82	4.38	3.46
0.29	1.79	2.41	1.60	1.59
2.97	3.56	4.47	1.39	2.29
2.19	2.26	0.85	3.04	1.83

7. Fifteen batteries are placed on a life test. It is believed
    that battery lifetime can be approximated by a Weibull distri-
    bution. Test this hypothesis by means of probability paper
    using the following data. If the model is appropriate, estimate
    the parameters.

Failure Times of Batteries (hr)

2.5	8.9	23.3	33.2	50.6
3.6	15.3	25.7	41.7	61.7
4.0	17.6	28.6	43.8	73.2

8. Use the data of Problem 2 to test for normality by means of a W test with $\alpha = 0.05$.

9. Use the data of Problem 3 to test for the exponential distribution by means of a W test with $\alpha = 0.05$. Assume the origin is unknown.

10. Repeat test of Problem 9 assuming that the origin $\mu_0 = 1$.

11. Use the data of Problem 5 to test for a lognormal distribution using a W test with $\alpha = 0.05$. Note that logs of data are normally distributed.

12. Show that the W statistic for testing for uniform distribution is given by

$$W = \frac{(y_n - y_1)^2}{s^2} \qquad y_1 \le y_2 \le \cdots \le y_n$$

noting that

$$m_i = \frac{1}{n+1}$$

$$v_{ij} = \frac{i(n - j + 1)}{(n + 1)^2(n + 2)}$$

and

$$V^{-1} = (n + 1)(n + 2) \begin{pmatrix} 2 & -1 & \cdots & & & \\ -1 & 2 & -1 & \cdots & & \bigcirc \\ 0 & -1 & 2 & \cdots & & \\ & \cdot & & \cdot & & \\ & & \cdot & & \cdot & \\ & & & \cdot & 2 & -1 \\ \bigcirc & & & & -1 & 2 \end{pmatrix}$$

6.3 DISTANCE TESTS

A large number of tests has been developed based on the idea that the cumulative distribution function $F(x)$ for any density $f(x)$ is uniformly distributed.

Theorem 6.3.1 The random variable $Z = F(x)$, where $F(x)$ is the distribution function of x, is uniformly distributed on (0,1).

Proof: Let $Z = F(x)$ be a continuous function whose derivative exists for all values of x where the function is defined. It is required to find the distribution of Z where X is a random variable with density function $f(x)$. Thus,

$$z = \int_{-\infty}^{x} f(y) \, dy = F(x)$$

Differentiating both sides, it follows that

$$\frac{dz}{dx} = F'(x) = f(x)$$

and since the rule for transforming variables is

$$g(z) = f(x)\left|\frac{dx}{dz}\right| \qquad \frac{dz}{dx} \neq 0$$

then

$$g(z) = f(x)\,\frac{1}{f(x)} = 1 \qquad 0 < z < 1$$

Hence, $Z = F(x)$ is uniformly distributed on [0, 1].

It follows that given a random sample of observations from $f(x)$, the distribution of the transformed values using $z = F(x)$ is uniformly distributed on the unit interval. This assumes that the data are a random sample drawn from the population described by $f(x)$; if not, the transformed data will not be uniformly distributed. Hence a test of the model $f(x)$ can be made by testing whether the transformed data are uniform irrespective of the form of the parent distribution as long as the continuity conditions are satisfied.

Therefore, to test that the data come from some hypothesized f(x) distribution, one need only assess whether the transformed data do come from a uniform distribution. The test is carried out by examining the difference between the transformed values $F(x_{(r)})$ and the fraction of the data less than or equal to each $x_{(r)}$, i.e., $r/n = F_n(x)$, where r is the order number. This latter expression is called the *empirical distribution function*, and the tests are called EDF tests or *distance tests*. The various types of tests arise from the different ways of comparing $F_n(x) - F(x_{(r)})$.

The distance tests can be simplified if one assumes that the parameters are known. These are called simple hypothesis tests. In such cases the distribution of the test statistic is independent of the form of f(x). The earliest version of these procedures was developed by Kolmogorov (1933) and Smirnov (1939); the test is based on the maximum discrepancy between $F(x_{(r)})$ and $F_n(x)$, namely,

$$D = \sqrt{n} \max_{1 < r < u} \left| \frac{r}{n} - F(x_{(r)}) \right| \tag{6.17}$$

The distribution of D, the Kolmogorov-Smirnov test statistic, is given in Table A.9. The power of this procedure is relatively low when compared to other procedures, and is mainly of interest from a mathematical point of view.

Example 6.3.1 Use the Kolmogorov-Smirnov test to assess whether the exponential with origin parameter $\mu = 0$ and scale parameter $\theta = 5$ is a reasonable model for the following set of data:

2.3	4.5
2.8	4.7
3.5	6.9
3.9	8.2
4.1	12.5

$F(x_{(1)}) = F(2.3) = 1 - e^{-2.3/5} = 0.37$ $F(x_{(4)}) = 0.54$

$F(x_{(2)}) = 0.43$ $F(x_{(5)}) = 0.56$

$F(x_{(3)}) = 0.50$ $F(x_{(6)}) = 0.59$

$$F(x_{(7)}) = 0.61 \qquad\qquad\qquad F(x_{(9)}) = 0.81$$

$$F(x_{(8)}) = 0.75 \qquad\qquad\qquad F(x_{(10)}) = 0.92$$

$$D = \sqrt{n} \; \max_{1 < r < n} \left| \frac{r}{n} - F(x_{(r)}) \right|$$

which occurs at the first value. Therefore,

$$D = \sqrt{10} \left| \frac{1}{10} - 0.37 \right| = \sqrt{10}(0.27) = 0.854$$

One sees, referring to Table A.9, for n = 10 the 5% value of D is 1.30, and hence the hypothesis of exponentiality is not rejected.

In an attempt to increase the power of the Kolmogorov-Smirnov procedure one might consider examining the discrepancy between $F_n(x)$ and $F(x_{(r)})$ over the entire interval rather than just looking at the maximum difference. Such a procedure was proposed by Cramér (1928) and Von Mises (1931). The Cramér-Von Mises statistic is defined as

$$W^2 = n \int_0^1 [F_n(x) - F(x)]^2 \, dF(x)$$

which can be simplified to

$$W^2 = \frac{1}{12n} + \frac{1}{n} \sum_{r=1}^{n} \left[F(x_r) - \frac{2r - 1}{2n} \right]^2 \tag{6.18}$$

The percentiles of W^2 are given in Anderson and Darling (1952). This procedure still has relatively poor power.

In an attempt to make the Cramér-Von Mises statistic more sensitive to discrepancies in the tails of the distribution, Anderson and Darling (1954) proposed the following statistic:

$$A^2 = n \int_0^1 [F_n(x) - F(x_{(r)})]^2 \, \frac{dF(x_{(r)})}{F(x_{(r)}) [1 - F(x_{(r)})]}$$

which can be simplified to

$$A^2 = -n - \frac{1}{n} \sum_{r=1}^{n} (2r - 1)\{\ln F(x_{(r)}) + \ln[1 - F(x_{(n-r+1)})]\}$$

$$(6.19)$$

When x is in the lower tail of the distribution, then $1/F(x)$ will be large, and when x is in the upper tail, then $1/[1 - F(x)]$ will be large; hence the difference between $F_n(x)$ and $F(x_{(r)})$ in the tails will receive a larger weight than the difference in the center of the distribution where $F(x) = 1 - F(x) = 0.5$. Tables for A^2 can be found in Anderson and Darling (1954).

The above simple hypothesis tests are of limited use since in most practical cases the parameters of the distribution are unknown. Also, the power of these simple hypothesis tests is relatively low compared to other procedures. A composite hypothesis (parameter unknown) version of these tests was developed by Lilliefors (1967) and Stephens (1974, 1978). When estimates of the parameters are used in place of known values, the distribution of the test statistic is no longer independent of the parent distribution $f(x)$ and hence a table of percentiles is needed for each null distribution. The approach presented here is due to Stephens (1974, 1978).

The test for the normal distribution is made using the variables

$$w_{(r)} = \frac{x_{(r)} - \bar{x}}{s} \qquad r = 1, 2, \ldots, n \qquad (6.20)$$

where $x_{(i)}$ is the ith-ordered variate such that $x_{(1)} < x_{(2)} < \cdots < x_{(n)}$ and

$$\bar{x} = \frac{1}{n} \sum_{r=1}^{n} x_r$$

$$s^2 = \frac{n \sum_{r=1}^{n} x_r^2 - \left(\sum_{r=1}^{n} x_r\right)^2}{n(n - 1)}$$

The transformed values are then

$$z_{(r)} = F(w_{(r)})$$

The $z_{(r)}$'s are used to compute either W^2 or A^2 using equation (6.18) or (6.19), and these statistics are modified as follows:

$$W_M^2 = W^2 \left(1.0 + \frac{0.5}{n} \right) \tag{6.18a}$$

or

$$A_M^2 = A^2 \left(1.0 + \frac{0.75}{n} + \frac{2.25}{n^2} \right) \tag{6.19a}$$

The percentiles of the modified statistics are given in Table A.10. Note that the tabulated values do not depend on the sample size. The tests are used as "upper tail." The hypothesis of normality is rejected with an α type I error if the test statistic exceeds the α percentile given in the table.

Example 6.3.2 Use the Anderson-Darling statistic to test, at the 5% level, the normality of the following data:

9.7	14.3	$\bar{x} = 14.55$
10.5	15.9	$s = 4.47$
11.1	16.2	
11.6	20.5	
12.5	23.2	

Use of (6.20) yields

$$w_{(1)} = \frac{9.7 - 14.55}{4.47} = -1.09 \qquad w_{(6)} = -0.06$$

$$w_{(2)} = -0.91 \qquad\qquad\qquad w_{(7)} = 0.30$$

$$w_{(3)} = -0.77 \qquad\qquad\qquad w_{(8)} = 0.37$$

$$w_{(4)} = -0.66 \qquad\qquad\qquad w_{(9)} = 1.33$$

$$w_{(5)} = -0.46 \qquad\qquad\qquad w_{(10)} = 1.94$$

Then if one uses Table A.1, the values $z_{(1)}, z_{(2)}, \ldots, z_{(10)}$ are obtained, where

$$z_{(1)} = F(w_{(1)}) = F(-1.09) = 0.14 \qquad z_{(6)} = 0.48$$

$$z_{(2)} = 0.18 \qquad\qquad\qquad\qquad z_{(7)} = 0.62$$

$$z_{(3)} = 0.22 \qquad\qquad\qquad\qquad z_{(8)} = 0.64$$

$$z_{(4)} = 0.25 \qquad\qquad\qquad\qquad z_{(9)} = 0.91$$

$$z_{(5)} = 0.32 \qquad\qquad\qquad\qquad z_{(10)} = 0.97$$

Equation (6.19) yields

$$A^2 = -10 + \frac{104.60}{10} = 0.460$$

and (6.19a) yields

$$A_M^2 = 0.460\left(1.0 + \frac{0.75}{n} + \frac{2.25}{100}\right)$$

$$= 0.505$$

From Table A.10 one sees that $A_{M0.05}^2 = 0.752$, and hence the hypothesis of normality is not rejected.

The test statistic for the exponential distribution requires that the origin parameter μ is known. The test is performed using

$$w_{(r)} = \frac{x_{(r)} - \mu}{\bar{x}} \tag{6.21}$$

The transformed variables are obtained from

$$z_{(r)} = F(w_{(r)}) = 1 - e^{-w_{(r)}} \tag{6.22}$$

The percentiles for the modified statistics are given in Table A.10, and the statistics are defined by

$$W_M^{2*} = W^2\left(1.0 + \frac{0.16}{n}\right) \tag{6.18b}$$

and

$$A_M^{2*} = A^2\left(1.0 + \frac{0.3}{n}\right) \tag{6.19b}$$

One benefit derived from using estimates of the parameters instead
of their known values is that the power of these procedures is
better than the power in the simple hypothesis case. The procedure
A_M^2 has the highest power for a wide range of alternate distributions and is almost as powerful as the regression tests. A major
advantage is that an extensive table of coefficients or percentiles
is not required to apply the test.

Example 6.3.3 Use the Cramér-Von Mises test at the 10% level of
significance to assess whether the exponential with $\mu = 0$ is a
reasonable model for the data shown below:

0.42	4.60	
1.32	5.18	
1.69	5.20	$\bar{x} = 3.92$
3.97	6.03	
4.05	6.77	

Equation (6.21) yields

$$w_{(1)} = \frac{0.42}{3.92} = 0.107 \qquad w_{(6)} = 1.173$$

$$w_{(2)} = 0.337 \qquad w_{(7)} = 1.321$$

$$w_{(3)} = 0.431 \qquad w_{(8)} = 1.327$$

$$w_{(4)} = 1.013 \qquad w_{(9)} = 1.538$$

$$w_{(5)} = 1.033 \qquad w_{(10)} = 1.727$$

Further, (6.22) gives

$$z_{(1)} = 0.101 \qquad z_{(6)} = 0.691$$

$$z_{(2)} = 0.286 \qquad z_{(7)} = 0.733$$

$$z_{(3)} = 0.350 \qquad z_{(8)} = 0.735$$

$$z_{(4)} = 0.637 \qquad z_{(9)} = 0.785$$

$$z_{(5)} = 0.644 \qquad z_{(10)} = 0.822$$

Then from (6.18) and (6.18b),

$$W^2 = \frac{1}{120} + \frac{1}{10}(0.1987) = 0.028$$

and

$$W_M^{2*} = 0.028(1 + 0.016) = 0.028$$

Table A.10 shows the upper 10% point of W_M^{2*} is 0.175; hence the hypothesis of an exponential model is not rejected.

Problems

1. Use the Kolmogorov-Smirnov procedure to test the hypothesis of normality using the data of Problem 2 in Section 6.2. Let $\alpha = 0.10$.

2. Use the Kolmogorov-Smirnov procedure to test the exponential hypothesis using the data of Problem 3 in Section 6.2. Let $\alpha = 0.10$ and assume $\mu = 0$.

3. Use the data of Problem 6 in Section 6.2 to test the hypothesis of an exponential distribution applying the A_M^2 statistic. Assume $\alpha = 0.15$, $\mu = 0$.

4. Take the logs of the data in Problem 5 in Section 6.2 and test for normality (i.e., lognormality of the untransformed values) using the W_M^2 statistic. Use $\alpha = 0.15$.

5. Use the A_M^2 statistic to test the hypothesis of normality using the data of Problem 2 in Section 6.2. Use $\alpha = 0.15$.

6. Use the W_M^2 statistic to test whether an exponential distribution can be used as a model for the following failure data. Assume $\mu = 0$ and use $\alpha = 0.10$:   0.75, 1.31, 1.62, 1.83, 1.99, 2.57, 3.14, 3.98, 4.76, 5.32, 5.44, 6.09, 8.04, 8.55, 9.78, 12.21, 14.63, 20.92.

6.4   CHI-SQUARE GOODNESS-OF-FIT TEST

Perhaps the oldest procedure for testing distributional assumptions is the chi-square test developed by Karl Pearson (1900). For this test the data are divided into k mutually exclusive and exhaustive

cells, the parameters estimated, and the actual number of observations in each cell is compared to the expected number under the
assumption that the hypothesized distribution is correct. The
resulting test statistic has an approximate chi-square distribution with k - 1 - t degrees of freedom, where k is the number of
cells and t is the number of parameters estimated. [See Cochran
(1952) for further details.] The advantages of this procedure are
the distribution of the test statistic is independent of the form
of the distribution being tested and the test requires only a table
of the chi-square distribution. The disadvantages are that for
continuous distributions the power of the test is relatively low
compared to other test procedures and the resulting decision to
accept or reject depends to some extent on the number of categories
into which the data were subdivided. The procedure is useful for
discrete models, for large samples where power differentials would
be small, or for problems where the measurements have been dis-
cretized such as in the case where the data came as pretabulated
counts. An example of this latter situation is where ball bearings
are screened by size and the only data available is the number in
each size interval, i.e., no actual measurements are made.

The steps for performing such a test are as follows:

1. If the data are not pretabulated or there is no natural
breakdown (such as for discrete data) to determine the number of
cells required and the cell boundaries. The choice of k, the number
of cells, should be kept at a reasonable size, i.e., not too large;
a good rule is $n/k \geq 5$. When n is large one can use the criterion
suggested by Williams (1950): Choose k as the closest integer to

$$k' = 4[0.75(n - 1)^2]^{1/5} \qquad (6.23)$$

The problem of the number of cells for the chi-square test has been
treated more recently by Dahiya and Gurland (1972). In selecting
the cell boundaries a good rule to follow that they advocated is
to choose the boundaries so that the probability of being in each

cell is the same. Hence, if there are k cells and if X_{Ui} and X_{Li}
are respectively the upper and lower bounds of the ith cell, then

$$pr\{X_{Li} \leq X \leq X_{Ui}\} = \frac{1}{k} \qquad \text{for } i = 1, 2, \ldots, k$$

In order to determine these boundaries one must estimate the parameters of the distribution. Thus, if $x_{(1)}$, $x_{(2)}$, $\ldots$, $x_{(k)}$ are
the upper cell boundaries, one chooses the $x_{(i)}$'s such that the
probability of being in any one cell is 1/k. If one sets

$$x_{(0)} = -\infty \qquad \text{and} \qquad x_{(k)} = \infty$$

then $x_{(i)}$ is chosen so that

$$pr\{X < x_{(i)}\} = \frac{i}{k} \qquad i = 1, 2, \ldots, k - 1 \qquad (6.24)$$

The lower cell boundaries can be specified as the first significant
number greater than the upper bound of the previous cell. Thus,
if the upper bound of the ith cell is 13.525, the lower bound of
the (1 + 1)st cell is 13.526.

2. The next step is to calculate the expected number in each
cell and obtain the count f_i of the number of observations in each
cell. In the case where the cells were chosen so that the probability of being in each cell is 1/k, then the expected number per
cell is

$$E_i = \frac{n}{k} \qquad \text{for all } i \qquad i = 1, 2, \ldots, k \qquad (6.25)$$

where n is the sample size. If the data were pretabulated, then it
is necessary first to calculate p_i, the probability of an observation falling into the ith cell; that is,

$$p_i = pr\{X < x_{(i)}\} - pr\{X < x_{(i-1)}\} \qquad i = 1, 2, \ldots, k$$
$$(6.26)$$

Then the expected number per cell is

$$E_i = np_i \qquad i = 1, 2, \ldots, k \qquad (6.27)$$

where n is the sample size. A requirement for using the form of
the chi-square test presented here is that the E_i are all at least
5. If this is not satisfied, it is necessary to combine the cells
with E_i's less than 5 with their neighbors until this criterion is
satisfied.

   3. The third step is to calculate the test statistic. For
pretabulated data, the statistic is generally written as

$$x^2 = \sum_{i=1}^{k} \frac{(f_i - E_i)^2}{E_i} \qquad (6.28)$$

where

   f_i is the observed frequency in the ith cell

   E_i is the expected number for the ith cell

   In the case where the expected number per cell is n/k,

$$\frac{k}{n} \sum_{i=1}^{k} f_i^2 - n \qquad (6.29)$$

where f_i is defined above, k is the number of cells, and n is the
sample size.

   4. The null hypothesis is rejected with a type I error of α
if the value of x^2 from (6.28) or (6.29) is greater than χ_{α}^2, a
chi-square variate with $k - 1 - t$ degrees of freedom where t is
the number of parameters estimated. Values of χ_{α}^2 are found in
Table A.2. This is an upper tail test since large departures from
the null hypothesis lead to large positive values of the chi-square
statistic (6.28) or (6.29).

Example 6.4.1 Calculate the upper cell boundaries for testing the
hypothesis of a normal distribution where k = 20, n = 100, $\bar{x}$ = 3.0,
and s = 1.2. Equation (6.24) gives

$$pr\{X < x_{(i)}\} = \frac{i}{20}$$

One determines by using Table A.1 the value of z that satisfied

$$\text{pr}\{Z < z_{i/20}\} = \frac{i}{20}$$

Since

$$z_{i/20} = \frac{x_{(i)} - \bar{x}^{\dagger}}{s}$$

one obtains

$$x_{(i)} = \bar{x} + z_{i/20}s \qquad i = 1, 2, \ldots, k - 1$$

where

$\bar{x}$ is the sample mean

s is the sample standard deviation

and the $z_{i/20}$ are obtained from Table A.1, the percentiles of the normal distribution. Thus, the upper bound of the first cell is

$$x_{(1)} = 3.0 - 1.645(1.2) = 1.026$$

since

$$\frac{i}{20} = 0.05$$

and

$$z_{0.05} = -1.645$$

Furthermore, the upper bound of the second cell is

$$x_{(2)} = 3.0 - 1.28(1.2) = 1.464$$

The corresponding lower bounds for the first two cells are $-\infty$ and 1.026.

[dagger]It is assumed that the sample size is large enough so that the normal approximation is valid.

Example 6.4.2   Calculate the E_i's for the previous example.   Since the probability of being in the ith cell is 0.05, then using (6.25), $E_i = (100)(0.05) = 5$ for all i.

Example 6.4.3   A test of pretabulated data for the assumption of a normal distribution is desired.   Determine the expected number for the first two cells given $\bar{x} = 10.0$, $s = 2.0$, $n = 100$, and the cell boundaries for the first two cells are $-\infty$ up to but not including 6.0 and up to but not including 7.0.   Equation (6.26) yields for the first cell

$$P_1 = pr\{x < 6.0\} - pr\{x < -\infty\}$$

$$= pr\left\{z < \frac{6.0 - 10.0}{2.0}\right\}$$

since

$$pr\{x < -\infty\} = 0 \qquad \text{and} \qquad z = \frac{x - \bar{x}}{s}$$

Use of Table A.1 yields $p_1 = 0.0228$.   Also,

$$P_2 = pr\{x < 7.0\} - pr\{x < 6.0\}$$

$$= pr\{z < -1.5\} - 0.0228 = 0.0668 - 0.0228 = 0.044$$

Use of (6.27) gives

$$E_1 = 100 \times 0.0228 = 2.28$$

$$E_2 = 100 \times 0.0440 = 4.40$$

Since both these expected values are less than 5, the first two cells are combined and their expected value is 6.68.

Example 6.4.4   Test the following data to see if the exponential distribution with origin parameter known to be zero can be used to model the system output.   The number of hours (to two decimal places) to exhaust a cartridge of CO_2 gas was recorded for 100 cartridges.   The total time for the 100 tests was 98.23 hr.   It is believed that time-to-exhaustion is exponentially distributed with a minimum value of zero (it is possible that no gas was put in the cartridge). Equation (6.23) yields

$$k' = 4\left[0.75(99)^2\right]^{1/5} = 23.7$$

However, since n/k must be 5 or more, set k = 20. The next step is to estimate the parameter λ. The raw data are not shown, but $\Sigma_{i=1}^{100} x_i = 98.23$ hr. Thus, using equation (3.10),

$$\hat{\lambda} = \frac{100}{98.23} = 1.018$$

The upper cell boundaries are obtained as follows. Equation (6.24) yields

$$pr\{x < x_i\} = \frac{i}{20}$$

and from (6.2),

$$x_p = \frac{-1}{\lambda} \ln(1 - p)$$

Thus,

$$x_{(1)} = -\frac{1}{1.018} \ln\left(1 - \frac{1}{20}\right) = 0.05$$

$$x_{(2)} = -\frac{1}{1.018} \ln\left(1 - \frac{2}{20}\right) = 0.10$$

and similarly for the others.

The cell boundaries and the frequencies are shown below.

Time to Exhaust Cartridges (hr)

Cell	Frequency	Cell	Frequency
0 - 0.05	1	0.69 - 0.78	7
0.06 - 0.10	3	0.79 - 0.90	9
0.11 - 0.16	3	0.91 - 1.03	15
0.17 - 0.22	4	1.04 - 1.18	10
0.23 - 0.28	2	1.19 - 1.36	6
0.29 - 0.35	5	1.37 - 1.58	4

Time to Exhaust Cartridges (hr)

Cell	Frequency	Cell	Frequency
0.36 - 0.42	4	1.59 - 1.86	2
0.43 - 0.50	5	1.87 - 2.26	1
0.51 - 0.59	6	2.27 - 2.94	2
0.60 - 0.68	6	Over 2.94	5

Equation (6.29) yields the value

$$x^2 = \frac{20}{100}(718) - 100 = 43.6$$

and using Table A.2 the $\chi^2_{0.05}$ value for $20 - 2 = 18$ degrees of freedom is 28.9. Since $x^2 > \chi^2_{0.05}$, the hypothesis that the data are exponentially distributed with an origin parameter of zero is rejected.

Problems

1. Using (6.28), show that for the equal probability case the chi-square test statistic reduces to (6.29).

2. Use the data of Example 6.4.4 to test for normality. Note that this data must be treated as pretabulated since the original observations are not given. Use $\alpha = 0.10$.

3. Use a chi-square test to test the hypothesis of normality using the data in Problem 2 in Section 6.2 Use $\alpha = 0.05$.

REFERENCES

Aitken, A. C. (1935). On least squares and linear combination of observations, *Proc. Roy. Soc. Edin. 55*, pp. 42-48.

Anderson, T. W., and Darling, D. A. (1952). Asymptotic theory of certain goodness of fit criteria based on stochastic processes, *An. Math. Stat. 23*, pp. 193-212.

Anderson, T. W., and Darling, D.A. (1954). A test for goodness of fit, *J. Amer. Stat. Assoc. 49*, pp. 765-769.

Andrews, D. F. (1971). A note on the selection of a data transformation, *Biometrika 58*, pp. 249-254.

Blom, G. (1958). *Statistical Estimates and Transformed Beta Values*, John Wiley and Sons, Inc., New York.

Cochran, W. G. (1952). The χ^2 test of goodness of fit, *An. Math. Stat. 23*, pp. 315-345.

Cramer, H. (1928). On the composition of elementary errors, *Skand. Aktuar. 11*, pp. 141-180.

Dahiya, R. C., and Gurland, J. (1972). Goodness of fit tests for the gamma and exponential distributions, *Technometrics 14*, pp. 791-801.

David, H. A. (1970). *Order Statistics*, John Wiley and Sons, Inc., New York.

Filliben, J. J. (1975). The probability plot correlation coefficient test for normality, *Technometrics 17*, pp. 111-117.

Fraser, D. A. S. (1967). Data transformations and their linear models, *An. Math. Stat. 38*, pp. 1455-1465.

Freund, J. E. (1971). *Mathematical Statistics*, Prentice-Hall, Inc., Englewood Cliffs, N. J., pp. 224-227.

Kendall, M. G., and Stuart, A. (1961). *The Advanced Theory of Statistics, Vol. 2*, Hafner Publishing Co., New York.

Kolmogorov, A. N. (1933). Sulla determinazione empirica di une legge di distribuzione, *Georn, dell' Instit. degli alt 4*, pp. 83-91.

LaBrecque, J. (1977). Goodness of fit tests based on non-linearity in probability plots, *Technometrics 19*, pp. 293-306.

Lilliefors, H. W. (1967). On the Kolmogorov-Smirnov tests for normality with mean and variance unknown, *J. Amer. Stat. Assoc. 62*, pp. 399-402.

Nelson, W. (1972). Theory and application of hazard plotting for censored failure data, *Technometrics 14*, pp. 945-966.

Pearson, K. (1900). On the theory of contingency and its relation to association and normal correlation, *Philos. Mag. 50*, pp. 157-175.

Shapiro, S. S., and Francia, R. S. (1972). An approximate analysis of variance test for normality, *J. Amer. Stat. Assoc. 67*, pp. 215-216.

Shapiro, S. S., and Wilk, M. B. (1965). Analysis of variance test for normality (complete samples), *Biometrika 52*, pp. 591-611.

Shapiro, S. S., and Wilk, M. B. (1972). An analysis of variance test for the exponential distribution, *Technometrics 14*, pp. 355-370.

Smirnov, N. V. (1939). Sur les scorto dela courbe de distribution empirique, *Rec. Math. (NS)(Mat Sborn) 6 (98)*, pp. 3-26.

Stephens, M. A. (1974). EDF statistics for goodness of fit and some comparisons, *J. Amer. Stat. Assoc. 69*, pp. 730-737.

Stephens, M. A. (1978). On the W test for exponentiality with
   origin known, *Technometrics 20*, pp. 33-35.

Stephens, M. A. (1978). Private communication.

Von Mises, R. (1931). *Wahrscheinlichkeitsrechnung*, Deuticke,
   Leipzig

Williams, C. A. (1950). On the choice of the number and width of
   classes for the chi-square test of goodness of fit, *J. Amer.
   Stat. Assoc. 45*, pp. 77-86.

Chapter 7

ANALYSIS OF SYSTEMS

7.1 INTRODUCTION

In this chapter several of the techniques described previously are
combined with two new procedures for the purpose of analyses of
complex systems. The term *system* again is being used in a very
general context and can be a piece of hardware, the clerical
operations necessary to process an application for a driver's
license, a statistical quantity that varies as a function of several
random variables, etc. The techniques in previous chapters have
been mainly limited to the use of a single random variable to
represent the output of a system. Specific models for continuous
and discrete mechanisms are described in Chapters 3 and 4 and
empirical models for fitting sets of data are covered in Chapter 5.
In this chapter one desires to estimate the distribution of the
ouptut of a system where the output can be described as a function
of several input or design variables which are random variables.
Thus, it is desired to estimate the distribution of the output
knowing the functional relationship between input and output and
the probability distribution of the input variables. Examples of
these types of problems are as follows:

1. The distribution of projected profit for the coming year is
    obtained by treating the uncertainties in estimates of sales,
    material costs, labor costs, interest expenses, etc., as random
    variables.

2. The tolerance of an electronic device is estimated from the
    tolerance of the components of the device.   The tolerance values
    of the components are treated as random variables.

3. The probability of a nuclear reactor exceeding a safety factor
    is estimated by treating the actual values taken on by the com-
    ponents and the uncertainty in knowledge of its parameters as
    random variables.

4. The distribution of the time for processing an application for
    an insurance policy is determined by treating the times for
    each clerical operation as random variables.

5. The null distribution of a new statistical procedure is esti-
    mated from the distribution of each of the random variables
    used in its calculation.

6. The reliability of a system (probability of operating for a
    given mission time) is obtained from the failure distribution
    of each of its components.

There are several ways of handling such problems depending on
the complexity of the system. If the relationship between input
and output is simple (linear, for example) and the probability distributions chosen to represent each of the input variables are such
that one can determine analytically the distribution of the output,
then techniques such as the transformation of variables and
moment-generating functions can be used. These techniques are described in most mathematical statistics texts and are useful for the
most simple of problems. See Hahn and Shapiro (1967) and Freund
(1971) for a further description of these procedures. Examples of
such relationships that can be obtained from these procedures are
as follows.

1. Linear combinations of n normal random variables, $\sum_{i=1}^{n} a_i X_i$, are
    normally distributed with mean

$$\mu = \sum_{i=1}^{n} a_i \mu_i$$

where

a_i's are constants

$$\mu_i = E(X_i) \qquad i = 1, 2, \ldots, n$$

and variance

$$\sigma^2 = \sum_{i=1}^{n} a_i^2 \, var(X_i) + 2 \sum_{i<j} \sum a_i a_j \, cov(X_i X_j)$$

2. The sum of n independent exponential variates (3.4) with the same parameter λ have a gamma distribution (3.34) with parameters λ and n.

3. The sum of the squares of n independent standard normal variables has a chi-square distribution (3.35a) with n degrees of freedom.

4. If the output

$$Y = \sum_{i=1}^{n} a_i X_i$$

is a linear function of n uncorrelated random variables with finite means μ_i and variances σ_i^2, then the central limit theorem (Theorem 2.7.1) states that if n is large enough, Y is normally distributed with mean

$$\mu = \sum_{i=1}^{n} a_i \mu_i$$

and variance

$$\sigma^2 = \sum_{i=1}^{n} a_i^2 \sigma_i^2$$

where a_i's are constants.

Therefore, the central limit theorem can be used to determine the distribution of system output and often holds for rather samll n. The magnitude necessary for approximate normality is dependent on the distributional form of each of the X_i's. Often four or five terms are sufficient to assure approximate normality. If the X_i's are all normal, then Y is normally distributed for any n.

Example 7.1.1 The time for a rat to get through a maze is a linear function of the time it takes to get through six submazes. Each submaze is different, and the time to "solve" each maze is independent of the time for the other five. Since this is linear system, the time to complete the maze is the sum of the times for each of the elements. Thus total time,

$$T = \sum_{i=1}^{6} T_i$$

is approximately normally distributed. If the means (in minutes) and variances for the six submazes are

$$\mu_1 = 5.2 \qquad \mu_4 = 45.0 \qquad \sigma_1^2 = 1.0 \qquad \sigma_4^2 = 3.5$$

$$\mu_2 = 15.3 \qquad \mu_5 = 23.6 \qquad \sigma_2^2 = 2.0 \qquad \sigma_5^2 = 4.6$$

$$\mu_3 = 7.9 \qquad \mu_6 = 17.5 \qquad \sigma_3^2 = 2.0 \qquad \sigma_6^2 = 3.0$$

then $\mu = 114.5$ min. and $\sigma^2 = 16.1$ min^2.

The major limitation of the central limit theorem is that it holds only for linear systems. Two other techniques which are commonly used for more complex systems are Monte Carlo simulation and propagation of system moments (sometimes called propagation of error). These two techniques are described in some detail in the following sections. The last section of this chapter contains a number of case studies that illustrate many of the techniques described in this book.

Problems

1. The time for a technician to prepare a specimen for use in an experiment is exponentially distributed with a mean of 2.0 hr. Find the probability that it will take more than 8 hr. to

prepare three specimens. Assume preparation times for each
are independent.

2. An electronic device is assembled in 10 discrete steps at 10
   different stations. The time for assembly is independent at
   each station and the average and standard deviation for each
   operation are shown below. Use the central limit theorem to
   determine the probability that an assembly of a unit will take
   more than 165 min.

Station	Mean assembly time (min)	Standard Deviation (min)
1	9.1	0.5
2	27.6	0.6
3	15.2	0.3
4	9.8	0.3
5	20.0	1.0
6	14.2	0.7
7	5.1	0.8
8	5.5	1.0
9	8.2	0.3
10	6.3	0.4

3. The score on a psychological exam is computed by combining the
   scores on each of three parts using the equation

   $$Y = 3 X_1 + 4 X_2 + X_3$$

   where X_i is the score on the ith part. Given that the X_i's
   are normally distributed with the following information

   $E(X_1) = 30.0$ $var(X_1) = 2.0$

   $E(X_2) = 20.0$ $var(X_2) = 2.0$

   $E(X_3) = 40.0$ $var(X_3) = 3.0$

   $cov(X_1 X_2) = 0.5$

   $cov(X_1 X_3) = 1.0$

   $cov(X_2 X_3) = 0.8$

find the probability that a student will score less than 200
on the exam.

4. The distance a depth charge misses its intended target is a
    function of the error in three dimensions: depth, latitude,
    and longitude.   If R^2 is the square of the miss distance, then

$$R^2 = X_1^2 + X_2^2 + X_3^2$$

where the X_i, i = 1, 2, 3, represent the three dimensions.
Assuming that the error in each dimension is normally distributed with a mean of zero, with unit standard deviation, and
independent of the other dimensional errors, find the probability that R is less than 3.0 units.   Hint: R less than 3.0
units is equivalent to R^2 less than 9.0.

5. A life insurance company processes an application for an
    insurance policy in an assembly-line fashion.   It passes
    through 8 clerical operations.   If each operation is independent
    of each other, and each has a mean processing time of 15 min.
    with a standard deviation of 2 min. find the probability that
    it will take more than 2-1/2 hrs. to process an application.

6. The number of defects in a television receiver is Poisson-
    distributed with a mean of 2 per set.   Find the probability that
    there are more than 8 defects in three sets.   Assume the
    number of defects in each set is independent.

7.2   MONTE CARLO SIMULATION

Monte Carlo, the capital of Monaco, is known for its gambling
operations where people play games of chance in hopes of winning
their fortunes.   The output of these games is determined by chance
factors.   In a similar vein, the technique known as Monte Carlo
simulation is based on a procedure that creates an output based
on playing a "game" whose outcome depends on a number of chance
factors.   In a simulation study the "game" is a functional or
mathematical representation of a system and the chance factors are

the random variables that are used to represent the elements of the system.   A simulation study usually involves the following steps:

1. Determine the functional and/or mathematical model that relates system output to its elements or inputs.
2. Assign to each design or input variable a probability distribution.
3. Generate random values for each of the input variables and using these with the functional model compute the value of the output.
4. Repeat the above step many times to get a data set representing system output.
5. Summarize the data set by fitting an empirical distribution and calculating desired percentiles and moments.

    The first step in this process requires basic knowledge of the system and is most often accomplished by the technical personnel with some assistance from the statistician.

Example 7.2.1   A container is manufactured that is in the shape of a cylinder.   It is desired to determine the distribution of the amount of fluid which can be put into the container.   Federal law specifies that 99.0% of the containers must exceed a given minimum volume.   From geometry the volume of a cylinder is

$$V = \pi r^2 h$$

where π is the ratio of the circumference to the diameter of a
          circle, i.e., 3.141592...
       r is the radius of the circular base
       h is the height of the cylinder

    The amount of fluid in the container also depends on how close the liquid level is to the top.   Therefore, the volume of the liquid is

$$Y = \pi r^2 h p$$

where p is a variable representing the percentage of the container

which is filled. In this model the volume of liquid is the output
variable and r, h, and p are the input variables. The above is,
of course, a very simplified illustration of how a functional
relationship is developed. In highly complex systems the relationship is more involved and perhaps can only be expressed by systems
of partial differential equations which may take minutes to solve
even on a high-speed computer.

Once the model is formulated it is necessary to select probability distributions to represent each variable in the equation.
The method of selection of the appropriate statistical model will
vary depending on what is known about each input variable. If there
are data available for the input variables, then the procedures of
Chapter 5 for fitting an empirical distribution or one of the models
of Chapter 3 or 4 may be used. The parameters of such models may
be estimated by the techniques previously described. The techniques
of Chapter 6 may be used to check the adequacy of the selected
model.

If there are no data available on the performance of an input
variable, then subjective procedures must be used. Several suggestions are given below for the selection process in such cases.

Case 1. Mechanism or Behavior Known
In some instances the performance of the input variable can be
matched to one of the underlying mechanisms described in Chapters
3 or 4 or can be related to similar mechanisms for which data are
available. Choice of a model in these cases is straightforward.
Hence, the use of a normal, exponential, extreme-value, etc. model
may be justified from a knowledge of the mechanism underlying the
input variable. The parameters of the selected model must be estimated from the subjective opinions of those familiar with it. Techniques such as matching of moments or percentiles (see Section 5.2)
are often used with these subjective estimates. For example, if
there are two parameters to be estimated, they can be obtained by
matching the opinions of the value of the mean and variance with the
corresponding expressions in terms of the parameters. An estimate

of the mean can be obtained by asking several project people what
they believe is the expected or average value for the variable.
Often this is the "design" or specified value. When several opinions are obtained, these can be averaged. A rough estimate of the
standard deviation can be obtained from

$$\sigma = \frac{1}{6}(U - L) \qquad\qquad (7.1)$$

where U is the value that has only a very small chance of being
           exceeded (less than 1.5 in a 1000)
       L is the value below which only 1.5 in a 1000 values will
           fall

Example 7.2.2 A gamma distribution (3.34) was chosen to model the
time to process an order based on the fact that it took a number
of subevents to occur before the order was processed and each of
these times was believed to be independently exponentially distributed with the same parameter. A survey of the managers indicated
that the average of their opinions for the expected time of processing was 3.5 hrs. and that there was a very small probability
that the job could be done in less than 2.0 hrs. or would take more
than 7.0 hrs. The method of matching of moments with the mean equal
to 3.5 and the variance estimated with the aid of (7.1) as

$$\sigma^2 = \frac{1}{36}(7.0 - 2.0)^2 = \frac{25}{36} = 0.694$$

yields by applying the first two equations in Example 5.2.3 with
$\varepsilon = 0$,

$$\lambda^* = \frac{3.5}{25/26} = 5.04$$

and

$$\eta^* = 5.04(3.5) = 17.64$$

Example 7.2.3 It is desired to model the input variable, sales, as
one of the factors in estimating income for the coming year. The
best guess (most likely) is that it will be \$12.5 million and that

the possible values are symmetrically distributed about this figure.
Further, in questioning sales management people, they are quite
certain that sales volume will not be less than $11.0 million nor
greater than $14.0 million. Since the distribution is symmetrical,
a normal model is used with μ = 12.5 and σ = (1/6)(14.0 - 11.0) =
0.5.

In some cases it is easier to use percentiles rather than
moments due to their computational simplicity and ease in obtaining
information on percentiles from project people. In such cases the
estimates can be obtained as the answer to the question: What
value would one expect to be exceeded $(1 - \alpha)100\%$ of the time? One
estimate is needed for each unknown parameter. The technique is
limited to those distributions whose percentiles can be expressed
in closed form or those distributions where tables of the percentiles are available.

Example 7.2.4 A Weibull distribution (3.23) was chosen to model
the failure time of an electronic device based on past experience
with similar devices. A survey of the engineers indicated that
there was only a 10% chance that the device would last longer than
100 hrs. and a 90% chance that it would survive 10 hrs. It was
possible for the device not to work; hence, the minimum value of
the distribution was taken to be zero. The percentile function for
the Weibull distribution is

$$x_p = \left\{ \frac{-1}{\lambda^*} \ln\left[1 - F(x_p)\right] \right\}^{1/\eta}$$

and hence setting $F(x_p)$ = 0.10 and 0.90, using $x_{0.10}$ = 10 and
$x_{0.90}$ = 100 yields the equations

$$10 = \left[\frac{-1}{\lambda^*} \ln(1 - 0.10) \right]^{1/\eta^*}$$

$$100 = \left[\frac{-1}{\lambda^*} \ln(1 - 0.90) \right]^{1/\eta^*}$$

Divide the second equation by the first to eliminate λ^* and solve
the resulting equation to obtain

$$\eta^* = 1.34$$

Substitution back into the first equation yields

$$\lambda^* = 0.0048$$

If one (two) of the parameters represents (represent) a terminal
point, then this (these) parameter(s) can be estimate by determining the maximum or minimum (or both) possible values. In Example
7.2.4 the starting point for the Weibull distribution was determined
by the minimum possible failure time.

Case 2. Mechanism or Behavior Unknown

If nothing is known about the mechanism, then a Johnson S_B or S_U
distribution (5.22) or (5.23) can be used to approximate the input
distribution. The first step would be to decide whether a bounded
distribution should be used. If one or both bounds are known and
these occur within a reasonable distance of the operating values of
the system, then the S_B distribution should be used. If two bounds
are specified, then the parameters of η and λ can be estimated from
(5.35) and (5.35a), where the percentiles $x_{1-\alpha}$ and x_α are obtained
by asking to what value of the random variable will be exceeded by
$100(1 - \alpha)\%$ of the observations and what value will be exceeded by
only $100\alpha\%$ of the observations? If the estimates of several people
are averaged, the desired estimates of $x_{1-\alpha}$ and x_α are obtained.
If only one terminal value is known then $x_{1-\alpha}$ and x_α must be
supplemented by the median and use made of equation (5.36) or
(5.36a). A response to the question, "What value has a 50-50 chance
of not being exceeded?" yields a subjective estimate of the median.

If the bounds are either unknown or quite distant from the
operating range of the system, then either an S_U or S_B is used
depending on the percentiles estimates obtained from project per-
sonnel. If one chooses four percentiles to satisfy use of the discriminant function described in Section 5.4, the subjective estimates

are used to calculate m, n, and p and an S_U is used if $mn/p^2 > 1$, S_B is used if $mn/p^2 < 1$, and the S_L if the equality should hold. Equation (5.37) is used for the S_B distribution and (5.43) is used for the S_U.

Example 7.2.5 The following information was collected from six engineers about the voltage output of cathode-ray tubes. It is desired to fit a Johnson distribution to these estimates. While the theoretical limits were 0 and ∞, these values were not realistic possibilities for an operating device. Therefore, no limits were assumed and the four percentiles 6.7%, 30.8%, 69.2%, and 93.3% that satisfy the relationship required for the discriminant function were selected. The engineers were asked what values of voltage would be exceeded approximately 7%, 30%, 70%, and 93% of the time. (Note that these percentages were rounded off since it makes no sense to distinguish 6.7% from 7% in a subjective estimate.) The answers for each of the six were averaged together to obtain subjective estimates

$$x_{0.067} = 119.6 = x_{-3z}$$

$$x_{0.308} = 125.2 = x_{-z}$$

$$x_{0.692} = 133.6 = x_{z}$$

$$x_{0.933} = 141.3 = x_{3z}$$

The values of m, n, and p are

m = 7.7

n = 5.6

p = 8.4

and the discriminant function is

$$\frac{(7.7)(5.6)}{(8.4)^2} = 0.61$$

Since this is less than 1 a Johnson S_B is used. Note that the value of z = 0.5 was used. Using equation (5.37) yields the following values for the parameters:

$\eta^* = 0.95$

$\gamma^* = 0.33$

$\lambda^* = 33.47$

$\epsilon^* = 115.38$

The voltage output can be modeled by a Johnson S_B distribution with the four parameters above.

Case 3.   Mechanism or Behavior Unknown - Simple Beta Distribution
A simple procedure that is sometimes used is to restrict the choice of a model to one of four beta distributions (3.67) depending on whether the model is symmetric or skewed. The four choices are as follows:

1. For symmetric cases where all values are felt to be equally likely, the uniform distribution is used:

$$f(x;a,b) = (b - a)^{-1} \qquad a \le x \le b \qquad\qquad (7.2)$$

2. For symmetric cases where central values are more likely, than tail values, the parabolic distribution is used:

$$f(x;a,b) = \frac{6}{(b - a)^3} (x - a)(b - x) \qquad a \le x \le b \qquad (7.3)$$

3. For cases where values are skewed to the right, the right triangular distribution is used:

$$f(x;a,b) = \frac{2}{(b - a)^2} (x - a) \qquad a \le x \le b \qquad (7.4)$$

4. For cases where values are skewed to the left, the left triangular distribution is used:

$$f(x;a,b) = \frac{2}{(b - a)^2} (b - x) \qquad a \le x \le b \qquad (7.5)$$

Once one of the four forms is selected, the distribution is determined by specifying the maximum and minimum possible values (b and a), usually those values which under normal conditions would not be exceeded.   Project people can be shown a simple sketch of each distribution and asked which one best fits their device.

Example 7.2.6   In a simulation study of an inventory system the purchasing manager indicated that the parabolic distribution best represents the time to receive an order after a request to purchase was made.   Due to paper handling and other red tape the minimum time to receive an order was 5 days.   He also felt that in no circumstance would the time exceed 15 days.   In the simulation a parabolic distribution, equation (7.3), was used with a = 5 and b = 15.

Case 4.   Mechanism or Behavior Unknown - Beta Distribution

A somewhat more flexible procedure than Case 3 is to use the beta distribution in general to represent the subjective knowledge about the variable.   Assume a model with a single mode ($\alpha > 1$ and $\beta > 1$), and use the following procedure with (3.67).   Three subjective estimates are obtained from those who are involved with the device. These estimates are obtained as responses to the questions:

1.   What is the expected or average value for the input variable?
     (Let this be x_m.)
2.   What is the highest value expected?   (Let this be b.)
3.   What is the lowest value expected? (Let this be a.)

From this information a beta distribution is fitted on the interval (a,b) with an expected value of x_m and a variance of $(1/36)(b - a)^2$.   The variance of the beta distribution in terms of its parameters is

$$\sigma^2 = \frac{\alpha\beta(b - a)^2}{(\alpha + \beta)^2(\alpha + \beta + 1)} \tag{7.6}$$

and its mean is

$$E(X) = \frac{\alpha}{\alpha + \beta}(b - a) + a \tag{7.7}$$

The values for α and β are obtained by setting equation (7.6) equal to $(1/36)(b - a)^2$ and (7.7) equal to x_m. The resulting estimates are

$$\alpha^* = 36(1 - y)\left[y(1 - y) - \frac{1}{36}\right]$$

and   (7.8)

$$\beta^* = \frac{y\alpha^*}{1 - y}$$

where

$$y = \frac{x_m - a}{b - a}$$

Example 7.2.7 Use a beta distribution to model the length of time required to prepare an order at a takeout restaurant during a given period of the day. The manager states that a minimum time is 30 mins. and the maximum is 75 mins. He would expect that the average order takes 45 mins. Using equation (7.8) with

$$a = 30 \qquad b = 75 \qquad x_m = 45$$

$$y = \frac{45 - 30}{75 - 30} = \frac{1}{3}$$

one obtains

$$\alpha^* = 36\left(1 - \frac{1}{3}\right)\left[\frac{1}{3}\left(1 - \frac{1}{3}\right) - \frac{1}{36}\right] = \frac{14}{3}$$

and

$$\beta^* = \frac{(1/3)(14/3)}{1 - 1/3} = \frac{7}{3}$$

Thus, the beta distribution with parameters $\alpha = 14/3$ and $\beta = 7/3$ defined on the interval $(30,75)$ is used to simulate the time to prepare an order.

The next step in the simulation is to generate the actual random sample values (deviates) from the selected distributions. There are available on almost all computers pseduo-random deviates from

the uniform distribution on the interval (0,1) and the standard
normal distribution ($\mu = 0$, $\sigma^2 = 1$). These deviates are denoted
pseudo-random because they are generated by an algorithm which
starting with a fixed number will always yield the same sequence of
deviates and hence are not truly random. However, these pseudo-
random deviates are adequate for most studies and pass most statistical tests for randomness. If one uses these, there are several
methods of generating random deviates from a chosen distribution.

The first technique can be used for those distributions whose
cumulative distribution function $F(x)$ can be expressed in closed
form. It was previously shown in Theorem 6.3.1 that $F(x)$ is uniformly distributed on the interval (0,1). Thus, random values of
$F(x)$ are generated using the uniform random deviates available on
the computer, and the percentile function, equation (5.5), is used
to generated the desired deviates, i.e.,

$$X = F^{-1}(p)$$

Example 7.2.8 Generate random deviates from the exponential distribution. Equation (5.5a) is the percentile function for exponential distribution. Substitution of a uniform random deviate U for
$p = F(x)$ yields the exponential deviate

$$X = -\frac{1}{\lambda} \ln (1 - U) \tag{7.9}$$

To add a location parameter ε, the equation

$$Y = X + \varepsilon \tag{7.9a}$$

is used where ε is known.

Example 7.2.9 Generate random deviates from the Weibull distribution (3.23). The distribution function for this distribution is

$$F(t) = 1 - e^{-\lambda t^{\eta}}$$

and hence its percentile function is

$$T = \left\{ \frac{-1}{\lambda} \ln[1 - F(T)] \right\}^{1/\eta} \tag{7.10}$$

Thus, generating a uniform deviate and substituting it for $F(T)$ in (7.10) yields the desired Weibull variate.

Another technique used to obtain random deviates is to utilize relationships between the uniform, normal, or other easily obtainable random deviates and the distribution of interest. The following are examples of such relationships:

1. *Lognormal*

The lognormal distribution (3.50) $f(y)$ can be obtained from the normal distribution $g(x)$ via the transformation $Y = e^X$. Using Theorem 5.4.1, we have

$$g(x) = \frac{1}{\sqrt{2\pi}\sigma} e^{-(1/2)[(x-\mu)/\sigma]^2}$$

and hence

$$f(y) = \frac{1}{\sqrt{2\pi}\sigma} e^{-(1/2)[(\ln y-\mu)/\sigma]^2} \left|\frac{dx}{dy}\right|$$

$$= \frac{1}{\sqrt{2\pi}\sigma y} e^{-(1/2)[(\ln y-\mu)/\sigma]^2}$$

$$y > 0, \quad \sigma > 0, \quad -\infty < \mu < \infty$$

Thus, lognormal deviates can be generated from normal deviates via the transformation $Y = e^X$. To generate a lognormal deviate with parameters μ and σ we set

$$Y = e^{\mu+\sigma X} \tag{7.11}$$

If the lognormal has a minimum value of ε, we use

$$Y = e^{\mu+\sigma X} + \varepsilon \tag{7.11a}$$

2. *Gamma Distribution - Integer Values of* η *(Erlangian)*

The gamma distribution (3.35) with integer values for η is often called the Erlangian distribution and can be derived as the sum of independent exponential variates with common scale parameter λ.

Theorem 7.2.1   If X_1, X_2, ..., X_η are independent exponential variates (3.4) with common parameter λ then $Y = X_1 + X_2 + \cdots + X_\eta$ has an Erlangian distribution with parameter λ and η.

Proof: The moment-generating function for X_i is (from Section 3.2, Problem 1).

$$m_{x_i}(\theta) = \frac{1}{1 - \theta/\lambda}$$

and

$$m_y(\theta) = \prod_{i=1}^{n} \frac{1}{1 - \theta/\lambda} = \left(\frac{1}{1 - \theta/\lambda}\right)^{\eta}$$

It is simple to check that this is the moment-generating function of a gamma distribution (see equation 3.34). Therefore, to generate a random deviate from an Erlangian distribution with parameters λ and η it is necessary to generate η independent exponential variates with parameter λ and add them together, i.e., using (7.9):

$$Y = -\lambda^{-1} \sum_{i=1}^{\eta} \ln (1 - U_i)$$

A more efficient form is to use

$$Y = -\lambda^{-1} \ln \prod_{i=1}^{\eta} U_i{}^{\dagger} \tag{7.12}$$

3.   Beta Distribution

The beta distribution (3.67) is related to the gamma distribution as demonstrated in the following theorem.

Theorem 7.2.2   Let Y and Z be independently distributed gamma variates with shape parameter γ and η, respectively, and scale parameters equal to 1.   Then

$$X = \frac{Y}{Y + Z}$$

---

†Note: if U_i is uniform on $(0,1)$, so is $(1 - U_i)$ and vice versa!

has a beta distribution with parameters γ and η.

Proof: Since Y and Z are independently distributed, their joint distribution is given by the product of their density functions:

$$f(y,z) = \frac{1}{\Gamma(\gamma)\Gamma(\eta)} \, y^{\gamma-1} z^{\eta-1} e^{-(y+z)} \qquad y > 0, \quad z > 0$$

The change of variable

$$X = \frac{Y}{Y + Z}$$

leads to

$$Z = Y\left(\frac{1 - X}{X}\right)$$

If this expression is substituted for Z, using results of Theorem 5.4.1 yields

$$g(y,x) = \frac{1}{\Gamma(\gamma)\Gamma(\eta)} \, y^{\gamma-1} y^{\eta-1} \left(\frac{1 - x}{x}\right)^{\eta-1} e^{-y/x} \frac{y}{x^2}$$

$$0 < x < 1 \qquad 0 < y < \infty$$

The density function for X is obtained by integrating out Y.   Thus,

$$h(x) = \frac{1}{\Gamma(\gamma)\Gamma(\eta)} \, \frac{(1 - x)^{\eta-1}}{x^{\eta+1}} \int_0^\infty y^{\gamma+\eta-1} e^{-y/x} \, dy$$

and making use of the fact that

$$\int_0^\infty z^{a-1} e^{-z/b} \, dz = \Gamma(a) \, b^a$$

it follows that

$$h(x) = \frac{\Gamma(\gamma + \eta)}{\Gamma(\gamma)\Gamma(\eta)} \, x^{\gamma-1}(1 - x)^{\eta-1} \qquad 0 < x < 1$$

which is beta density.

Thus, beta variates on a unit interval with parameter values of η and γ can be generated from gamma variates.   To obtain a beta variate on the interval (a,b), let

$$X = \left(\frac{Y}{Y + Z}\right)(b - a) + a \qquad\qquad (7.13)$$

where Y has a gamma distribution with parameter γ, and Z has a gamma distribution with parameter η.

4. *Binomial Distribution*

A binomial random deviate (4.1) with parameters n and p can be generated from uniform deviates as follows. Generate n uniform random deviates and count the number above p. This number has a binomial distribution with parameters of n and p. If n is large and p is not small, the relationship between the normal and binomial can be used. Generate a normal deviate with mean equal to np and variance equal to np(1 - p). Round the generated deviate to a whole number, i.e., any number between $11.5 \leqslant x < 12.5$ should be rounded to 12.

5. *Poisson Distribution*

A Poisson random deviate (4.13) can be generated from the fact that it represents the number of occurrences in an interval where the rate of occurrence, λ per interval, is constant (see Section 4.3). The distribution of the time to the occurrence of an event is exponential with parameter λt, where t is the length of an interval. Thus, the Poisson distribution represents the number of occurrences of an exponential variate in a fixed interval (can be set to 1). Therefore, the Poisson variate k is the number of occurrences of an exponential variate with parameter λ in the time interval (0,1) and can be represented by

$$Y = k$$

where k is the smallest integer such that

$$\sum_{i=1}^{k+1} -\frac{1}{\lambda} \ln(1 - U_i) > 1$$

From (7.9) it is known that $-\lambda^{-1} \ln(1 - U)$ is an exponential variate. Hence, sum up k + 1 of these until a total of 1 is exceeded. It is possible to avoid computing the $\ln(1 - U)$ function, since

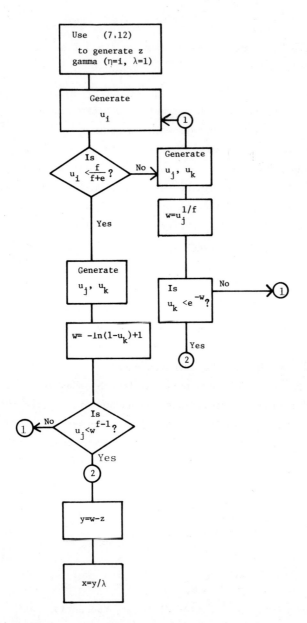

Figure 7.1 Rejection technique for generating gamma variates with noninteger shape parameter.

the criterion for determining k can be written

$$\sum_{i=1}^{k+1} -\frac{1}{\lambda}\ln(1 - U_i) = \frac{-1}{\lambda}\ln\left[\prod_{i=1}^{k+1} (1 - U_i)\right] > 1$$

Now multiplying by negative λ and exponentiating both sides yields

$$\prod_{i=1}^{k+1} (1 - U_i) < e^{-\lambda} \tag{7.14}$$

Thus, k is the smallest integer that satisfies (7.14).

There are several other techniques which are used for generating random variates, but discussion of them is beyond the scope of this test. Some of the most common are the rejection technique, composition technique [see National Bureau of Standards (1964)], and a set of highly efficient procedures suggested by Marsaglia and Bray (1964). Techniques for generating random variables are given below for most of the distributions covered in this book. It has been assumed that standard normal ($\mu = 0$, $\sigma^2 = 1$) and uniform variates (0,1) (Z_i and U_i) are available on a computer or from a table. A synopsis of the techniques for generating most of the distributions discussed in this book is now given.

1. Exponential equation (7.9).

2. Weibull equation (7.10).

3. Lognormal equation (7.11).

4. Erlangian equation (7.12).

5. Gamma distribution (η, λ) noninteger shape parameter: A combination of the method used to get (7.12) and the rejection method suggested by McGrath and Irving (1975) is used. Divide the shape parameter η into two parts: i = integer part and f equal to the fractional part. Thus, $\eta = if$. Let Z = the gamma variate generated using (7.12) with $\eta = i$ and $\lambda = 1$. The rejection method operates as follows. Generate u_i and determine if u_i is greater than $f/(f + e)$, where e is the base of natural logs. If it is greater than this ratio, generate two new uniform deviates u_j and u_k. Let $w = u_j^{1/f}$ and check to see if $u_k < e^{-w}$. If it is not,

reject the process and go back and start over, generating a new value of u_i. If $u_k < e^{-w}$, then let $y = w - z$ and the desired gamma variate is

$$X = \frac{Y}{\lambda} \tag{7.15}$$

If u_i is less than $f/(f + e)$, generate u_j and u_k and let $w = 1 - \ln(1 - u_k)$. If $u_j > w^{f-1}$, reject the process and start over, generating a new value of u_i. If $u_j < w^{f-1}$, then let $y = w - z$ and use (7.15). A diagram of this process is shown in Figure 7.1.

    6. Beta distribution equation (7.13).

    7. Binomial distribution (n,p): Let

$$Y_i = \begin{cases} 0 & \text{if } U_i \leq p \\ 1 & \text{if } U_i > p \end{cases} \qquad i = 1, 2, \ldots, n \tag{7.16}$$

Then

$$X = \sum_{i=1}^{n} Y_i$$

is binomial.

    8. Poisson (λ): Use equation (7.14) to determine k, the Poisson variable.

    9. Rayleigh (λ): The Rayleigh distribution is a special case of the Weibull with $\gamma = 2$; hence, equation (7.10) is used with $\gamma = 2$.

    10. Kodlin's distribution (λ, γ): Kodlin's distribution (3.68) has a distribution function that can be expressed in closed form, and hence random deviates can be generated from its percentile function (see Problem 1):

$$X = \frac{\lambda}{\gamma} \left\{ \left[1 - \frac{2\gamma}{\lambda^2} \ln(1 - U) \right] - 1 \right\} \tag{7.17}$$

    11. Pareto (α, β): The Pareto distribution (3.69) has a distribution function which can be expressed in closed form, and hence random deviates can be generated from its percentile function (see Problem 2):

$$X = \frac{1}{\beta} \left[(1 - U)^{-1/\alpha} - 1 \right] \tag{7.18}$$

12. Type I extreme-value distribution of maximum value (α, β): The distribution function of the type I extreme-value distribution (3.56a) is expressible in closed form; hence, the percentile function is used to generate random variates:

$$X = -\beta \ln[-\ln(1 - U)] + \alpha \tag{7.19}$$

13. Extreme-value distribution minimum value (α, β): The density function (3.56) is directly integrable, and hence the percentile function is used to generate random variables:

$$X = \beta \ln[-\ln(1 - U)] + \alpha \tag{7.20}$$

14. Johnson system (see Chapter 5, equations 5.34, 5.39, and 5.41).

15. Normal distribution (μ, σ):

$$X = \sigma Z + \mu \tag{7.21}$$

16. Uniform (a, b):

$$X = (b - a)U + a \tag{7.22}$$

17. Geometric distribution (θ), $0 < \theta < 1$: The geometric distribution (4.36) has a simple distribution function

$$F(x) = \sum_{i=1}^{x} \theta(1 - \theta)^{i-1}$$

Therefore, one generates a uniform deviate U and finds the smallest value of integer x such that $\sum_{i=1}^{x} \theta(1 - \theta)^{i-1}$ exceeds U.

18. Negative binomial (n, θ): The negative binomial (4.37) can be derived as a sum of n independently distributed geometric random variables with common parameter θ. Hence, its random deviates can be obtained by first generating geometric random variates X_i with parameter θ described above and then letting

$$Y = \sum_{i=1}^{n} X_i \tag{7.22a}$$

The next steps in a simulation study are to generate the desired random variates, plug the numbers into the function describing
the system, and calculate the system output. This process is
repeated many times in order to obtain the desired simulation output. The number of repetitions depends on the purpose of the simulation and the types of statements that will be made as a result
of the study. Often the purpose of a simulation study is to determine the percentage of the output that is beyond a stated limit.
In such studies the output can be divided into two parts, those
above and below the limits. In these situations the data can be
treated as a sample from a binomial population for which it is
desired to estimate θ, the proportion beyond the limit (consider
success as any value beyond the limit). The procedure to determine the number of simulation runs is exactly the same as that of
determining the sample size needed to estimate the parameter, θ of
the binomial distribution within an error band of $\pm D$ using a confidence interval of $(1 - \alpha)100\%$.

This procedure requires an initial estimate of θ say θ', and
n can be obtained from the equation

$$n = \left[\frac{\theta'(1 - \theta')}{D^2} z^2_{1-\alpha/2} \right] + 1 \qquad (7.23)$$

where θ' is an initial estimate of θ

   D   is 1/2 the width of the error band

   $z_{1-\alpha/2}$ is the normal deviate corresponding to a confidence
            coefficient of $1 - \alpha$

   [a]   is the largest integer less than or equal to a

This formula is derived in most elementary textbooks and
assumes that $n\theta$ and $n(1 - \theta)$ are greater than 5, which will be
satisfied in most cases. In the case where no guess can be made
concerning the value of θ, a value of 1/2, which maximizes n, may
be used or a small number of trials m can be run and the value of
θ' estimated from $\theta' = x/m$, where x is the number exceeding the
limit and m is the number of trials.

Example 7.2.10 Determine the number of simulation runs necessary
in order to estimate the percentage of times an electronic heater
will exceed the specified maximum temperature limit for the device.
It is felt that only 1% of the units will exceed this limit and it
is desired to estimate this percentage within ±0.005 with a 99% confidence interval.

For this problem,

$$\theta' = 0.01 \qquad D = 0.005 \qquad \alpha = 0.01 \qquad z_{1-\alpha/2} = 2.58$$

Using equation (7.23),

$$n = \left[\frac{0.01(0.99)(2.58)^2}{(0.005)^2} \right] + 1 = 2636$$

Thus, a total of 2636 simulation runs are required.

In some simulation runs the measurement of interest is the
average output value. In these cases the number of simulation runs
required is obtained from

$$n = \left[\frac{z_{1-\alpha/2}^2 \; \sigma'^2}{D^2} \right] + 1 \qquad\qquad (7.24)$$

where σ'^2 is an a priori estimate of the population variance and
D and $z_{1-\alpha/2}$ are defined in (7.23). Here again the value of σ'^2
can be estimated from a small number of simulation runs by

$$\sigma'^2 = \sum_{i=1}^{m} \frac{(x_i - \bar{x})^2}{m - 1}$$

However, m should be at least 30; otherwise there are some theoretical problems in applying (7.24).

Example 7.2.11 Determine the number of simulation runs necessary
to estimate the gross profit of a corporation where it is desired
to estimate the average within ±$5000 with a 95% confidence inter-
val and σ' is estimated to be $20,000. For this problem,

$$\sigma' = 20{,}000 \qquad D = 5000 \qquad z_{1-\alpha/2} = 1.96$$

Equation (7.24) yields

$$n = \left[\frac{(1.96)^2 (20{,}000)^2}{(5000)^2} \right] + 1 = 62$$

Thus, a total of 62 simulations runs is required. Equation (7.24) is based on the normal distribution, but since by the central limit theorem the distribution of $\bar{x}$ is approximately normal for even moderately sized samples, the approximation will usually be adequate.

Once the simulation is completed the techniques of Chapter 5 can be used to find an empirical approximation, then the desired percentiles may be obtained, the percentage observations beyond or between given limits may be calculated, and moments may be estimated.

Problems

1. Derive the formula for generating random variables from Kodlin's distribution (7.17).

2. Derive the formula for generating random variables from the Pareto distribution (7.18).

3. Derive the formula for generating random variables from the simple beta distributions (7.3), (7.4), and (7.5), without using gamma variates. Why is the formula for the parabolic distribution not useful?

4. Show that the sum of independent geometric variates can be used to generate variates from the negative binomial distribution.

5. A two-parameter Weibull distribution will be used in a simulation study to model the time-to-death of a given insect in the study of an ecosystem involving the wildlife in a given area. Scientists believe that 90% of the insects live at least 15 days and that only 10% of them live past 60 days. From this information determine the equation for generating the random variable, lifetime of the insect.

6. Past studies have indicated that the annual salary of a group
   under study has a lognormal distribution with a minimum value
   of zero (unemployed). This information is one part of a simulation study. If the average income of the group is $20,000
   per year and the variance is $10,000, determine the parameters
   of the lognormal.

7. Determine the parameters of the appropriate "simple" beta
   distributions (7.2), (7.3), (7.4), and (7.5) for the following
   cases:

   (a) The most probable value for the time it takes to repair an
       electronic component is 5.0 mins. (the time to change the
       circuit board). However, it could take as long as 15 mins.
       if the failure were not located in the circuit board.

   (b) The best estimate for the company sales forecast for the
       next year is $650,000. It is felt that even under the
       worst economic conditions it will be no less than $500,000,
       and if the factory produces at full capacity, the most it
       could produce is $800,000.

   (c) The start of a drag race occurs when a green light goes
       on in the starting signal. The light could go on anytime
       within 10 secs. of the time when the cars are in position
       to start.

   (d) A piece of equipment has a lifetime such that the longer
       it is in service the higher the probability of expiring
       in the next instant of time. Due to this feature any
       unit which is in service for 100 hr. is considered to
       have failed and is replaced even though it is still
       operational.

8. In the simulation study of a hospital one of the components
   under study was the operating room and the variable of interest
   was the time to perform an operation. From a study of the
   past records it was found that the longest operation took 2.5
   hrs. and the shortest was 0.2 hrs. Also, the average operation
   took 0.75 hrs. Determine the parameters of the beta distribution to use in the simulation.

9. Write a computer program to generate the distribution of

    $Y = XZ$

    where X is an exponential variate with mean of 5.0 and Z has a
    Rayleigh distribution with $\lambda = 1$. Determine the sample size
    needed if we wish to estimate the percentage of the values of
    Y that exceed 10 within $\pm 1\%$ with a 95% confidence level.

10. Write a computer program to generate gamma random variates
    with $\eta = 2.3$ and $\lambda = 2$.

11. How many simulations are required to estimate the mean output
    of a system to within ± 0.1 unit with a 95% confidence interval
    if the variance of the output is 3.0 units?

12. If a simulation were run with a total of 1000 outputs, within
    what accuracy can we estimate the mean using a 99% confidence
    interval if the variance was 7.0 units?

13. A trial of 100 simulations indicated that 3% of the values
    generated exceeded the critical limit for the system. How
    many more runs are needed to be 99% sure that this figure is
    in error by no more than $\pm 0.25\%$?

7.3 PROPAGATION OF MOMENTS

The technique known as propagation of moments or propagation of error
offers another method of estimating system output based on the
knowledge of the functional relationship between input and output
and information about the distribution or moments of the input variables. The method consists of linearizing the functional relationship that describes system output by means of a multivariate Taylor
series expansion about the mean values of the input variables,
obtaining an expression of the first four moments of the linearized
function, evaluating these four moments by inserting values of the
first four moments of the input variables in the expressions
obtained, and then fitting an empirical distribution to the resulting moments. The first four moments of the input variables are
obtained by assigning probability distributions of each of these

variables, as was done in the case of performing a simulation, or,
if data are available, using the first four sample moments. A
decision must be made as to the precision desired, which affects
the number of terms kept in the Taylor series expansion. Most
often the quadratic model given below is adequate and sometimes
even a linear model is sufficient; the accuracy will depend on how
nonlinear the function is.

   Nonlinear multivariable functions can be linearized by expanding them in a Taylor series. If $f(x_1, x_2, \ldots, x_n)$ is such a
function of random variables $X_1, X_2, \ldots, X_n$ with mean $E(X_1) = \mu_1$,
$E(X_2) = \mu_2, \ldots, E(X_n) = \mu_n$, and the function exists at the point
$(\mu_1, \mu_2, \ldots, \mu_n)$, then the Taylor series expansion of $f(x_1, x_2, \ldots, x_n)$ about its mean values is given by

$$z = z(x_1, x_2, \ldots, x_n) = f(\mu_1, \mu_2, \ldots, \mu_n) + \sum_{i=1}^{n} \frac{\partial f^*}{\partial x_i} (x_i - \mu_i)$$

$$+ \frac{1}{2} \sum_{i=1}^{n} \frac{\partial^2 f^*}{\partial x_i^2} (x_i - \mu_i)^2 + \sum_{i<j}^{n-1} \sum^{n} \frac{\partial^2 f^*}{\partial x_i \partial x_j} (x_i - \mu_i)(x_j - \mu_j) \tag{7.25}$$

and terms of higher order; where $\partial f^*/\partial x_i$ is the partial derivative
of $f(x_1, x_2, \ldots, x_n)$ with respect to x_i evaluated at the mean value
of the x_i's, i.e., at $x_1 = \mu_1, x_2 = \mu_2, \ldots, x_n = \mu_n$.

   In some cases, often referred to as the propagation of error,
the series is truncated after the second term of (7.25); however,
with the advent of high-speed computers the complete quadratic model
can be easily used. Since (7.25) is a linear expression, it is a
straightforward task to obtain $E(Z)$, and $E[Z - E(Z)]^k$ for $k = 2, 3$,
and 4.

   The approximate mean z is obtained from (7.25) by taking the
expected value of both sides, yielding

$$E(Z) = f(\mu_1, \mu_2, \ldots, \mu_n) + \sum_{i=1}^{n} \frac{\partial f^*}{\partial x_i} E(X_i - \mu_i) + \frac{1}{2} \sum_{i=1}^{n} \frac{\partial^2 f^*}{\partial x_i^2}$$

$$\times E(X_i - \mu_i)^2 + \sum_{i<j}^{n-1} \sum^{n} \frac{\partial^2 f^*}{\partial x_i \partial x_j} E[(X_i - \mu_i)(X_j - \mu_i)]$$

plus higher-order terms. Noting that $E(X_i - \mu_i) = 0$ and dropping
the higher-order terms yields

$$E(Z) = f(\mu_1, \mu_2, \ldots, \mu_n) + \frac{1}{2} \sum_{i=1}^{n} \frac{\partial^2 f^*}{\partial x_i^2} \, \mathrm{var}(X_i)$$

$$+ \sum_{i < j}^{n-1} \sum^{n} \frac{\partial^2 f^*}{\partial x_i \partial x_j} \, \mathrm{cov}(X_i X_j) \qquad (7.26)$$

If the input variables, X_1, X_2, $\ldots$, X_n are mutually independent so
that $\mathrm{cov}(X_i X_j) = 0$ for all $i \neq j$, then (7.26) simplifies to

$$E(Z) = f(\mu_1, \mu_2, \ldots, \mu_n) + \frac{1}{2} \sum_{i=1}^{n} \frac{\partial^2 f^*}{\partial x_i^2} \, \mathrm{var}(X_i) \qquad (7.27)$$

Example 7.3.1 The relationship between the actual volume of liquid
in a cylindrical container and the physical dimensions was discussed
in Example 7.2.1. The technique of propagation of moments can be
used to obtain the average volume of the liquid in the container.
The system equation is

$$Z = \pi r^2 hp$$

where r, h, and p are independent random variables. Assume r and
h are normal variables. The specified tolerance for the radius is
6.00 ± 0.6 in. and is 10.5 ± 0.3 in. for the height. These can be
converted into moments by setting the expected value of each random
variable equal to the specified value and the standard deviation
equal to 1/6 the tolerance spread. Thus

    $E(r) = 6.00$ $E(h) = 10.5$

    $\mathrm{var}(r) = 0.04$ $\mathrm{var}(h) = 0.01$

Assume that data are obtained on p, the proportion of volume containing liquid, from a similar vessel and the mean and variance are,

respectively,

$$E(p) = 0.98 \text{ and } var(p) = 0.04$$

The required partial derivatives are

$$\frac{\partial z}{\partial r} = 2\pi rhp \qquad \frac{\partial^2 z}{\partial r^2} = 2\pi hp \qquad \frac{\partial^2 z*}{\partial r^2} = 2\pi(10.5)(0.98) = 64.65$$

$$\frac{\partial^2 z}{\partial h^2} = 0 \qquad \frac{\partial^2 z}{\partial p^2} = 0$$

and from (7.27),

$$E(Z) = \pi(6.00)^2(10.5)(0.98) + \frac{1}{2} 64.65(0.04)$$

$$= 1163.77 + 1.29 = 1165.06 \text{ in}^3$$

If the radial and height dimensions were correlated, for example, if they were machined at one pass so that if one were smaller than the nominal value, the other also tended to be smaller, then (7.26) would be used with the covariance term. Assume $cov(r,h) = 0.001$ and all other covariances are zero. Then (7.26) yields, with $\partial^2 z*/\partial r \partial h = 36.95$,

$$E(Z) = 1165.06 + (36.95)(0.001) = 1165.10$$

When the higher moments of Z are calculated it simplifies notation if the origin is shifted and the notation is modified. Thus, one defines Z [as suggested by Cox (1979)] as

$$Z = b_0 + \sum_{i=1}^{n} b_i(X_i - \mu_i) + \sum_{i=1}^{n} b_{ii}(X_i - \mu_i)^2$$

$$+ \sum_{i<j}^{n-1} \sum^{n} b_{ij}(X_i - \mu_i)(X_j - \mu_j) \qquad (7.28)$$

and sets $Y = Z - b_0$. Now the moments of Y can be obtained directly as $E(Y)$, $var(Y)$, $\mu_3(Y)$, and $\mu_4(Y)$ and except for the expected value term they are identical to the corresponding values for Z. To get

E(Z) one merely computes

$$E(Z) = E(Y) + b_0 \tag{7.29}$$

In this notation,

$$b_0 = f(\mu_1, \mu_2, \ldots, \mu_n)$$

$$b_i = \frac{\partial f*}{\partial x_i} \qquad b_{ii} = \frac{1}{2}\frac{\partial^2 f*}{\partial x_i^2} \qquad b_{ij} = \frac{\partial^2 f*}{\partial x_i \partial x_j}$$

In what follows the moments of Y, $E(Y^k)$, $k = 1$, 2, 3, and 4, are obtained under the assumptions that the input variables X_i's are mutually independent and that all terms in (7.28) are included. A partial treatment for the dependent case can be found in Hahn and Shapiro (1967). Now from (7.27) and (7.29),

$$E(Y) = E(Z) - b_0 = \sum_{i=1}^{n} b_{ii} \, var(X_i) \tag{7.27a}$$

$$E(Y^2) = E\left[\sum_{i=1}^{n} b_i (X_i - \mu_i) + \sum_{i=1}^{n} b_{ii}(X_i - \mu_i)^2 \right.$$

$$\left. + \sum_{i<j}^{n-1} \sum^{n} b_{ij}(X_i - \mu_i)(X_j - \mu_j) \right]^2$$

Terms of the form $E\left[(X_i - \mu_i)(X_j - \mu_j)^k\right]$, $i \neq j$, drop out since from the independence property this equals $E(X_i - \mu_i) \, E(X_j - \mu_j)^k$ and hence zero. Thus,

$$E(Y^2) = \sum_{i=1}^{n} \left[b_i^2 \, var(X_i) + 2b_i b_{ii}\mu_3(X_i) + b_{ii}^2\mu_4(X_i) \right]$$

$$+ \sum_{i<j}^{n-1} \sum^{n} \left(2b_{ii}b_{jj} + b_{ij}^2 \right) var(X_i)var(X_j) \tag{7.30}$$

In many problems equation (7.30) is truncated to include only powers up to third order. In that case,

$$E(Y^2) = \sum_{i=1}^{n} \left[b_i^2 \, var(X_i) + 2b_i b_{ii} \mu_3(X_i) \right]$$ (7.30a)

[Note: $var(Y_i) \, var(X_j)$ is a fourth-order term.]

The variance of Y is obtained from $E(Y^2) - [E(Y)]^2$

Example 7.3.2 Recompute the mean of Z for Example 7.3.1 and then
obtain the variance of Z. From (7.27a),

$$E(Y) = \frac{1}{2} (64.65)(0.04) = 1.29$$

and hence

$$E(Z) = b_0 + E(Y) = 1163.77 + 1.29 = 1165.06$$

as before. In order to calculate the second moment the values of
partial derivatives are required as well as some more information
about the distribution or moments of the X_i's, that is r, h, and p:

$$\frac{\partial z}{\partial r} = 2\pi rhp = 387.92 \qquad \frac{\partial z}{\partial p} = \pi r^2 h = 1187.52$$

$$\frac{\partial z}{\partial h} = \pi r^2 p = 110.84 \qquad \frac{\partial^2 z}{\partial r \partial h} = 2\pi rp = 36.95$$

$$\frac{\partial^2 z}{\partial r \partial p} = 2\pi rh = 395.84$$

$$\frac{\partial^2 z}{\partial h \partial p} = \pi r^2 = 113.10$$

The variables r and h are normally distributed; hence, $\mu_3(r) = \mu_3(h) = 0$. From the data the value of $\mu_3(p) = 0.0000042$. Also,
$\mu_4(r) = 3\sigma^4 = 0.0048$, $\mu_4(h) = 0.0003$, and $\mu_4(p) = 5.7 \times 10^{-7}$. By
(7.30), omitting zero value terms gives

$$E(Y^2) = (387.92)^2(0.04) + (110.84)^2(0.01)$$

$$+ (1187.52)^2(0.0004) + \left(\frac{64.65}{4}\right)^2(0.0048)$$

$$+ (36.95)^2(0.04)(0.01) + (395.84)^2(0.04)(0.0004)$$

$$+ (113.10)^2(0.01)(0.0004)$$

$$= 6714.33$$

Hence,

$$\text{var}(Y) = \text{var}(Z) = 6714.33 - (1.29)^2 = 6712.67$$

The third moment of Y can be obtained by cubing (7.28) after subtraction of b_0, taking expected values, and dropping all terms containing $E(X_i - \mu_i)$. Thus,

$$E(Y^3) = \sum_{i=1}^{n} [b_i^3 \mu_3(X_i) + b_{ii}^3 \mu_6(X_i) + 3b_i^2 b_{ii} \mu_4(X_i)$$

$$+ 3b_i b_{ii}^2 \mu_5(X_i)] + \sum_{i}^{n-1} \sum_{j}^{n} [b_{ij}^3 \mu_3(X_i)\mu_3(X_j)$$
$$\qquad \qquad \qquad \qquad i < j$$

$$+ 6b_i b_j b_{ij} \,\text{var}(X_i)\,\text{var}(X_j) + 6b_{ii} b_{jj} b_{ij} \mu_3(X_i)\mu_3(X_j)]$$

$$+ \sum_{i}^{n} \sum_{j}^{n} [3b_{ii}^2 b_{jj} \mu_4(X_i)\,\text{var}(X_j) + 6b_i b_{jj} b_{ij}\,\text{var}(X_i)\mu_3(X_j)$$
$$\quad i \neq j$$

$$+ 3b_{ii} b_j^2 \,\text{var}(X_i)\,\text{var}(X_j) + 6b_i b_{ii} b_{jj} \mu_3(X_i)\,\text{var}(X_j)$$

$$+ 3b_i b_{ij}^2 \mu_3(X_i)\,\text{var}(X_j) + 3b_{ii} b_{ij}^2 \mu_4(X_i)\,\text{var}(X_j)]$$

$$+ \sum_{i=1}^{n-2} \sum_{j=i+1}^{n-1} \sum_{k=j+1}^{n} [6b_{ii} b_{jj} b_{kk} + 6b_{ij} b_{ik} b_{jk} + 3b_{ii} b_{jk}^2$$

$$+ 3b_{jj} b_{ik}^2 + 3b_{kk} b_{ij}^2]\,\text{var}(X_i)\,\text{var}(X_j)\,\text{var}(X_k) \qquad (7.31)$$

This equation can be simplified if all terms of fourth order or higher are omitted. This results in

$$E(Y^3) = \sum_{i=1}^{n} b_i^3 \mu_3(X_i) \qquad (7.31a)$$

which is often adequate.

Following along the same lines, $E(Y^4)$ can be obtained. This is shown as equation (B.1) in Appendix B. If terms in $E(Y^4)$ of fifth order and higher are dropped, (B.1) reduces to

$$E(Y^4) = \sum_{i=1}^{n} b_i^4 \mu_4(X_i) + \sum_{i<j}^{n-1} \sum^{n} 6b_i^2 b_j^2 \, \text{var}(X_i) \, \text{var}(X_j) \qquad (7.32)$$

The third and fourth moments about the mean of Y can be obtained from

$$\mu_3(Y) = E(Y^3) - 3E(Y^2)E(Y) + 2[E(Y)]^3$$

and

$$\mu_4(Y) = E(Y^4) - 4E(Y^3)E(Y) + 6E(Y^2)[E(Y)]^2 - 3[E(Y)]^4$$

The complete second-order model requires knowing the central moments of up to order 8 for each of the input variables. These can be obtained from a knowledge of the distribution or be estimated from data. (Estimates of high-order moments will often be poor.)

A complete second-order expansion requires some involved calculations. However, there is a computer program entitled SOERP[†] that performs the complete moment calculations. In order to use this program it is necessary to convert the input variables X_i to standard form, that is, use

$$W_i = \frac{X_i - \mu_i}{\sigma(X_i)} \qquad (7.33)$$

where $\sigma(X_i)$ is the standard deviation of X_i. Using this form, (7.28) is written as

---

[†]SOERP is available from the Argonne Code Center, Building 203, Room C230, 9700 South Cass Ave., Argonne, Ill. 60439. It was prepared by N. Cox and C. Miller.

$$Z = b_0 + \sum_{i=1}^{n} b_i \sigma(X_i) \left[\frac{X_i - \mu_i}{\sigma(X_i)}\right] + \sum_{i=1}^{n} b_{ii} \sigma^2(X_i) \left[\frac{X_i - \mu_i}{\sigma(X_i)}\right]^2$$

$$+ \sum_{i < j}^{n-1} \sum_{}^{n} b_{ij} \sigma(X_i) \sigma(X_j) \left[\frac{X_i - \mu_i}{\sigma(X_i)}\right] \left[\frac{X_j - \mu_j}{\sigma(X_j)}\right]$$

At this point define $a_i = b_i \sigma_i$, $a_{ii} = b_{ii} \sigma_i^2$, and $a_{ij} = b_{ij} \sigma_i \sigma_j$, so that

$$Z = b_0 + \sum_{i=1}^{n} a_i W_i + \sum_{i=1}^{n} a_{ii} W_i^2 + \sum_{i < j}^{n-1} \sum_{}^{n} a_{ij} W_i W_j \qquad (7.34)$$

Equations (7.27a), (7.30), (7.31), and (B.1) can be used with the a's replacing the b's and the W's replacing the X's. This is the input required for SOERP. The advantage of using the standardized variables W_i's is that their moments do not depend on location and scale parameters and hence are always the same for each shape-free parameter distribution. Therefore, the moments of W_i for a normal distribution random variable are always the same, i.e., the values for the standardized variate.

$$\mu = 0, \qquad \mu_2 = 1, \qquad \mu_3 = 0, \qquad \mu_4 = 3, \qquad \mu_5 = 0,$$

$$\mu_6 = 15, \qquad \mu_7 = 0, \qquad \mu_8 = 105 \qquad (7.35)$$

and for the exponential distribution the moments of W_i are always

$$\mu = 1, \qquad \mu_2 = 1, \qquad \mu_3 = 2, \qquad \mu_4 = 9, \qquad \mu_5 = 44,$$

$$\mu_6 = 265, \qquad \mu_7 = 1854, \qquad \mu_8 = 14833. \qquad (7.35a)$$

Example 7.3.3 The following example appears in Volk (1969) and was used by Cox (1979). It illustrates the use of SOERP with a simple model. The equation for the volumetric gas flow measured by an orifice meter can be written as

$$Q = C \sqrt{\left\{\frac{HP}{M}\left(\frac{520}{t + 460}\right)\right\}^{1/2}}$$

where Q = volumetric flowrate, standard ft^3/hr

C = a known constant depending on orifice size and units of measurement = 38.4

H = pressure differential, in. water

M = molecular weight of gas mixture

P = absolute pressure, psia

t = temperature degree $^{\circ}F$

It appeared that all four variables were symmetrically distributed about their nominal or design values and were mutually independent, and that a normal model described adequately the distribution for each variable. The means (nominal value) and standard deviations (1/6 the tolerance spread) were as follows:

Variable	Mean	Standard deviation
H	64 in. water	0.5
M	16	0.1
P	361 psia	2.0
t	165°F	0.5

A Taylor series expansion (7.34) of Q yields the coefficients shown below. For example, the coefficient a_H is obtained from

$$b_H = \frac{Q^*}{H} = \frac{C}{2}\left\{\frac{P}{HM}\left(\frac{520}{t+460}\right)\right\}^{1/2} = \frac{38.4}{2}\left[\frac{(361)(520)}{(64)(16)(625)}\right]^{1/2}$$

$$= 10.40$$

$$a_H = b_H\sigma_H = 10.40(0.5) = 5.20$$

Coefficients:

$b_0 = 1331$ $a_{HM} = -0.015$ $a_{Pt} = -0.001$

$a_H = 5.2$ $a_{HP} = 0.01$ $a_{HH} = -0.010$

$a_M = -4.16$ $a_{Ht} = -0.0025$ $a_{MM} = 0.0195$

$a_P = 3.68$ $a_{MP} = -0.012$ $a_{PP} = -0.0051$

$a_t = -0.53$ $a_{Mt} = 0.0015$ $a_{tt} = 0.0003$

These coefficients are used as input to the SOERP program along with the standardized first eight moments of each variable. Since all four distributions are Gaussian, the moments are the same as given in (7.35). The input cards contain the following information which is shown in Table 7.1:

The first card contains the title of the problem.

The second card specifies number of independent variables   n.

The third card contains the linear coefficients a_i.

The fourth card contains the quadratic coefficients a_{ii}.

The next n cards (four in this example) contain the central moments

of the input variable W_i in order from 2 to 8.

The last cards contain the cross-product terms and a designation

indicating which terms they are.

The first two numbers on the cards indicate the two variables i and

j, and this is followed by the value of a_{ij}.

Table 7.1   Volumetric Gas Flow Analysis

Card#								
1	Volumetric Gas Flow Analysis							
2	4							
3	5.2	−4.16	3.68	−0.53				
4	−0.01	0.0195	−0.0051	0.0003				
5	0.0	1.0	0.0	3.0	0.0	15.0	0.0	105.0
6	0.0	1.0	0.0	3.0	0.0	15.0	0.0	105.0
7	0.0	1.0	0.0	3.0	0.0	15.0	0.0	105.0
8	0.0	1.0	0.0	3.0	0.0	15.0	0.0	105.0
9	1	2	−0.015					
10	1	3	0.01					
11	1	4	−0.0025					
12	2	3	−0.012					
13	2	4	0.0015					
14	3	4	−0.001					

The output for the problem gives the mean, the first four moments, the first four central moments, $\sqrt{\beta_1}$, and β_2. The actual mean of Q is obtained by adding b_0 to the value indicated. The values obtained for this problem are

$$E(Q) = 1331 \qquad \sqrt{\beta_1} = 0.009$$

$$\text{var}(Q) = 57.6 \qquad \beta_2 = 3.000$$

The values $\sqrt{\beta_1}$ and β_2 of 0.009 and 3.000 obtained for this problem indicate that Q can be assumed to be normally distributed. Therefore, the distribution used to approximate Q is normal with a mean of 1331 and a standard deviation of 7.59. If it were of interest to determine the probability of obtaining a flowrate of less than 1300, if follows that

$$\text{pr}\{(Q < 1300)\} = \text{pr}\{Z < \frac{(1300 - 1331)}{7.59}\} = \text{pr}\{Z < -4.08\} = 0.000023$$

The technique of propagation of moments also supplies the analyst with information in regard to which variables are the major contributors to the total variation. Since each variable is in standardized form, the coefficient a_i indicates the relative effect of a change in the variability of one component on the total variance. Thus, the variable with the largest a_i has the greatest effect on total variability. The expression for the variance is obtained from $E(Y^2)$ (7.30) with a_i's and W_i's substituted for b_i's and X_i's; however, in most cases the simplified version (7.30a) is sufficient. Note that in many case (7.30a) is the approximate expression for the variance, i.e., when the $E(Y)$ is small relative to b_0.

Example 7.3.4 From the data in Example 7.3.3 find the major contributors to the variance. Equation (7.30a) yields

$$\text{var}(Y) = \sum_{i=1}^{4} a_i^2 \, \text{var}(W_i) = \sum_{i=1}^{4} a_i^2$$

since $\mu_3(W_i) = 0$ for $i = 1, 2, 3,$ and 4, $\text{var}(W_i) = 1$, and $E(Y) \approx 0$.

$$\text{var}(Y) = (5.2)^2 + (4.16)^2 + (3.68)^2 + (0.53)^2 = 58.2$$

(Note that omission of the truncated terms increased the variance
from 56.7 to 58.2). The first term, due to H, accounts for over
46% of the variance (27/58.2), while the second term M accounts
for less than 30% of the variance. If it were desired to reduce
the variability of Q, then the possibility of reducing the variability of H should be considered first. A cost analysis may be
performed to determine the least expensive procedure for reducing
variability.

Problems

(In all problems in this section use the truncated form of the
equations unless the SOERP program is available.)

1. Determine the third and fourth moments of the volume of liquid
    in Example 7.3.1.

2. Use the technique of Chapter 5 and the results of Problem 1,
    to select and then fit an approximating distribution to the
    volume of the liquid in Example 7.3.1. Find the probability
    that the volume of liquid will exceed 1400 in$.^3$.

3. Use the information of Example 7.3.1 to write the expression
    for volume of liquid in the standardized form required for the
    SOERP program.

4. The amount of garbage collected in a day is the sum of the
    amount collected by six different methods. The distribution
    of the amount collected by each method is given below. Find
    the distribution of the total by the method of moments and the
    central limit theorem. Compare the probability obtained by
    each method of getting a total amount of garbage for a single
    day greater than twice the expected amount

$$Y = \sum_{i=1}^{6} X_i \text{ where } X_1 \text{ is normal} \qquad \mu = 5, \ \sigma^2 = 9$$

$$X_2 \text{ is exponential} \qquad \lambda = 0.4$$

$$X_3 \text{ is lognormal} \qquad \mu = 0.6, \ \sigma^2 = 0.6$$

$$X_4 \text{ is Weibull} \qquad \lambda = 0.2, \ \gamma = 3$$

$$X_5 \text{ is gamma} \qquad \lambda = 4, \ \eta = 6$$

$$X_6 \text{ is Rayleigh} \qquad \lambda = 1/4$$

If you had to choose one of the sources as the major contributor to the variability in the amount of garbage collected which one would you choose?

5. The relationship of the output torque of a motor, y, and five design variables is given by

$$y = x_1^2 + \frac{x_2 x_3^2}{x_4} = e^{-x_5}$$

Find the distribution of Y given the following information about X_1, X_2, X_3, X_4, and X_5:

Variables	Mean	Variance	3	4
X_1	1	0.5	0.1	1.0
X_2	2	1	-1	3.0
X_3	2	0.5	0	2.0
X_4	5	2	0	16
X_5	1	0.2	0	8

Assume X_1, X_2, X_3, X_4, and X_5 are all statistically independent.

6. Find the moments of Y based on the following model (assume all high-order product moments are zero).

$$Y = \frac{X_1 X_2}{X_3} \text{ where } X_1 \text{ is normal} \qquad \mu = 4, \ \sigma^2 = 1$$

$$X_2 \text{ is normal} \qquad \mu = 6, \ \sigma^2 = 2$$

$$X_3 \text{ is exponential} \qquad \lambda = 1$$

and

$$\text{cov}(X_1 X_2) = 1$$

$$\text{cov}(X_1 X_3) = 0$$

$$\text{cov}(X_2 X_3) = 0$$

7. Use the propagation technique to obtain the moments of the distribution of the following function:

$$y = x_1 x_2^2 e^{x_3} \quad \text{where} \quad X_1 \text{ is normal} \qquad \mu = 1, \sigma = 2$$

$$X_2 \text{ is normal} \qquad \mu = 2, \sigma = 2$$

$$X_3 \text{ is exponential} \qquad \lambda = 2$$

and all three are mutually independent. Find the probability that Y is greater than 55.0.

8. In Problem 5, what percentage of the total variability in Y is due to variable X_3?

7.4 CASE STUDIES

In the remainder of this chapter several case studies are discussed. Some of these have been scaled down from real problems for ease in presentation.

Example 7.4.1 A sales forecast is being prepared for a company. The president wishes to estimate the sales volume for next year and determine the probability that it will either exceed $5.1 million or be less than $3.6 million. The company has seven regional offices. The vice-president of marketing requests each regional sales manager to submit three forecasts to him. The first is their best estimate of the volume expected next year under normal conditions. The second estimate is the most optimistic estimate of the volume assuming everything goes smoothly and conditions for demand increase at the most optimistic rate. The third is the most pessimistic estimate of volume assuming the worst possible conditions.

The system model is simply

$$T = \sum_{i=1}^{7} T_i$$

where T_i is the forecast for the ith region. Since T is linear and contains seven terms of approximately equal weight which are independent of each other,[†] the central limit theorem is applied. Hence, T is approximately normally distributed with

$$E(T) = \sum_{i=1}^{7} E(T_i)$$

$$var(T) = \sum_{i=1}^{7} var(T_i)$$

The reports from the seven regional managers are as follows:

Region	Best estimate	Optimistic	Pessimistic	1/6 Range
1	0.4	0.6	0.2	0.067
2	0.7	1.2	0.4	0.133
3	0.5	0.8	0.3	0.083
4	0.7	0.9	0.6	0.050
5	0.6	0.9	0.4	0.083
6	1.3	1.7	0.8	0.150
7	0.3	0.5	0.2	0.050

Estimating the standard deviation as 1/6 the difference between the maximum and minimum forecasts, the following input is obtained:

$$E(T_1) = 0.4 \qquad var(T_1) = 0.0044$$

$$E(T_2) = 0.7 \qquad var(T_2) = 0.0178$$

$$E(T_3) = 0.5 \qquad var(T_3) = 0.0069$$

$$E(T_4) = 0.7 \qquad var(T_4) = 0.0025$$

$$E(T_5) = 0.6 \qquad var(T_5) = 0.0069$$

$$E(T_6) = 1.3 \qquad var(T_6) = 0.0225$$

$$E(T_7) = 0.3 \qquad var(T_7) = 0.0025$$

Thus,

[†]Regional sales managers do not communicate with each other.

E(T) = \$4.5 million

var(T) = \$0.0635 million

and T is normally distributed. The two desired probabilities are
then obtained as

$pr\{T > 5.1\} = pr\{Z > 2.38\} = 0.0087$

$pr\{T < 3.6\} = pr\{Z < -3.57\} = 0.0002$

Thus, the probability of exceeding \$5.1 million is a little less
than 0.01, and the probability of sales dropping below \$3.6 million
is negligible.

Example 7.4.2 An antimissile device is being designed and it is
necessary to determine the distribution of its "miss" distance,
i.e., the distance from the actual target to the detonation point,
in order to design the size of the explosive charge needed so that
the probability of a kill is above some minimum value. The device
works as follows. Based on radar returns the antimissile device
locks on a target and tracks the missile. The device's computer
determines the speed of the missile and programs an intercept
course. This course is set into the device's guidance system and
it is launched. When the device reaches the predetermined position,
it is detonated. There are four components to the intercept location: height, longitude, latitude, and time. The objective is to
have the missile detonate at the right location at the right time.
Therefore, a miss "distance" consists of four components. The
system has been designed so that on the average the miss distance
in all four dimension is zero, that is, the aiming device is
unbiased. Experimentation has shown that the aim error standard
deviation in each dimension if 4 units (feet for the distance variables and tenths of a second for the time variable).

The system model or output variable is the radial distance

$$w = \sqrt{x^2 + y^2 + z^2 + t^2}$$

where x, y, and z are the distance variables and t is the time

variable. A discussion with the system designers has indicated
that the variables X, Y, Z, and T are mutually independent. In
prior error analyses the assumption that each variable is normally
distributed gave good results; hence, this assumption is used for
this problem. Therefore, X, Y, Z, and T are independent normal
random variables with zero means (since the average deviation from
target in each dimension is zero) and variance equal to 16 units.
It is required to find the distribution of W, defined above.

This problem can be solved by transformation of variables,
but it is somewhat easier to make use of the fact that a linear
combination of the squares of n independent, standard normal random
variables has a chi-square distribution with n degrees of freedom
(gamma distribution with scale parameter equal to 1/2 and shape
parameter equal to n/2). Thus,

$$V = \left(\frac{X}{4}\right)^2 + \left(\frac{Y}{4}\right)^2 + \left(\frac{Z}{4}\right)^2 + \left(\frac{T}{4}\right)^2 = \frac{1}{16}\left[X^2 + Y^2 + Z^2 + T^2\right]$$

has a chi-square distribution with four degrees of freedom,

$$f(v) = \frac{1}{4\Gamma(2)}\, v e^{-v/2} = \frac{1}{4}\, v e^{-v/2} \qquad v \geq 0$$

since X/4, Y/4, Z/4, and T/4 are standard normal variables. Now
if $W = 4V^{1/2}$ so that

$$W^2 = 16V = X^2 + Y^2 + Z^2 + T^2$$

then g(w) is the desired distribution. Since

$$dv = \frac{2w\,dw}{16} \qquad v = \frac{w^2}{16}$$

then Theorem 5.4.1 yields

$$g(w) = \frac{1}{512}\, w^3 e^{-w^2/32} \qquad w \geq 0$$

For any given value of W, say w, it is now possible to obtain
pr{W < w}. For example, if a sufficient charge is placed in the

missile so that the kill distance is 8 ft, then the probability of
a kill is given by

$$F(8) = \frac{1}{512} \int_0^8 w^3 e^{-w^2/32} \, dw = 0.594$$

The value of this integral can be determined by integration by parts
or it can be transformed to a gamma integral and looked up in a
table of the incomplete gamma function.

Example 7.4.3 Monte Carlo simulations are usually used on very
large problems where other techniques would be too time-consuming
or cumbersome. However, it is not feasible to describe such a
system in a book such as this. Therefore, this case study confines
itself to a small segment of an overall system in order to illustrate the technique. The large system is a nuclear reactor and the
the project is to simulate the maximum clad temperature. One factor
in the simulation is the gap conductance, which is given by

$$h_g = \frac{k_g}{t_g + 5.448\left[\dfrac{\mu}{P}\left(\dfrac{T}{M}\right)^{1/2}\right]} + h_r$$

where k_g = the gas thermal conductivity (helium)

$\quad\ \ t_g$ = gap thickness

$\quad\ \ \mu$ = viscosity, g/cm-sec

$\quad\ \ P$ = pressure, lb/in^2

$\quad\ \ T$ = temperature, $^\circ$K

$\quad\ \ M$ = molecular weight of helium

$\quad\ \ h_r$ = radiant heat transfer coefficient

The object is to determine the distribution of h_g. The numbers
following have all been synthesized and bear no reality to the real
problem.

Following the procedure outlined for Monte Carlo simulation,
it is first necessary to define the system equation. This is

obtained directly from the laws of physics and is given above.
The second step is to assign the random variables in this equation
their appropriate statistical models:

k_g: The gas thermal conductivity of helium is given as 0.0049.
However, it is known that this may vary between 0.0043 and
0.0054. While the distribution is unknown, it was believed
by the engineers to be symmetrical with values around 0.0049
more likely. A parabolic distribution was selected (equation
7.3) with a = 0.0043 and b = 0.0054. Since the parabolic
distribution is a beta variable with $\alpha = \beta = 2$, one uses
equation (7.13) to generate the random variable. This requires in turn that two gamma variates X and Y with parameters
α and β be generated from equation (7.12). Thus, X and Y are
obtained from the sum of α and β independent exponential variates, respectively, with $\alpha = \beta = 2$, and $\lambda = 1$.

t_g: Gap thickness is a dimensional quantity and its variation
represents the variability in the manufacturing process. A
normal distribution is used to represent the tolerance for
this quantity. The tolerance for the dimension is 0.025 ±
0.001; hence, set $\mu = 0.025$ and $\sigma = 0.00033$ in equation (7.21)
since $\mu \pm 3\sigma$ contains 99.8% of the area under the normal
curve.

μ: The viscosity is again a physical property of the material
and is believed to be 0.000039; however, it may vary between
0.000030 and 0.000048. The distribution is unknown and any
value between the extremes is equally likely; hence, a uniform random variable, equation (7.22), with a = 0.000030 and
b = 0.000048, is used to generate the viscosities.

P: The pressure was expected to have a distribution which was
skewed to the right with a minimum value of 450, and it is
believed that a lognormal distribution adequately represents
this phenomenon. The average pressure is designed to be 500,
and a variance of 100 was common in past applications. The
parameters of the lognormal may be obtained by the techniques

of matching of moments described in Section 5.2. Let ε = 450,

$$500 = e^{\mu + (1/2)\sigma^2} + \varepsilon$$

and

$$100 = ^{2\mu + \sigma^2}(e^{\sigma^2} - 1)$$

then

$$50 = e^{\mu + (1/2)\sigma^2}$$

Square this quantity, thus

$$2500 = e^{2\mu + \sigma^2}$$

Divide into the second equation,

$$\frac{100}{2500} = e^{\sigma^2} - 1$$

$$\sigma^2 = 0.039$$

Now

$$\ln 50 = \mu + \frac{1}{2}(0.039)$$

$$\mu = 3.893$$

Therefore, equation (7.11a) is used with μ = 3.893, σ = 0.197, and ε = 450.

T: The temperature is expected to be symmetrically distributed about 1600 with a standard deviation of 30°. Since temperatures should cluster about the 1600 reading and tail off rapidly, a normal distribution with μ = 1600 and σ = 30 is used in equation (7.21).

M: The molecular weight of helium will be treated as a constant equal to 4.0. The variability from this figure is so small that we need not consider it a random variable.

h_r: The radiant heat transfer coefficient has a minimum value of
0.6, which is its most likely value, and the probabilities of
higher values decrease from this point.   The average value is
0.80.   An exponential model can be used to represent this
variable and $\lambda = (0.8 - 0.6)^{-1} = 5$ is used in equation (7.9a)
with $\varepsilon = 0.6$.

Once the distributions are identified and the method of generating the random deviates are specified, the system can be pro-
grammed.   The next step is to determined the number of simulation
runs.   A sufficient number of simulations is required so that the
percentage of the times the limit of 1.4 is exceeded may be estimated with a 95% confidence interval with an error band of ± 0.01.
It is believed that theta is roughly 1%.   Equation (7.23) with

$$\theta' = 0.01 \qquad D = 0.01 \qquad z_{1-\alpha/2} = 1.96$$

$$n = \left\lceil \frac{0.01(0.99)}{(0.01)^2} (1.96)^2 \right\rceil + 1 = 381$$

The results of simulation are shown in Table 7.2.   From these
results an estimate of 0.047 is obtained for the probability of
exceeding 1.4.   The mean of h_g is 0.991; the variance is 0.0364.

Example 7.4.4   The next example is one that arises in genetics and
ecology.   Assume that a population characteristic is controlled by
a gene that can take on one of two forms (A or a) called alleles.
In a mating the offspring obtains one gene from each parent,
resulting in three possible genotypes (AA, Aa, or aa).   The proportions of these three types in the initial population are denoted
as p^2, 2pq, and q^2, where $p = pr\{A\}$ and $q = pr\{a\}$.   Furthermore,
assume that there are n environments in which the population exists
and in each of these one allele is favored over the other.   The
gene frequency is a given environment increases or decreases depending on whether the gene is favored.   After selection let c_i be
the proportion of the total surviving population in each of the
environments, $\sum_{i=1}^{n} c_i = 1$.   Li (1955) has suggested the following

Table 7.2   Results of Simulation for Example 7.4.3

Values of h_g	Frequency
0.750 - 0.799	19
0.800 - 0.849	72
0.850 - 0.899	66
0.900 - 0.949	48
0.950 - 0.999	40
1.000 - 1.049	31
1.050 - 1.099	25
1.100 - 1.149	17
1.150 - 1.199	14
1.200 - 1.249	11
1.250 - 1.299	6
1.300 - 1.349	8
1.350 - 1.399	6
1.400 - 1.449	4
1.450 - 1.499	3
1.500 - 1.549	4
1.550 - 1.599	1
1.600 - 1.649	2
1.650 - 1.699	1
1.700 - 1.749	1
1.750 - 1.799	1
1.800 - 1.849	1
Total	381

Mean   = 0.991

Variance = 0.0364

model to represent the change in the genetic composition of the
population for the case where at the time of reproduction the sur-
vivors leave the environment they are living in and mate at random
in the entire population:

$$\Delta q = pq \sum_{i=1}^{n} c_i \frac{(1 - W_i) \, p - (1 - V_i)q}{W_i p^2 + 2pq + V_i q^2}$$

where $p = 1 - q$ and c_i are defined above; W_i is the ability of fitness of genotype AA to survive in environment i, relative to Aa;
and V_i is the ability or fitness of genotype aa to survive in
environment i, relative to Aa. The quantity Δq is a random variable depending on the random variables p, c_i, W_i, and V_i and can
be studied by means of Monte Carlo simulation. Discussion with
the ecologist yields the following information that is used to
select the random variables. There were $n = 3$ environments. The
random variable p lies between 0 and 1 and in some cases the value
can be close to the lower bound. Therefore, the S_B distribution was
selected. The scientists believed that there was only a 1/20
chance that p would be below 0.10 and only a 1/20 chance that it
would exceed 0.20. Equations (5.35) and (5.35a) with $\varepsilon = 0$, and
$\lambda = 1$ and $z_{1-\alpha/2} = 1.645$ yield the values of the parameters of the
Johnson S_B distribution as

$$\eta = \frac{2(1.645)}{\ln (0.20)(0.90/(0.10)(0.80)} = 4.057$$

$$\gamma = 1.645 - 4.057 \ln\left(\frac{0.20}{0.80}\right) = 7.269$$

Random variates are generated from (5.39).

The value of $q = 1 - p$.

The c_i, i = 1, 2, 3 are the proportion among the total survi-
vors in each of the three environments. The scientists felt that
c_i could be any value between 0 and 1 subject to the constraint
that the c_i's sum to unity. Thus, a uniform distribution is used
to generate the c_i's as follows. First c_1 is generated from a

uniform $(0,1)$ distribution. Next c_2 is generated from a uniform $(0, 1 - c_1)$ distribution using equation (7.22) and finally $c_3 = 1 - c_1 - c_2$. The fitness coefficients W_i for AA and V_i for aa are measures relative to the fitness of the heterozygote Aa, which is given a fitness measure of 1.0. Thus, a W_i value of 0.8 means that in environment i the genotype AA has 20% less survival chance that the genotype Aa, and a V_i value of 1.2 means that genotype aa has a 20% higher survival chance than Aa. The W_i's and V_i's are all symmetric random variables, and even though they are bounded below by zero, the actual values are never close to the boundary. Therefore, a normal distribution was used to represent the fitness coefficients. The scientists believe that variability of the V_i and W_i (i = 1, 2, 3) was the same, that the range of possible values was ±0.10 units of the mean. Thus $\sigma = 0.033$. They specified the following means:

$$W_1 = 1.2 \qquad V_1 = 0.9$$
$$W_2 = 0.6 \qquad V_2 = 0.8$$
$$W_3 = 0.7 \qquad V_3 = 1.1$$

The number of simulations was determined from the desire to estimate the mean of Δq within ±0.005 unit with a confidence interval of 95%. The variance of Δq was estimated by running four simulations and calculating σ^2 from these runs. This value was 0.0377. Therefore, the number of simulation runs required using (7.24) is

$$n = \left\lceil \frac{(1.96)^2 (0.0377)^2}{(0.005)^2} \right\rceil + 1 = [218.2] + 1 = 219$$

The output of this simulation is shown in Table 7.3.

Example 7.4.5 A problem of interest today discussed by Duguay (1977) is to determine the cost of generating electricity from solar energy. One formula used in the calculation is

$$C = \frac{r(c_1 + c_2)}{K E_1 E_2}$$

Table 7.3   Results of Simulation for Example 7.4.4

Values of Δq	Frequency
$-0.0240 - (-0.0221)$	1
$-0.0220 - (-0.0201)$	4
$-0.0200 - (-0.0181)$	11
$-0.0180 - (-0.0161)$	9
$-0.0160 - (-0.0141)$	15
$-0.0140 - (-0.0121)$	28
$-0.0120 - (-0.0101)$	19
$-0.0100 - (-0.0081)$	28
$-0.0080 - (-0.0061)$	20
$-0.0060 - (-0.0041)$	18
$-0.0040 - (-0.0021)$	16
$-0.0020 - (-0.0001)$	11
$0.0000 - 0.0019$	13
$0.0020 - 0.0039$	9
$0.0040 - 0.0059$	7
$0.0060 - 0.0079$	2
$0.0080 - 0.0099$	5
$0.0100 - 0.0119$	2
$0.0120 - 0.0139$	1
$0.0140 - 0.0159$	0
$0.0160 - 0.0179$	1
Total	220

Mean $= -0.0074$

Variance $= 0.000057$

where C = cost in dollars per gigajoule (GJ)

r = cost of invested capital

c_1 = cost of $300cm^2$ of silicon solar cells (per $1\ m^2$ of concentrator)

c_2 = cost per m^2 of DC-to-AC inverter

K = amount of direct solar radiation in $1\ m^2$ cross section

E_1 = collector efficiency

E_2 = overall efficiency for converting sunlight to AC electricity

The problem is to find the distribution of the cost using propagation of moments based on the following information. The random variables are independent and for illustrative purposes it is assumed that

r is uniformly distributed on $(0.10,0.20)$
c_1 is normally distributed with $\mu = 50$, $\sigma^2 = 9$
c_2 is normally distributed with $\mu = 15$, $\sigma^2 = 1$
K is normally distributed with $\mu = 6.05$, $\sigma^2 = 0.8$
E_1 has a beta distribution on $(0,1)$ with $\alpha = 2.0$ and $\beta = 0.8$
E_2 has a beta distribution on $(0,1)$ with $\alpha = 15.0$ and $\beta = 45.0$

For the purposes of this example the truncated expressions (7.27a), (7.30a), (7.31a) and (7.32) are used and C is expressed in terms of variables X_i with coefficients b_i. Thus, the problem requires the first four moments of each of the variables. The following are the necessary input values for this analysis:

Variable	Mean	Variance	Third central moment	Fourth central moment
r	0.15	0.00083	0.0	0.0000012
c_1	50.0	9.0	0.0	243.0
c_2	15.0	1.0	0.0	3.0
K	6.05	0.8	0.0	1.92
E_1	0.71	0.054	−0.0097	0.0079
E_2	0.10	0.0018	0.000056	0.000012

The b_i's are obtained next:

$$b_1 = \frac{\partial C}{\partial r} = \frac{c_1 + c_2}{KE_1E_2} = \frac{50 + 15}{6.05(0.71)(0.10)} = 151.32$$

$$b_2 = \frac{\partial C}{\partial c_1} = \frac{r}{KE_1E_2} = \frac{0.15}{0.43} = 0.35$$

$$b_3 = \frac{\partial C}{\partial c_2} = 0.35 \qquad b_4 = \frac{\partial C}{\partial K} = \frac{-r(c_1 + c_2)}{K^2 E_1 E_2} = -3.75$$

$$b_5 = \frac{\partial C}{\partial E_1} = \frac{-r(c_1 + c_2)}{KE_1^2 E_2} = -31.97$$

$$b_6 = \frac{\partial C}{\partial E_2} = \frac{-r(c_1 + c_2)}{KE_1 E_2^2} = -226.98$$

$$b_{11} = 0 \qquad b_{22} = 0 \qquad b_{33} = 0$$

$$b_{44} = \frac{1}{2} \frac{2r(c_1 + c_2)}{K^3 E_1 E_2} = 0.62$$

$$b_{55} = \frac{1}{2} \frac{2r(c_1 + c_2)}{KE_1^3 E_2} = 45.03 \qquad b_{66} = \frac{1}{2} \frac{2r(c_1 + c_2)}{KE_1 E_2^3} = 2269.80$$

$$b_0 = 22.70$$

If

$$Y = C - b_0$$

one obtains from (7.27a)

$$E(Y) = 0.62(0.8) + 45.03(0.054) + 2269.8(0.0018) = 7.01$$

$$E(C) = 22.70 + 7.01 = 29.71$$

Using (7.30a),

$$E(Y^2) = (151.32)^2(0.00083) + (0.35)^2(9.0)$$
$$+ (0.35)^2(1.0) + (3.75)^2(0.8) + (31.97)^2(0.054)$$
$$+ (226.98)^2(0.0018) + 2(-31.97)(45.03)(-0.0097)$$
$$+ 2(-226.98)(2269.80)(0.000056)$$
$$= 149.63$$

and

$$\text{var}(C) = \text{var}(Y) = 149.63 - 49.14 = 100.49$$

Use (7.31a). Hence,

$$E(Y^3) = (-31.97)^3(-0.0097) + (-226.98)^3(0.000056)$$
$$= -337.91$$
$$\mu_3(C) = \mu_3(Y) = -337.91 - 3(149.63)(7.01) + 2(7.01)^3$$
$$= -2795.68$$
$$\sqrt{\beta_1} = -\frac{-2795.68}{(100.49)^{3/2}} = 2.78$$

Use (7.32). Hence,

$$E(Y^4) = (151.32)^4(0.0000012) + (0.35)^4(243) + (0.35)^4(3.0)$$
$$+ \cdots + 6(31.97)^2(225.98)^2(0.054)(0.0018)$$
$$= 101,273.61$$
$$\mu_4(C) = \mu_4(Y) = 101,273.61 - 4(-337.91)(7.01) + 6(149.63)(7.01)^2$$
$$- 3[7.01]^4$$
$$\beta_2 = \frac{147,621.36}{(100.49)^2} = 14.62$$

These values of $\sqrt{\beta_1}$ and β_2 indicate that a Pearson type III (gamma distribution) can be used to approximate the cost of solar energy. Note from Example 5.2.2 that the relationship of β_2 and β_1 for

for a gamma distribution is

$$\beta_2 = 3\left(1 + \frac{\beta_1}{2}\right)$$

In the above problem setting,

$$\beta_2 = 3\left[1 + \frac{(2.78)^2}{2}\right] = 14.59$$

which is close to the value obtained of 14.62. The technique of
matching moments shows the parameters of the approximating gamma
distribution can be obtained as in Example 5.2.3 with $\varepsilon = 0$:

$$29.71 = \frac{\eta^*}{\lambda^*}$$

$$100.49 = \frac{\eta^*}{\lambda^{*2}}$$

and hence,

$$\lambda^* = 0.296$$

$$\eta^* = 8.78$$

Thus, questions such as the probability of the cost of electricity
exceeding \$35/GJ may now be answered. It is interesting to note
that if one just inserts average values in the cost function, the
estimate of the average cost would be \$22.70/GJ, which is about
\$7/GJ too low and disregards the inherent variable in the cost.
The term contributing the most to the variability is E_2. In order
to reduce the variability in estimating C, it is necessary to have
a more precise estimate on the overall efficiency for converting
collected sunlight to AC electricity.

Problems

1. Redo Example 7.4.1 assuming that the sales managers in regions
    1, 2, and 3 will discuss the economic situation and that their
    estimates will be correlated with $\rho = 0.70$. See (2.67) for
    definition of ρ.

2. Redo Example 7.4.2 using the propagation-of-moment technique.

3. Write the computer program for doing the Monte Carlo simulation study in Example 7.4.3.

4. In Example 7.4.3 determine the parameters to use for P is instead of a lognormal, a gamma distribution with minimum value of 450 is to be used.

5. Write the computer program for doing the Monte Carlo simulation study in Example 7.4.4.

6. Evaluate the effect on C in Example 7.4.5 if r has a parabolic distribution on the interval (0.10, 0.20)

7. Evaluate the effect on C in Example 7.4.5 if C_2 has an exponential distribution with $\lambda = 1/15$.

REFERENCES

Cox, N. D. (1979).   Tolerance analysis by computer, *J. Quali. Tech.* *11*, pp. 80-87.

Duguay, M. A. (1977).   Solar electricity: The hybrid system approach, *Amer. Scientist 65*, pp. 422-427.

Freund, J. E. (1971).   *Mathematical Statistics*, Prentice-Hall, Inc., Englewood Cliffs, N.J.,   pp. 182-189.

Hahn, G. J. and Shapiro, S. S. (1967), *Statistical Models in Engineering*, John Wiley and Sons, Inc., New York.

Li, C. C. (1955).   *Population Genetics*, University of Chicago Press, Chicago p. 266.

McGrath, E. J., and Irving, D. C. (1975).   Techniques for efficient Monte Carlo simulation, Vol. 2, Tech. Report ORNL-RSIC-38, Radiation Shielding Information Center, Oak Ridge, Tenn.

Marsaglia, G., and Bray, T. A. (1964).   A convenient method for generating normal variables, *SIAM Rev 6*, pp. 260-264.

National Bureau of Standards (1964).   Handbook of mathematical functions: Numerical methods, *Appl. Math Ser. 55*, pp. 949-953.

Volk, W. (1969).   *Applied Statistics for Engineers*, McGraw-Hill Book Company, New York, pp. 157-158.

Appendix A

TABLES

Table A.1 Standard Cumulative Normal Distribution

$$F(y) = (2\pi)^{-1/2} \int_{-\infty}^{y} e^{-z^2/2} \, dz$$

y	·00	·01	·02	·03	·04	·05	·06	·07	·08	·09
·0	·5000	·5040	·5080	·5120	·5160	·5199	·5239	·5279	·5319	·5359
·1	·5398	·5438	·5478	·5517	·5557	·5596	·5636	·5675	·5714	·5753
·2	·5793	·5832	·5871	·5910	·5948	·5987	·6026	·6064	·6103	·6141
·3	·6179	·6217	·6255	·6293	·6331	·6368	·6406	·6443	·6480	·6517
·4	·6554	·6591	·6628	·6664	·6700	·6736	·6772	·6808	·6844	·6879
·5	·6915	·6950	·6985	·7019	·7054	·7088	·7123	·7157	·7190	·7224
·6	·7257	·7291	·7324	·7357	·7389	·7422	·7454	·7486	·7517	·7549
·7	·7580	·7611	·7642	·7673	·7703	·7734	·7764	·7794	·7823	·7852
·8	·7881	·7910	·7939	·7967	·7995	·8023	·8051	·8078	·8106	·8133
·9	·8159	·8186	·8212	·8238	·8264	·8289	·8315	·8340	·8365	·8389
1·0	·8413	·8438	·8461	·8485	·8508	·8531	·8554	·8577	·8599	·8621
1·1	·8643	·8665	·8686	·8708	·8729	·8749	·8770	·8790	·8810	·8830
1·2	·8849	·8869	·8888	·8907	·8925	·8944	·8962	·8980	·8997	·90147
1·3	·90320	·90490	·90658	·90824	·90988	·91149	·91309	·91466	·91621	·91774
1·4	·91924	·92073	·92220	·92364	·92507	·92647	·92785·	·92922	·93056	·93189
1·5	·93319	·93448	·93574	·93699	·93822	·93943	·94062	·94179	·94295	·94408
1·6	·94520	·94630	·94738	·94845	·94950	·95053	·95154	·95254	·95352	·95449
1·7	·95543	·95637	·95728	·95818	·95907	·95994	·96080	·96164	·96246	·96327
1·8	·96407	·96485	·96562	·96638	·96712	·96784	·96856	·96926	·96995	·97062
1·9	·97128	·97193	·97257	·97320	·97381	·97441	·97500	·97558	·97615	·97670
2·0	·97725	·97778	·97831	·97882	·97932	·97982	·98030	·98077	·98124	·98169
2·1	·98214	·98257	·98300	·98341	·98382	·98422	·98461	·98500	·98537	·98574
2·2	·98610	·98645	·98679	·98713	·98745	·98778	·98809	·98840	·98870	·98899
2·3	·98928	·98956	·98983	·9^{2}0097	·9^{2}0358	·9^{2}0613	·9^{2}0863	·9^{2}1106	·9^{2}1344	·9^{2}1576
2·4	·9^{2}1802	·9^{2}2024	·9^{2}2240	·9^{2}2451	·9^{2}2656	·9^{2}2857	·9^{2}3053	·9^{2}3244	·9^{2}3431	·9^{2}3613
2·5	·9^{2}3790	·9^{2}3963	·9^{2}4132	·9^{2}4297	·9^{2}4457	·9^{2}4614	·9^{2}4766	·9^{2}4915	·9^{2}5060	·9^{2}5201
2·6	·9^{2}5339	·9^{2}5473	·9^{2}5604	·9^{2}5731	·9^{2}5855	·9^{2}5975	·9^{2}6093	·9^{2}6207	·9^{2}6319	·9^{2}6427
2·7	·9^{2}6533	·9^{2}6636	·9^{2}6736	·9^{2}6833	·9^{2}6928	·9^{2}7020	·9^{2}7110	·9^{2}7197	·9^{2}7282	·9^{2}7365
2·8	·9^{2}7445	·9^{2}7523	·9^{2}7599	·9^{2}7673	·9^{2}7744	·9^{2}7814	·9^{2}7882	·9^{2}7948	·9^{2}8012	·9^{2}8074
2·9	·9^{2}8134	·9^{2}8193	·9^{2}8250	·9^{2}8305	·9^{2}8359	·9^{2}8411	·9^{2}8462	·9^{2}8511	·9^{2}8559	·9^{2}8605
3·0	·9^{2}8650	·9^{2}8694	·9^{2}8736	·9^{2}8777	·9^{2}8817	·9^{2}8856	·9^{2}8893	·9^{2}8930	·9^{2}8965	·9^{2}8999
3·1	·9^{3}0324	·9^{3}0646	·9^{3}0957	·9^{3}1260	·9^{3}1553	·9^{3}1836	·9^{3}2112	·9^{3}2378	·9^{3}2636	·9^{3}2886
3·2	·9^{3}3129	·9^{3}3363	·9^{3}3590	·9^{3}3810	·9^{3}4024	·9^{3}4230	·9^{3}4429	·9^{3}4623	·9^{3}4810	·9^{3}4991
3·3	·9^{3}5166	·9^{3}5335	·9^{3}5499	·9^{3}5658	·9^{3}5811	·9^{3}5959	·9^{3}6103	·9^{3}6242	·9^{3}6376	·9^{3}6505
3·4	·9^{3}6631	·9^{3}6752	·9^{3}6869	·9^{3}6982	·9^{3}7091	·9^{3}7197	·9^{3}7299	·9^{3}7398	·9^{3}7493	·9^{3}7585
3·5	·9^{3}7674	·9^{3}7759	·9^{3}7842	·9^{3}7922	·9^{3}7999	·9^{3}8074	·9^{3}8146	·9^{3}8215	·9^{3}8282	·9^{3}8347
3·6	·9^{3}8409	·9^{3}8469	·9^{3}8527	·9^{3}8583	·9^{3}8637	·9^{3}8689	·9^{3}8739	·9^{3}8787	·9^{3}8834	·9^{3}8879
3·7	·9^{3}8922	·9^{3}8964	·9^{4}0039	·9^{4}0426	·9^{4}0799	·9^{4}1158	·9^{4}1504	·9^{4}1838	·9^{4}2159	·9^{4}2468
3·8	·9^{4}2765	·9^{4}3052	·9^{4}3327	·9^{4}3593	·9^{4}3848	·9^{4}4094	·9^{4}4331	·9^{4}4558	·9^{4}4777	·9^{4}4988
3·9	·9^{4}5190	·9^{4}5385	·9^{4}5573	·9^{4}5753	·9^{4}5926	·9^{4}6092	·9^{4}6253	·9^{4}6406	·9^{4}6554	·9^{4}6696
4·0	·9^{4}6833	·9^{4}6964	·9^{4}7090	·9^{4}7211	·9^{4}7327	·9^{4}7439	·9^{4}7546	·9^{4}7649	·9^{4}7748	·9^{4}7843
4·1	·9^{4}7934	·9^{4}8022	·9^{4}8106	·9^{4}8186	·9^{4}8263	·9^{4}8338	·9^{4}8409	·9^{4}8477	·9^{4}8542	·9^{4}8605
4·2	·9^{4}8665	·9^{4}8723	·9^{4}8778	·9^{4}8832	·9^{4}8882	·9^{4}8931	·9^{4}8978	·9^{5}0226	·9^{5}0655	·9^{5}1066
4·3	·9^{5}1460	·9^{5}1837	·9^{5}2199	·9^{5}2545	·9^{5}2876	·9^{5}3193	·9^{5}3497	·9^{5}3788	·9^{5}4066	·9^{5}4332
4·4	·9^{5}4587	·9^{5}4831	·9^{5}5065	·9^{5}5288	·9^{5}5502	·9^{5}5706	·9^{5}5902	·9^{5}6089	·9^{5}6268	·9^{5}6439
4·5	·9^{5}6602	·9^{5}6759	·9^{5}6908	·9^{5}7051	·9^{5}7187	·9^{5}7318	·9^{5}7442	·9^{5}7561	·9^{5}7675	·9^{5}7784
4·6	·9^{5}7888	·9^{5}7987	·9^{5}8081	·9^{5}8172	·9^{5}8258	·9^{5}8340	·9^{5}8419	·9^{5}8494	·9^{5}8566	·9^{5}8634
4·7	·9^{5}8699	·9^{5}8761	·9^{5}8821	·9^{5}8877	·9^{5}8931	·9^{5}8983	·9^{6}0320	·9^{6}0789	·9^{6}1235	·9^{6}1661
4·8	·9^{6}2067	·9^{6}2453	·9^{6}2822	·9^{6}3173	·9^{6}3508	·9^{6}3827	·9^{6}4131	·9^{6}4420	·9^{6}4696	·9^{6}4958
4·9	·9^{6}5208	·9^{6}5446	·9^{6}5673	·9^{6}5889	·9^{6}6094	·9^{6}6289	·9^{6}6475	·9^{6}6652	·9^{6}6821	·9^{6}6981

Source: A. Hald, *Statistical Tables and Formulas*, John Wiley and Sons, Inc., New York, 1952 (Table II).

Table A.2 Percentiles of the χ^2 Distribution

Degrees of Freedom	0.90	0.95	0.975	0.990
1	2.71	3.84	5.02	6.63
2	4.61	5.99	7.38	9.21
3	6.25	7.81	9.35	11.3
4	7.78	9.49	11.1	13.3
5	9.24	11.1	12.8	15.1
6	10.6	12.6	14.4	16.8
7	12.0	14.1	16.0	18.5
8	13.4	15.5	17.5	20.1
9	14.7	16.9	19.0	21.7
10	16.0	18.3	20.5	23.2
11	17.3	19.7	21.9	24.7
12	18.5	21.0	23.3	26.2
13	19.8	22.4	24.7	27.7
14	21.1	23.7	26.1	29.1
15	22.3	25.0	27.5	30.6
16	23.5	26.3	28.8	32.0
17	24.8	27.6	30.2	33.4
18	26.0	28.9	31.5	34.8
19	27.2	30.1	32.9	36.2
20	28.4	31.4	34.2	37.6
21	29.6	32.7	35.5	38.9
22	30.8	33.9	36.8	40.3
23	32.0	35.2	38.1	41.6
24	33.2	36.4	39.4	43.0
25	34.4	37.7	40.6	44.3
26	35.6	38.9	41.9	45.6
27	36.7	40.1	43.2	47.0
28	37.9	41.3	44.5	48.3
29	39.1	42.6	45.7	49.6
30	40.3	43.8	47.0	50.9
35	46.1	49.8	53.2	57.3
40	51.8	55.8	59.3	63.7
45	57.5	61.7	65.4	70.0
50	63.2	67.5	71.4	76.2
75	91.1	96.2	100.8	106.4
100	118.5	124.3	129.6	135.6

Source: Abridged from A. Hald, *Statistical Tables and Formulas*, John Wiley and Sons, Inc., New York, 1952, Table V.

Table A.3 Lambda Parameters for Given Values of Skewness(α_3) and Kurtosis(α_4) when $\mu = 0$ and $\sigma = 1$

α_4	$\alpha_3 = 0.0$				$\alpha_3 = 0.05$				$\alpha_3 = 0.10$			
	LAM 1	LAM 2	LAM 3	LAM 4	LAM 1	LAM 2	LAM 3	LAM 4	LAM 1	LAM 2	LAM 3	LAM 4
1.8	.0	.5774	1.0000	1.0000	-1.703	.2861	.0000	.9502*	-1.678	.2835	.0000*	.9071*
2.0	.0	.4952	.5843	.5843	-1.229	.3122	.0505	.7603	-1.271	.3028	.0412	.7373
2.2	.0	.4197	.4092	.4092	-.802	.3314	.1128	.5802	-.872	.3177	.0941	.5700
2.4	.0	.3533	.3032	.3032	-.375	.3328	.1876	.3941	-.515	.3164	.1477	.4116
2.6	.0	.2949	.2303	.2303	-.143	.2924	.1973	.2605	-.269	.2863	.1678	.2831
2.8	.0	.2433	.1765	.1765	-.083	.2429	.1625	.1903	-.164	.2417	.1486	.2033
3.0	.0	.1974	.1349	.1349	-.059	.1975	.1276	.1425	-.117	.1977	.1205	.1503
3.2	.0	.1563	.1016	.1016	-.046	.1565	.0974	.1061	-.092	.1572	.0936	.1111
3.4	.0	.1191	.0742	.0742	-.038	.1194	.0718	.0770	-.076	.1203	.0698	.0803
3.6	.0	.0852	.0512	.0512	-.033	.0856	.0499	.0530	-.065	.0866	.0490	.0552
3.8	.0	.0545	.0317	.0317	-.027	.0548	.0311	.0327	-.057	.0558	-.0308	.0342
4.0	.0	.0262	.0148	.0148	-.026	.0264	.0146	.0153	-.049	.0276	-.0149	.0163
4.1	.0	.0128	.7140+	.7140+	-.024	.0132	.7184+	.7504+	-.048	.0142	.7606+	.8302+
4.2	.0	-.0659+	-.0363+	-.0363+	-.024	-.0704+	.0380+	-.0397+	-.046	.1440+	.0762+	-.0828+
4.3	.0	-.0123	-.6706+	-.6706+	-.022	-.0120	-.6386+	-.6643+	-.044	-.0109	-.5703+	-.6174+
4.4	.0	-.0241	-.0130	-.0130	-.022	-.0238	-.0126	-.0131	-.041	-.0227	-.0118	-.0127
4.6	.0	-.0466	-.0246	-.0246	-.019	-.0462	-.0240	-.0248	-.037	-.0452	-.0231	-.0247
4.8	.0	-.0676	-.0350	-.0350	-.018	-.0671	-.0342	-.0354	-.036	-.0661	-.0332	-.0354
5.0	.0	-.0870	-.0443	-.0443	-.017	-.0867	-.0435	-.0448	-.033	-.0857	-.0424	-.0450
5.2	.0	-.1053	-.0528	-.0528	-.016	-.1050	-.0519	-.0534	-.032	-.1040	-.0507	-.0537
5.4	.0	-.1227	-.0606	-.0606	-.015	-.1222	-.0596	-.0612	-.030	-.1213	-.0584	-.0616
5.6	.0	-.1389	-.0677	-.0677	-.014	-.1386	-.0667	-.0684	-.028	-.1375	-.0654	-.0688
5.8	.0	-.1541	-.0742	-.0742	-.014	-.1538	-.0731	-.0750	-.027	-.1530	-.0719	-.0755
6.0	.0	-.1686	-.0802	-.0802	-.013	-.1682	-.0791	-.0810	-.027	-.1674	-.0778	-.0816
6.2	.0	-.1823	-.0858	-.0858	-.012	-.1820	-.0847	-.0866	-.025	-.1811	-.0834	-.0872
6.4	.0	-.1954	-.0910	-.0910	-.012	-.1950	-.0899	-.0918	-.024	-.1943	-.0886	-.0925
6.6	.0	-.2077	-.0958	-.0958	-.011	-.2074	-.0947	-.0967	-.023	-.2066	-.0934	-.0973
6.8	.0	-.2194	-.1003	-.1003	-.011	-.2192	-.0992	-.1012	-.022	-.2184	-.0979	-.1019
7.0	.0	-.2306	-.1045	-.1045	-.011	-.2303	-.1034	-.1054	-.022	-.2297	-.1021	-.1062
7.2	.0	-.2414	-.1085	-.1085	-.010	-.2411	-.1074	-.1094	-.021	-.2405	-.1061	-.1102
7.4	.0	-.2518	-.1123	-.1123	-.010	-.2515	-.1112	-.1132	-.020	-.2507	-.1099	-.1139
7.6	.0	-.2615	-.1158	-.1158	-.979+	-.2613	-.1147	-.1167	-.020	-.2606	-.1134	-.1175
7.8	.0	-.2709	-.1191	-.1191	-.999+	-.2707	-.1180	-.1201	-.020	-.2699	-.1167	-.1208
8.0	.0	-.2800	-.1223	-.1223	-.928+	-.2797	-.1212	-.1232	-.019	-.2791	-.1199	-.1240
8.2	.0	-.2887	-.1253	-.1253	-.906+	-.2884	-.1242	-.1262	-.019	-.2878	-.1229	-.1270
8.4	.0	-.2969	-.1281	-.1281	-.931+	-.2968	-.1270	-.1291	-.018	-.2961	-.1258	-.1298
8.6	.0	-.3050	-.1308	-.1308	-.912+	-.3048	-.1297	-.1318	-.017	-.3041	-.1285	-.1325
8.8	.0	-.3128	-.1334	-.1334	-.852+	-.3125	-.1323	-.1343	-.017	-.3119	-.1311	-.1351
9.0	.0	-.3203	-.1359	-.1359	-.837+	-.3201	-.1348	-.1368	-.017	-.3193	-.1335	-.1376

Source: Ramberg et al., *Technometrics 21*, 1979, pp. 201-214.

$\alpha_3 = 0.15$

α_4	LAM 1	LAM 2	LAM 3	LAM 4
1.8	-1.655	.2811	.0000*	.8700*
2.0	-1.323	.2934	.0314	.7204
2.2	-.940	.3056	.0782	.5623
2.4	-.617	.3031	.1215	.4194
2.6	-.376	.2791	.1435	.2994
2.8	-.244	.2397	.1350	.2156
3.0	-.177	.1980	.1135	.1586
3.2	-.138	.1584	.0901	.1167
3.4	-.114	.1219	.0682	.0843
3.6	-.098	.0884	.0485	.0581
3.8	-.086	.0577	.0310	.0363
4.0	-.076	.0294	.0155	.0178
4.1	-.073	.0160	.8378*	.9564*
4.2	-.069	.3217*	.1667*	.1890*
4.3	-.066	.9113*	.4680*	.5278*
4.4	-.063	-.0210	-.0107	-.0120
4.6	-.055	-.0435	-.0218	-.0242
4.8	-.051	-.0644	-.0318	-.0351
5.0	-.048	-.0842	-.0410	-.0449
5.2	-.045	-.1025	-.0493	-.0537
5.4	-.045	-.1198	-.0569	-.0617
5.6	-.043	-.1361	-.0639	-.0690
5.8	-.040	-.1514	-.0703	-.0757
6.0	-.038	-.1660	-.0763	-.0819
6.2	-.037	-.1798	-.0819	-.0876
6.4	-.035	-.1928	-.0870	-.0929
6.6	-.034	-.2053	-.0919	-.0978
6.8	-.033	-.2172	-.0964	-.1024
7.0	-.032	-.2284	-.1006	-.1067
7.2	-.031	-.2392	-.1046	-.1107
7.4	-.031	-.2496	-.1084	-.1145
7.6	-.030	-.2593	-.1119	-.1180
7.8	-.029	-.2688	-.1153	-.1214
8.0	-.028	-.2780	-.1185	-.1246
8.2	-.027	-.2866	-.1215	-.1276
8.4	-.027	-.2948	-.1243	-.1304
8.6	-.027	-.3031	-.1271	-.1332
8.8	-.026	-.3108	-.1297	-.1357
9.0	-.025	-.3183	-.1322	-.1382

$\alpha_3 = 0.20$

α_4	LAM 1	LAM 2	LAM 3	LAM 4
2.0	-1.387	.2841	.0212	.7090
2.2	-1.011	.2947	.0638	.5571
2.4	-.706	.2919	.1013	.4246
2.6	-.471	.2718	.1233	.3120
2.8	-.322	.2374	.1221	.2273
3.0	-.237	.1983	.1065	.1672
3.2	-.184	.1599	.0866	.1230
3.4	-.154	.1240	.0667	.0889
3.6	-.132	.0908	.0482	.0615
3.8	-.116	.0601	.0314	.0389
4.0	-.103	.0318	.0164	.0198
4.1	-.097	.0185	.9467*	.0113
4.2	-.093	.5707*	-.2894*	-.3429*
4.3	-.089	-.6641*	-.3342*	-.3929*
4.4	-.085	-.0185	-.9261*	-.0108
4.6	-.079	-.0410	-.0202	-.0233
4.8	-.074	-.0622	-.0302	-.0345
5.0	-.069	-.0818	-.0392	-.0444
5.2	-.065	-.1003	-.0475	-.0534
5.4	-.061	-.1176	-.0551	-.0615
5.6	-.058	-.1339	-.0621	-.0689
5.8	-.055	-.1494	-.0686	-.0757
6.0	-.053	-.1639	-.0745	-.0819
6.2	-.051	-.1778	-.0801	-.0877
6.4	-.049	-.1909	-.0853	-.0930
6.6	-.047	-.2034	-.0901	-.0980
6.8	-.045	-.2153	-.0947	-.1026
7.0	-.044	-.2265	-.0989	-.1069
7.2	-.043	-.2374	-.1029	-.1110
7.4	-.041	-.2477	-.1067	-.1148
7.6	-.040	-.2577	-.1103	-.1184
7.8	-.038	-.2671	-.1136	-.1218
8.0	-.037	-.2762	-.1168	-.1250
8.2	-.036	-.2850	-.1199	-.1280
8.4	-.035	-.2935	-.1228	-.1309
8.6	-.035	-.3014	-.1255	-.1336
8.8	-.034	-.3092	-.1281	-.1362
9.0	-.034	-.3168	-.1306	-.1387
9.2	-.034	-.3241	-.1330	-.1411

$\alpha_3 = 0.25$

α_4	LAM 1	LAM 2	LAM 3	LAM 4
2.0	-1.465	.2748	.0105	.7034
2.2	-1.084	.2847	.0506	.5548
2.4	-.790	.2820	.0843	.4294
2.6	-.558	.2650	.1062	.3226
2.8	-.398	.2349	.1099	.2385
3.0	-.298	.1987	.0996	.1763
3.2	-.237	.1619	.0831	.1300
3.4	-.196	.1266	.0653	.0942
3.6	-.167	.0937	.0481	.0656
3.8	-.147	.0632	.0321	.0421
4.0	-.131	.0351	.0176	.0224
4.1	-.126	.0217	.0108	.0136
4.2	-.118	.8889*	.4408*	.5467*
4.3	-.113	-.3476*	-.1713*	-.2103*
4.4	-.108	-.0154	-.7540*	-.9175*
4.6	-.099	-.0380	-.0184	-.0220
4.8	-.094	-.0591	-.0282	-.0334
5.0	-.087	-.0790	-.0373	-.0436
5.2	-.082	-.0974	-.0455	-.0527
5.4	-.077	-.1149	-.0531	-.0610
5.6	-.073	-.1312	-.0601	-.0685
5.8	-.070	-.1467	-.0665	-.0754
6.0	-.067	-.1613	-.0725	-.0817
6.2	-.064	-.1753	-.0781	-.0876
6.4	-.062	-.1885	-.0833	-.0930
6.6	-.059	-.2010	-.0882	-.0980
6.8	-.058	-.2129	-.0927	-.1027
7.0	-.055	-.2242	-.0970	-.1070
7.2	-.054	-.2350	-.1010	-.1111
7.4	-.052	-.2455	-.1048	-.1150
7.6	-.051	-.2554	-.1084	-.1186
7.8	-.049	-.2649	-.1118	-.1220
8.0	-.048	-.2742	-.1151	-.1252
8.2	-.047	-.2829	-.1181	-.1283
8.4	-.046	-.2914	-.1210	-.1312
8.6	-.044	-.2995	-.1238	-.1339
8.8	-.044	-.3072	-.1264	-.1365
9.0	-.043	-.3147	-.1289	-.1390
9.2	-.042	-.3220	-.1313	-.1414

Table A.3 Continued

$\alpha_3 = 0.30$

α_4	LAM 1	LAM 2	LAM 3	LAM 4
2.0	-1.550	.2660	.0000	.7020
2.2	-1.164	.2755	.0380	.5556
2.4	-.871	.2733	.0695	.4348
2.6	-.642	.2586	.0911	.3324
2.8	-.474	.2323	.0983	.2495
3.0	-.362	.1991	.0925	.1859
3.2	-.288	.1641	.0796	.1377
3.4	-.239	.1298	.0640	.1003
3.6	-.204	.0973	.0481	.0704
3.8	-.179	.0671	.0330	.0460
4.0	-.160	.0389	.0190	.0255
4.2	-.144	.0127	.6175+	.8035+
4.3	-.138	.0789+	.0380+	.0489+
4.4	-.131	-.0116	-.5554+	-.7057+
4.5	-.129	-.0231	-.0110	-.0203
4.6	-.121	-.0343	-.0163	-.0319
4.8	-.113	-.0554	-.0260	-.0423
5.0	-.105	-.0752	-.0350	-.0517
5.2	-.100	-.0939	-.0432	-.0601
5.4	-.094	-.1114	-.0508	-.0684
5.6	-.089	-.1279	-.0578	-.0678
5.8	-.081	-.1435	-.0643	-.0748
6.0	-.082	-.1582	-.0703	-.0812
6.2	-.078	-.1722	-.0759	-.0872
6.4	-.075	-.1854	-.0811	-.0927
6.6	-.072	-.1979	-.0860	-.0977
6.8	-.069	-.2100	-.0906	-.1025
7.0	-.067	-.2214	-.0949	-.1069
7.2	-.065	-.2325	-.0990	-.1111
7.4	-.063	-.2427	-.1028	-.1149
7.6	-.061	-.2528	-.1064	-.1186
7.8	-.060	-.2623	-.1098	-.1220
8.0	-.058	-.2716	-.1131	-.1253
8.2	-.056	-.2805	-.1162	-.1284
8.4	-.055	-.2889	-.1191	-.1313
8.6	-.054	-.2971	-.1219	-.1341
8.8	-.053	-.3050	-.1246	-.1367
9.0	-.052	-.3125	-.1271	-.1392
9.2	-.051	-.3197	-.1295	-.1416

$\alpha_3 = 0.35$

α_4	LAM 1	LAM 2	LAM 3	LAM 4
2.0	-1.539	.2639	.0000*	.6836*
2.2	-1.252	.2668	.0256	.5599
2.4	-.955	.2653	.0559	.4415
2.6	-.724	.2528	.0775	.3423
2.8	-.550	.2298	.0873	.2606
3.0	-.427	.1996	.0854	.1961
3.2	-.343	.1665	.0758	.1462
3.4	-.285	.1333	.0625	.1072
3.6	-.243	.1014	.0482	.0760
3.8	-.213	.0714	.0340	.0505
4.0	-.191	.0434	.0206	.0293
4.2	-.172	.0173	.8158+	.0112
4.3	-.163	.4870+	.2293+	.3090+
4.4	-.156	-.7105+	-.3332+	-.4431+
4.5	-.151	-.0187	-.8723+	-.0115
4.6	-.142	-.0298	-.0139	-.0180
4.8	-.132	-.0511	-.0236	-.0300
5.0	-.124	-.0710	-.0325	-.0407
5.2	-.117	-.0898	-.0407	-.0503
5.4	-.110	-.1074	-.0483	-.0589
5.6	-.105	-.1240	-.0553	-.0668
5.8	-.096	-.1396	-.0618	-.0739
6.0	-.091	-.1545	-.0678	-.0805
6.2	-.088	-.1685	-.0735	-.0865
6.4	-.085	-.1818	-.0789	-.0921
6.6	-.082	-.1945	-.0836	-.0973
6.8	-.079	-.2067	-.0883	-.1021
7.0	-.077	-.2181	-.0926	-.1066
7.2	-.074	-.2291	-.0967	-.1108
7.4	-.072	-.2396	-.1006	-.1147
7.6	-.072	-.2496	-.1042	-.1184
7.8	-.070	-.2593	-.1077	-.1219
8.0	-.068	-.2685	-.1109	-.1252
8.2	-.066	-.2775	-.1141	-.1283
8.4	-.065	-.2860	-.1170	-.1313
8.6	-.064	-.2942	-.1198	-.1341
8.8	-.062	-.3020	-.1225	-.1367
9.0	-.060	-.3096	-.1251	-.1392
9.2	-.059	-.3172	-.1276	-.1417

$\alpha_3 = 0.40$

α_4	LIM 1	LAM 2	LAM 3	LAM 4
2.2	-1.354	.2582	.0129	.5683
2.4	-1.043	.2580	.0430	.4500
2.6	-.808	.2473	.0648	.3527
2.8	-.627	.2273	.0767	.2720
3.0	-.494	.2000	.0782	.2069
3.2	-.400	.1690	.0718	.1555
3.4	-.333	.1371	.0609	.1149
3.6	-.284	.1060	.0482	.0824
3.8	-.248	.0764	.0351	.0558
4.0	-.222	.0485	.0223	.0337
4.2	-.200	.0224	.0103	.0149
4.3	-.190	.0100	.4597+	.6521+
4.4	-.182	-.0397+	-.0182+	-.0254**
4.5	-.174	-.0136	-.6204+	-.8533+
4.6	-.166	-.0248	-.0113	-.0153
4.8	-.155	-.0462	-.0209	-.0277
5.0	-.146	-.0662	-.0297	-.0387
5.2	-.136	-.0850	-.0379	-.0485
5.4	-.129	-.1027	-.0455	-.0574
5.6	-.122	-.1194	-.0525	-.0654
5.8	-.115	-.1352	-.0591	-.0727
6.0	-.111	-.1501	-.0651	-.0794
6.2	-.106	-.1643	-.0708	-.0856
6.4	-.102	-.1778	-.0761	-.0913
6.6	-.098	-.1906	-.0811	-.0966
6.8	-.094	-.2026	-.0857	-.1014
7.0	-.091	-.2142	-.0901	-.1060
7.2	-.089	-.2253	-.0942	-.1103
7.4	-.086	-.2359	-.0981	-.1143
7.6	-.083	-.2459	-.1018	-.1180
7.8	-.081	-.2558	-.1053	-.1216
8.0	-.079	-.2650	-.1086	-.1249
8.2	-.077	-.2741	-.1118	-.1281
8.4	-.075	-.2827	-.1148	-.1311
8.6	-.073	-.2908	-.1176	-.1339
8.8	-.072	-.2988	-.1203	-.1366
9.0	-.070	-.3064	-.1229	-.1391
9.2	-.069	-.3139	-.1254	-.1416
9.4	-.067	-.3210	-.1278	-.1439

$\alpha_3 = 0.45$

α_4	LAM 1	LAM 2	LAM 3	LAM 4
2.2	-1.471	.2500	.0000	.5812
2.4	-1.138	.2511	.0305	.4608
2.6	-.894	.2424	.0528	.3641
2.8	-.707	.2248	.0663	.2840
3.0	-.565	.2003	.0707	.2184
3.2	-.460	.1716	.0674	.1657
3.4	-.384	.1412	.0590	.1236
3.6	-.329	.1110	.0480	.0897
3.8	-.287	.0818	.0361	.0619
4.0	-.255	.0542	.0241	.0388
4.2	-.230	.0282	-.0126	.0193
4.3	-.221	.0158	.7041+	-.0106
4.4	-.208	-.0102+	-.1833+	-.2691+
4.5	-.200	-.7861+	-.8501+	-.5065+
4.6	-.192	-.0191	-.0180	-.0249
4.8	-.178	-.0406	-.0268	-.0362
5.0	-.165	-.0607	-.0349	-.0464
5.2	-.157	-.0796	-.0425	-.0555
5.4	-.147	-.0975	-.0495	-.0637
5.6	-.140	-.1142	-.0530	-.0675
5.8	-.132	-.1302	-.0561	-.0712
6.0	-.127	-.1453	-.0622	-.0781
6.2	-.121	-.1595	-.0679	-.0844
6.4	-.116	-.1731	-.0733	-.0902
6.6	-.112	-.1860	-.0783	-.0956
6.8	-.108	-.1983	-.0830	-.1006
7.0	-.104	-.2098	-.0874	-.1052
7.2	-.101	-.2211	-.0916	-.1096
7.4	-.097	-.2316	-.0955	-.1136
7.6	-.095	-.2419	-.0992	-.1175
7.8	-.092	-.2518	-.1028	-.1211
8.0	-.090	-.2611	-.1061	-.1245
8.2	-.088	-.2702	-.1093	-.1277
8.4	-.085	-.2789	-.1124	-.1307
8.6	-.084	-.2871	-.1152	-.1336
8.8	-.081	-.2952	-.1180	-.1363
9.0	-.080	-.3029	-.1206	-.1389
9.2	-.078	-.3102	-.1231	-.1413
9.4	-.076	-.3176	-.1256	-.1437

$\alpha_3 = 0.50$

α_4	LAM 1	LAM 2	LAM 3	LAM 4
2.4	-1.245	.2445	.0178	.4748
2.6	-.987	.2376	.0410	.3770
2.8	-.790	.2225	.0561	.2969
3.0	-.639	.2006	.0630	.2307
3.2	-.525	.1742	.0625	.1768
3.4	-.440	.1454	.0566	.1332
3.6	-.376	.1163	.0476	.0979
3.8	-.329	.0877	.0369	.0689
4.0	-.290	.0604	.0259	.0447
4.2	-.262	.0345	.0149	.0243
4.3	-.248	.0221	.9582+	.0152
4.4	-.238	.0101	.4383+	-.6815+
4.5	-.228	-.1612+	-.0700+	-.1066+
4.6	-.219	-.0128+	-.5570+	-.8334+
4.8	-.202	-.0344	-.0236	-.0216
5.0	-.188	-.0546	-.0317	-.0333
5.2	-.177	-.0737	-.0393	-.0438
5.4	-.167	-.0917	-.0464	-.0537
5.6	-.157	-.1087	-.0529	-.0617
5.8	-.150	-.1246	-.0560	-.0694
6.0	-.142	-.1398	-.0591	-.0764
6.2	-.137	-.1542	-.0648	-.0829
6.4	-.131	-.1679	-.0702	-.0889
6.6	-.126	-.1809	-.0753	-.0945
6.8	-.122	-.1933	-.0800	-.0995
7.0	-.117	-.2050	-.0845	-.1042
7.2	-.114	-.2163	-.0887	-.1087
7.4	-.110	-.2270	-.0927	-.1128
7.6	-.107	-.2374	-.0965	-.1167
7.8	-.104	-.2473	-.1001	-.1204
8.0	-.101	-.2567	-.1035	-.1238
8.2	-.098	-.2659	-.1067	-.1271
8.4	-.095	-.2745	-.1098	-.1301
8.6	-.094	-.2830	-.1127	-.1331
8.8	-.091	-.2910	-.1155	-.1358
9.0	-.089	-.2986	-.1181	-.1384
9.2	-.088	-.3064	-.1207	-.1410
9.4	-.086	-.3134	-.1231	-.1433
9.6	-.084	-.3206	-.1255	-.1456

$\alpha_3 = 0.55$

α_4	LAM 1	LAM 2	LAM 3	LAM 4
2.4	-1.370	.2379	.4463+	.4931
2.6	-1.087	.2331	.0292	.3920
2.8	-.878	.2202	.0459	.3109
3.0	-.716	.2009	.0551	.2440
3.2	-.593	.1767	.0572	.1889
3.4	-.499	.1497	.0538	.1438
3.6	-.428	.1217	.0467	.1070
3.8	-.372	.0940	.0376	.0767
4.0	-.330	.0670	.0275	.0514
4.2	-.298	.0413	.0172	.0301
4.4	-.265	.0170	.7149+	-.0118
4.5	-.257	-.5355+	.2258+	-.3644+
4.6	-.247	-.5954+	.2515+	-.3975+
4.7	-.237	-.0169	-.7160+	-.0111
4.8	-.227	-.0276	-.0117	-.0178
5.0	-.213	-.0480	-.0203	-.0300
5.2	-.200	-.0671	-.0283	-.0408
5.4	-.187	-.0852	-.0359	-.0505
5.6	-.177	-.1024	-.0430	-.0593
5.8	-.169	-.1184	-.0495	-.0672
6.0	-.161	-.1338	-.0557	-.0745
6.2	-.153	-.1483	-.0615	-.0811
6.4	-.147	-.1620	-.0669	-.0872
6.6	-.141	-.1753	-.0721	-.0929
6.8	-.136	-.1878	-.0769	-.0981
7.0	-.131	-.1997	-.0814	-.1030
7.2	-.127	-.2111	-.0857	-.1075
7.4	-.123	-.2218	-.0897	-.1117
7.6	-.119	-.2322	-.0935	-.1157
7.8	-.115	-.2422	-.0972	-.1194
8.0	-.113	-.2519	-.1006	-.1230
8.2	-.110	-.2610	-.1039	-.1263
8.4	-.107	-.2698	-.1070	-.1294
8.6	-.104	-.2784	-.1100	-.1324
8.8	-.102	-.2864	-.1128	-.1352
9.0	-.100	-.2943	-.1155	-.1379
9.2	-.097	-.3019	-.1181	-.1404
9.4	-.095	-.3092	-.1206	-.1428
9.6	-.094	-.3164	-.1230	-.1452

Table A.3 Continued

$\alpha_3 = 0.60$

α_4	LAM 1	LAM 2	LAM 3	LAM 4
2.4	-1.411	.2347	.0000*	.4951*
2.6	-1.198	.2286	.0171	.4098
2.8	-.972	.2180	.0355	.3265
3.0	-.800	.2009	.0467	.2583
3.2	-.665	.1791	.0514	.2020
3.4	-.562	.1539	.0504	.1554
3.6	-.482	.1273	.0454	.1171
3.8	-.420	.1005	.0379	.0854
4.0	-.372	.0740	.0289	.0589
4.2	-.335	.0486	.0194	.0366
4.4	-.302	.0244	.9911+	.0175
4.5	-.289	.0128	.5215+	.8965+
4.6	-.277	-.1492+	.0611+	.1025+
4.7	-.266	-.9531+	-.8326+	-.6425+
4.8	-.256	-.0202	-.0134	-.0261
5.0	-.238	-.0407	-.0168	-.0373
5.2	-.222	-.0600	-.0248	-.0474
5.4	-.209	-.0782	-.0323	-.0565
5.6	-.197	-.0956	-.0394	-.0647
5.8	-.187	-.1118	-.0460	
6.0	-.179	-.1273	-.0522	-.0722
6.2	-.171	-.1419	-.0580	-.0790
6.4	-.163	-.1559	-.0635	-.0853
6.6	-.157	-.1691	-.0686	-.0911
6.8	-.151	-.1818	-.0735	-.0965
7.0	-.146	-.1938	-.0781	-.1015
7.2	-.141	-.2052	-.0824	-.1061
7.4	-.137	-.2163	-.0865	-.1105
7.6	-.132	-.2267	-.0904	-.1145
7.8	-.128	-.2368	-.0941	-.1183
8.0	-.124	-.2465	-.0976	-.1219
8.2	-.121	-.2557	-.1009	-.1253
8.4	-.118	-.2647	-.1041	-.1285
8.6	-.115	-.2732	-.1071	-.1315
8.8	-.113	-.2815	-.1100	-.1344
9.0	-.110	-.2894	-.1127	-.1371
9.2	-.108	-.2970	-.1153	-.1397
9.4	-.105	-.3045	-.1179	-.1422
9.6	-.103	-.3116	-.1203	-.1445

$\alpha_3 = 0.65$

α_4	LAM 1	LAM 2	LAM 3	LAM 4
2.6	-1.329	.2240	.3908+	.4318
2.8	-1.076	.2157	.0246	.3443
3.0	-.889	.2010	.0380	.2742
3.2	-.744	.1812	.0449	.2162
3.4	-.630	.1582	.0464	.1682
3.6	-.542	.1330	.0435	.1283
3.8	-.472	.1072	.0377	.0952
4.0	-.418	.0813	.0300	.0674
4.2	-.374	.0564	.0215	.0400
4.4	-.338	.0323	.0126	.0239
4.5	-.324	-.0207	.8137+	-.0150
4.6	-.310	-.9399+	.3719+	.6660+
4.7	-.297	-.1593+	-.0634+	-.1106+
4.8	-.285	-.0123	-.4921+	-.8391+
5.0	-.265	-.0328	-.0132	-.0216
5.2	-.248	-.0524	-.0211	-.0334
5.4	-.231	-.0707	-.0286	-.0438
5.6	-.219	-.0880	-.0356	-.0532
5.8	-.208	-.1046	-.0422	-.0618
6.0	-.198	-.1201	-.0484	-.0695
6.2	-.189	-.1350	-.0543	-.0766
6.4	-.181	-.1491	-.0598	-.0831
6.6	-.174	-.1625	-.0650	-.0891
6.8	-.167	-.1753	-.0700	-.0946
7.0	-.161	-.1874	-.0746	-.0997
7.2	-.155	-.1991	-.0790	-.1045
7.4	-.150	-.2100	-.0831	-.1089
7.6	-.145	-.2208	-.0871	-.1131
7.8	-.141	-.2309	-.0908	-.1170
8.0	-.137	-.2407	-.0944	-.1207
8.2	-.134	-.2501	-.0977	-.1242
8.4	-.130	-.2591	-.1010	-.1274
8.6	-.127	-.2677	-.1040	-.1305
8.8	-.124	-.2761	-.1069	-.1335
9.0	-.121	-.2840	-.1097	-.1362
9.2	-.119	-.2919	-.1124	-.1389
9.4	-.116	-.2994	-.1150	-.1414
9.6	-.114	-.3065	-.1174	-.1438
9.8	-.112	-.3136	-.1198	-.1461

$\alpha_3 = 0.70$

α_4	LAM 1	LAM 2	LAM 3	LAM 4
2.6	-1.368	.2217	.0000*	.4353*
2.8	-1.194	.2132	.0130	.3651
3.0	-.987	.2008	.0286	.2918
3.2	-.828	.1833	.0378	.2319
3.4	-.704	.1621	.0416	.1821
3.6	-.606	.1385	.0409	.1406
3.8	-.529	.1139	.0369	.1060
4.0	-.467	.0889	.0307	.0768
4.2	-.419	.0643	.0232	.0522
4.4	-.379	.0406	.0151	.0312
4.5	-.344	.0178	.6767+	.0130
4.6	-.331	.6799+	.2607+	.4872+
4.7	-.317	-.3917+	-.1512+	-.2750+
4.8	-.305	-.0144	-.9565+	-.9893+
5.0	-.294	-.0245	-.0173	-.0166
5.2	-.276	-.0441	-.0247	-.0289
5.4	-.257	-.0626	-.0317	-.0398
5.6	-.243	-.0802	-.0383	-.0496
5.8	-.229	-.0967	-.0445	-.0584
6.0	-.219	-.1125	-.0504	-.0665
6.2	-.209	-.1275	-.0560	-.0738
6.4	-.199	-.1417	-.0613	-.0805
6.6	-.191	-.1554	-.0662	-.0867
6.8	-.184	-.1682	-.0709	-.0924
7.0	-.177	-.1805	-.0754	-.0977
7.2	-.170	-.1923	-.0796	-.1026
7.4	-.165	-.2036	-.0836	-.1072
7.6	-.160	-.2144	-.0874	-.1115
7.8	-.155	-.2246	-.0910	-.1155
8.0	-.151	-.2346	-.0944	-.1193
8.2	-.147	-.2439	-.0977	-.1228
8.4	-.143	-.2532	-.1008	-.1262
8.6	-.139	-.2618	-.1038	-.1293
8.8	-.136	-.2703	-.1066	-.1323
9.0	-.133	-.2784	-.1093	-.1352
9.2	-.130	-.2862	-.1119	-.1379
9.4	-.127	-.2937	-.1144	-.1404
9.6	-.125	-.3011	-.1168	-.1429
9.8	-.122	-.3081		-.1452

$$\alpha_3 = 0.75$$

α_4	LAM 1	LAM 2	LAM 3	LAM 4
2.8	-1.334	.2104	.0000	.3903
3.0	-1.097	.2003	.0183	.3119
3.2	-.921	.1850	.0299	.2492
3.4	-.785	.1658	.0360	.1974
3.6	-.677	.1440	.0375	.1542
3.8	-.590	.1206	.0355	.1179
4.0	-.521	.0966	.0309	.0873
4.2	-.466	.0726	.0246	.0614
4.4	-.419	.0492	.0174	.0392
4.6	-.384	.0266	.9663*	.0202
4.7	-.367	.0156	.5749+	.0116
4.8	-.352	.4940*	.1833+	.3583+
4.9	-.339	-.5509*	-.2061+	-.3916+
5.0	-.324	-.0157	-.5915+	-.0109
5.2	-.306	-.0353	-.0207	-.0238
5.4	-.284	-.0539	-.0276	-.0352
5.6	-.268	-.0716	-.0342	-.0454
5.8	-.254	-.0884	-.0405	-.0547
6.0	-.240	-.1044	-.0464	-.0630
6.2	-.229	-.1195	-.0520	-.0706
6.4	-.219	-.1339	-.0573	-.0776
6.6	-.209	-.1476	-.0623	-.0840
6.8	-.201	-.1607	-.0670	-.0899
7.0	-.194	-.1731	-.0715	-.0954
7.2	-.188	-.1851	-.0758	-.1005
7.4	-.181	-.1964	-.0799	-.1052
7.6	-.175	-.2074	-.0837	-.1096
7.8	-.170	-.2177	-.0874	-.1137
8.0	-.165	-.2278	-.0909	-.1176
8.2	-.160	-.2375	-.0942	-.1213
8.4	-.156	-.2466	-.0974	-.1247
8.6	-.152	-.2554	-.1004	-.1279
8.8	-.148	-.2640	-.1033	-.1310
9.0	-.142	-.2722	-.1061	-.1339
9.2	-.138	-.2802	-.1088	-.1367
9.4	-.135	-.2879	-.1113	-.1393
9.6	-.133	-.2952	-.1137	-.1418
9.8	-.130	-.3023	-.1161	-.1442
10.0	-.130	-.3093	-.1161	-.1465

$$\alpha_3 = 0.80$$

α_4	LAM 1	LAM 2	LAM 3	LAM 4
3.0	-1.225	.1996	.6847+	.3356
3.2	-1.025	.1864	.0211	.2687
3.4	-.874	.1692	.0295	.2143
3.6	-.754	.1492	.0333	.1691
3.8	-.657	.1272	.0333	.1310
4.0	-.582	.1042	.0303	.0989
4.2	-.515	.0810	.0254	.0716
4.4	-.468	.0580	.0192	.0482
4.6	-.425	.0357	.0123	.0281
4.8	-.392	.0142	.5035*	.0107
4.9	-.375	.3770+	.1352+	.2770+
5.0	-.361	-.6291+	-.2278+	-.4531+
5.1	-.349	-.0164	-.5981+	-.0116
5.2	-.335	-.0261	-.9598+	-.0181
5.4	-.313	-.0449	-.0167	-.0301
5.6	-.295	-.0626	-.0235	-.0408
5.8	-.279	-.0795	-.0300	-.0504
6.0	-.264	-.0958	-.0363	-.0592
6.2	-.251	-.1110	-.0422	-.0671
6.4	-.240	-.1255	-.0478	-.0743
6.6	-.230	-.1394	-.0531	-.0810
6.8	-.220	-.1527	-.0582	-.0871
7.0	-.212	-.1653	-.0630	-.0928
7.2	-.204	-.1774	-.0676	-.0980
7.4	-.197	-.1889	-.0719	-.1029
7.6	-.191	-.2000	-.0760	-.1075
7.8	-.185	-.2104	-.0799	-.1117
8.0	-.180	-.2205	-.0836	-.1157
8.2	-.174	-.2304	-.0872	-.1195
8.4	-.169	-.2397	-.0906	-.1230
8.6	-.166	-.2488	-.0938	-.1264
8.8	-.161	-.2574	-.0969	-.1295
9.0	-.157	-.2658	-.0999	-.1325
9.2	-.154	-.2737	-.1027	-.1353
9.4	-.150	-.2815	-.1054	-.1380
9.6	-.147	-.2890	-.1080	-.1406
9.8	-.144	-.2962	-.1105	-.1430
10.0	-.141	-.3033	-.1129	-.1454
10.2	-.139	-.3100	-.1152	-.1476

$$\alpha_3 = 0.85$$

α_4	LAM 1	LAM 2	LAM 3	LAM 4
3.0	-1.303	.1985	.0000*	.3488
3.2	-1.145	.1875	.0110	.2912
3.4	-.973	.1723	.0220	.2332
3.6	-.838	.1541	.0281	.1855
3.8	-.732	.1336	.0301	.1455
4.0	-.645	.1179	.0291	.1117
4.2	-.577	.0895	.0256	.0829
4.4	-.519	.0671	.0206	.0582
4.6	-.472	.0451	.0146	.0370
4.8	-.430	.0238	.8001+	.0185
4.9	-.413	.0134	.4581+	.0102
5.0	-.398	.3503+	.1211+	.2612+
5.1	-.383	-.6701+	-.2345+	-.4896+
5.2	-.370	-.0165	-.5808+	-.0118
5.4	-.344	-.0353	.0127	.0244
5.6	-.324	-.0531	-.0193	-.0356
5.8	-.305	-.0703	-.0258	-.0457
6.0	-.290	-.0864	-.0319	-.0548
6.2	-.275	-.1019	-.0378	-.0631
6.4	-.262	-.1168	-.0435	-.0707
6.6	-.251	-.1307	-.0488	-.0776
6.8	-.241	-.1442	-.0539	-.0840
7.0	-.231	-.1570	-.0588	-.0899
7.2	-.223	-.1692	-.0634	-.0953
7.4	-.215	-.1809	-.0678	-.1004
7.6	-.207	-.1921	-.0720	-.1051
7.8	-.201	-.2028	-.0759	-.1095
8.0	-.195	-.2130	-.0797	-.1136
8.2	-.190	-.2229	-.0833	-.1175
8.4	-.184	-.2324	-.0868	-.1211
8.6	-.179	-.2416	-.0901	-.1246
8.8	-.175	-.2503	-.0932	-.1278
9.0	-.171	-.2587	-.0962	-.1309
9.2	-.167	-.2669	-.0991	-.1338
9.4	-.163	-.2748	-.1019	-.1366
9.6	-.159	-.2823	-.1045	-.1392
9.8	-.156	-.2897	-.1071	-.1417
10.0	-.153	-.2967	-.1095	-.1441
10.2	-.150	-.3037	-.1119	-.1464

Table A.3 Continued

$\alpha_3 = 0.90$

α_4	LAM 1	LAM 2	LAM 3	LAM 4
3.2	-1.277	.1880	.0000	.3160
3.4	-1.085	.1751	.0133	.2548
3.6	-.933	.1586	.0218	.2039
3.8	-.814	.1397	.0260	.1615
4.0	-.717	.1193	.0269	.1258
4.2	-.639	.0979	.0251	.0953
4.4	-.575	.0762	.0214	.0693
4.6	-.522	.0547	.0164	.0468
4.8	-.478	.0337	.0106	.0273
5.0	-.439	.0132	.4328+	.0102
5.1	-.422	-.3339+	.1111+	.2526+
5.2	-.407	-.6388+	-.215+	-.4735+
5.3	-.394	-.0759	-.5428+	-.0116
5.4	-.379	-.0252	-.8694+	-.0180
5.6	-.353	-.0432	-.0152	-.0298*
5.8	-.334	-.0605	-.0215	-.0405
6.0	-.317	-.0768	-.0275	-.0500
6.2	-.301	-.0924	-.0334	-.0587
6.4	-.287	-.1073	-.0390	-.0666
6.6	-.273	-.1215	-.0444	-.0738
6.8	-.262	-.1352	-.0495	-.0805
7.0	-.252	-.1481	-.0544	-.0866
7.2	-.242	-.1606	-.0591	-.0923
7.4	-.233	-.1723	-.0635	-.0975
7.6	-.225	-.1838	-.0678	-.1024
7.8	-.218	-.1947	-.0718	-.1070
8.0	-.212	-.2051	-.0756	-.1113
8.2	-.205	-.2151	-.0793	-.1153
8.4	-.199	-.2246	-.0828	-.1190
8.6	-.194	-.2340	-.0862	-.1226
8.8	-.189	-.2428	-.0894	-.1259
9.0	-.185	-.2514	-.0924	-.1291
9.2	-.180	-.2597	-.0954	-.1321
9.4	-.176	-.2676	-.0982	-.1349
9.6	-.172	-.2753	-.1009	-.1376
9.8	-.168	-.2827	-.1035	-.1402
10.0	-.165	-.2900	-.1060	-.1427
10.2	-.162	-.2969	-.1084	-.1450
10.4	-.159	-.3035	-.1107	-.1472

$\alpha_3 = 1.00$

α_4	LAM 1	LAM 2	LAM 3	LAM 4
3.4	-1.253	.1772	.0000*	.2854**
3.6	-1.169	.1664	.4828+	.2490
3.8	-1.010	.1509	.0141	.1996
4.0	-.886	.1333	.0193	.1588
4.2	-.787	.1142	.0212	.1244
4.4	-.706	.0943	.0206	.0950
4.6	-.638	.0741	.0182	.0697
4.8	-.581	.0539	.0144	.0477
5.0	-.533	.0340	.9695+	.0285
5.2	-.492	.0146	.4383+	.0117
5.3	-.474	.5192+	.1584+	.4061+
5.4	-.445	-.C317+	-.0101**	-.0242**
5.5	-.442	-.0132	.4176+	-.9946+
5.6	-.429	-.0222	.7097+	-.0164
5.8	-.403	-.0395	-.0129	-.0282*
6.0	-.358	-.0562	-.0187	-.0388
6.2	-.341	-.0721	-.0244	-.0484
6.4	-.325	-.0873	-.0299	-.0571
6.6	-.309	-.1019	-.0352	-.0651
6.8		-.1158	-.0404	-.0723
7.0	-.297	-.1291	-.0453	-.0790
7.2	-.285	-.1419	-.0500	-.0852
7.4	-.275	-.1540	-.0545	-.0909
7.6	-.265	-.1658	-.0589	-.0962
7.8	-.256	-.1769	-.0630	-.1011
8.0	-.248	-.1878	-.0670	-.1058
8.2	-.241	-.1980	-.0707	-.1101
8.4	-.233	-.2079	-.0744	-.1141
8.6	-.227	-.2174	-.0778	-.1179
8.8	-.220	-.2267	-.0812	-.1215
9.0	-.215	-.2356	-.0844	-.1249
9.2	-.210	-.2440	-.0874	-.1281
9.4	-.204	-.2522	-.0904	-.1311
9.6	-.200	-.2602	-.0932	-.1340
9.8	-.195	-.2678	-.0959	-.1367
10.0	-.191	-.2752	-.0985	-.1393
10.2	-.187	-.2824	-.1010	-.1418
10.4	-.184	-.2893	-.1034	-.1442
10.6	-.180	-.2959	-.1057	-.1464

$\alpha_3 = 1.10$

α_4	LAM 1	LAM 2	LAM 3	LAM 4
3.8	-1.215	.1582	.0000*	.2379
4.0	-1.108	.1459	.6035+	.2013
4.2	-.974	.1294	.0125	.1607
4.4	-.869	.1117	.0157	.1267
4.6	-.781	.0932	.0165	.0977
4.8	-.708	.0743	.0154	.0727
5.0	-.647	.0552	.0128	.0508
5.2	-.596	.0365	.9168+	.0318
5.4	-.552	.0181	.4839+	.0150
5.5	-.532	.9038+	.2484+	.7342+
5.6	-.517	-.0997+	-.0279+	-.0795+
5.7	-.497	-.8629+	-.2479+	-.6726+
5.8	-.481	-.0173	-.5046+	-.0132
6.0	-.451	-.0340	-.0103	-.0251
6.2	-.427	-.0501	-.0155	-.0358
6.4	-.403	-.0656	-.0208	-.0455
6.6	-.384	-.0805	-.0259	-.0544
6.8	-.366	-.0947	-.0309	-.0624
7.0	-.350	-.1084	-.0358	-.0698
7.2	-.335	-.1214	-.0405	-.0766
7.4	-.322	-.1341	-.0451	-.0829
7.6	-.311	-.1460	-.0494	-.0887
7.8	-.299	-.1577	-.0537	-.0941
8.0	-.289	-.1687	-.0577	-.0991
8.2	-.280	-.1794	-.0616	-.1038
8.4	-.271	-.1896	-.0653	-.1082
8.6	-.263	-.1994	-.0689	-.1123
8.8	-.256	-.2090	-.0724	-.1162
9.0	-.249	-.2180	-.0757	-.1198
9.2	-.242	-.2267	-.0788	-.1232
9.4	-.236	-.2353	-.0819	-.1265
9.6	-.231	-.2435	-.0848	-.1296
9.8	-.226	-.2513	-.0876	-.1325
10.0	-.221	-.2590	-.0903	-.1353
10.2	-.216	-.2664	-.0930	-.1379
10.4	-.211	-.2735	-.0955	-.1404
10.6	-.207	-.2804	-.0979	-.1428
10.8	-.203	-.2870	-.1002	-.1451
11.0	-.199	-.2936	-.1025	-.1473

α₃ = 1.20

α₄	LAM 1	LAM 2	LAM 3	LAM 4
4.2	-1.183	.1407	.0000*	.1997
4.4	-1.083	.1278	.5096+	.1675
4.6	-.965	.1113	.9968+	.1329
4.8	-.870	.0941	.0122	.1036
5.0	-.792	.0764	.0124	.0784
5.2	-.723	.0586	.0112	.0565
5.4	-.668	.0408	.8705+	.0372
5.6	-.619	.0233	.5411+	.0202
5.7	-.597	.0146	.3525+	.0124
5.8	-.577	.6088+	.1515+	.5050+
5.9	-.558	-.2319+	-.0594+	-.1884+
6.0	-.562	-.0962+	-.0245+	-.0784+
6.2	-.508	-.0268	-.7343+	-.0206
6.4	-.481	-.0024	-.0120	-.0315
6.6	-.452	-.0575	-.0168	-.0414
6.8	-.432	-.0719	-.0215	-.0504
7.0	-.412	-.0860	-.0262	-.0587
7.2	-.394	-.0993	-.0308	-.0662
7.4	-.378	-.1123	-.0353	-.0732
7.6	-.362	-.1247	-.0397	-.0796
7.8	-.349	-.1366	-.0439	-.0856
8.0	-.337	-.1480	-.0480	-.0911
8.2	-.325	-.1589	-.0519	-.0962
8.4	-.314	-.1695	-.0558	-.1010
8.6	-.305	-.1796	-.0594	-.1055
8.8	-.296	-.1896	-.0630	-.1098
9.0	-.287	-.1990	-.0664	-.1137
9.2	-.280	-.2082	-.0697	-.1175
9.4	-.273	-.2168	-.0728	-.1210
9.6	-.265	-.2253	-.0759	-.1243
9.8	-.259	-.2335	-.0788	-.1275
10.0	-.254	-.2414	-.0816	-.1305
10.2	-.248	-.2490	-.0843	-.1333
10.4	-.242	-.2564	-.0870	-.1360
10.6	-.237	-.2636	-.0895	-.1386
10.8	-.233	-.2704	-.0919	-.1410
11.0	-.228	-.2772	-.0943	-.1434
11.2	-.224	-.2837	-.0966	-.1456
11.4	-.220	-.2901	-.0988	-.1478

α₃ = 1.30

α₄	LAM 1	LAM 2	LAM 3	LAM 4
4.6	-1.156	.1244	.0000*	.1679
4.8	-1.084	.1129	.3174+	.1435
5.0	-.975	.0968	.7225+	.1130
5.2	-.886	.0802	.9035+	.0870
5.4	-.812	.0634	.9148+	.0645
5.6	-.749	.0466	.7959+	.0447
5.8	-.695	.0300	.5783+	.0273
6.0	-.604	.0286+	.6619$+	.0239+
6.1	-.617	.0446+	.0100+	-.0375+*
6.2	-.616	-.0526+	-.0118+	-.0402+
6.3	-.589	-.0104	-.2450+	-.8504+
6.4	-.572	-.0182	-.4399+	-.0146
6.6	-.539	-.0333	-.8469+	-.0258
6.8	-.510	-.0480	-.0127	-.0360
7.0	-.485	-.0622	-.0170	-.0453
7.2	-.463	-.0758	-.0213	-.0538
7.4	-.442	-.0890	-.0256	-.0616
7.6	-.424	-.1017	-.0298	-.0688
7.8	-.407	-.1140	-.0340	-.0754
8.0	-.392	-.1258	-.0380	-.0816
8.2	-.378	-.1372	-.0420	-.0873
8.4	-.365	-.1480	-.0458	-.0926
8.6	-.353	-.1584	-.0495	-.0975
8.8	-.342	-.1687	-.0531	-.1022
9.0	-.332	-.1784	-.0566	-.1065
9.2	-.322	-.1878	-.0600	-.1106
9.4	-.314	-.1969	-.0632	-.1145
9.6	-.305	-.2057	-.0664	-.1181
9.8	-.298	-.2141	-.0694	-.1215
10.0	-.291	-.2223	-.0723	-.1248
10.2	-.284	-.2304	-.0752	-.1279
10.4	-.277	-.2379	-.0779	-.1308
10.6	-.272	-.2453	-.0805	-.1336
10.8	-.266	-.2525	-.0831	-.1362
11.0	-.261	-.2595	-.0855	-.1388*
11.2	-.256	-.2662	-.0879	-.1412
11.4	-.251	-.2728	-.0902	-.1435
11.6	-.246	-.2792	-.0925	-.1457
11.8	-.242	-.2852	-.0946	-.1478

α₃ = 1.40

α₄	LAM 1	LAM 2	LAM 3	LAM 4
5.0	-1.132	.1092	.0000*	.1411
5.2	-1.106	.1011	.0787+	.1268
5.4	-1.001	.0855	.4546+	.0991
5.6	-.916	.0697	.6296+	.0754
5.8	-.844	.0538	.6530+	.0547
6.0	-.782	.0379	.5603+	.0365
6.2	-.729	.0222	.3785+	.0204
6.3	-.706	.0145	.2611+	+.0130
6.4	-.683	.6822+	.1292+	-.5987+
6.5	-.660	-.0244+	-.0244+	-.1052+
6.6	-.643	-.8266+	-.1702+	-.6968+
6.8	-.607	-.0230	-.5060+	.0187
7.0	-.575	-.0373	-.8670+	.0293
7.2	-.547	-.0510	.0124	.0389
7.4	-.521	-.0645	.0163	.0478
7.6	-.498	-.0775	.0202	.0559
7.8	-.475	-.0900	.0242	.0633*
8.0	-.458	-.1020	.0280	.0702
8.2	-.440	-.1137	.0319	.0766
8.4	-.423	-.1250	.0357	.0825*
8.6	-.410	-.1358	-.0393	-.0881*
8.8	-.395	-.1463	-.0430	-.0932
9.0	-.383	-.1564	-.0465	-.0980
9.2	-.372	-.1662	-.0499	-.1026
9.4	-.361	-.1756	-.0532	-.1068
9.6	-.351	-.1846	-.0564	-.1108
9.8	-.342	-.1935	-.0595	-.1146
10.0	-.333	-.2018	-.0625	-.1181
10.2	-.325	-.2102	-.0655	-.1215
10.4	-.317	-.2181	-.0683	-.1247
10.6	-.310	-.2257	-.0710	-.1277
10.8	-.303	-.2332	-.0737	-.1306
11.0	-.297	-.2405	-.0762	-.1334*
11.2	-.291	-.2475	-.0787	-.1360
11.4	-.285	-.2542	-.0811	-.1385*
11.6	-.279	-.2609	-.0835	-.1409
11.8	-.274	-.2671	-.0857	-.1431
12.0	-.269	-.2734	-.0879	-.1453
12.2	-.265	-.2794	-.0900	-.1474

Table A.3 Continued

α₃ = 1.50

α₄	LAM 1	LAM 2	LAM 3	LAM 4
5.4	-1.112	.0951	.0000*	.1182
5.6	-1.103	.0886	.0000*	.1083
5.8	-1.042	.0773	.1949+	.0899
6.0	-.957	.0622	.3907+	.0677
6.2	-.885	.0471	.4441+	.0483
6.4	-.824	.0321	.3885+	.0313
6.6	-.688	.0566+	.0104*	.0494**
6.7	-.747	.9962+	.1538+	.9059+
6.8	-.714	-.0290+	.4897$	-.0256+
6.9	-.704	-.4446+	-.0768+	-.3882+
7.0	-.684	.0115	-.2088+	.9875+
7.2	-.647	-.0254	-.4989+	.0210
7.4	-.615	-.0390	-.8156+	.0312
7.6	-.585	-.0520	-.0115	-.0404
7.8	-.558	-.0648	-.0150	-.0489
8.0	-.536	-.0767	-.0184	-.0565
8.2	-.514	-.0891	-.0221	-.0640
8.4	-.494	-.1007	-.0257	-.0707
8.6	-.476	-.1118	-.0292	-.0769
8.8	-.459	-.1225	-.0327	-.0826
9.0	-.443	-.1330	-.0362	-.0880
9.2	-.429	-.1431	-.0396	-.0931
9.4	-.416	-.1528	-.0429	-.0978
9.6	-.404	-.1622	-.0461	-.1022
9.8	-.392	-.1713	-.0493	-.1064
10.0	-.382	-.1803	-.0524	-.1104
10.2	-.372	-.1887	-.0553	-.1141
10.4	-.363	-.1969	-.0582	-.1176
10.6	-.354	-.2049	-.0611	-.1209
10.8	-.346	-.2127	-.0638	-.1241
11.0	-.338	-.2202	-.0665	-.1271
11.2	-.331	-.2273	-.0690	-.1299
11.4	-.325	-.2339	-.0713	-.1325
11.6	-.317	-.2414	-.0740	-.1353
11.8	-.311	-.2478	-.0763	-.1377
12.0	-.305	-.2544	-.0786	-.1401
12.2	-.300	-.2607	-.0808	-.1424
12.4	-.295	-.2662	-.0827	-.1444
12.6	-.289	-.2726	-.0851	-.1466

α₃ = 1.60

α₄	LAM 1	LAM 2	LAM 3	LAM 4
6.0	-1.086	.0757	.0000*	.0896
6.2	-1.078	.0698	.0000	.0814
6.4	-1.011	.0573	.1699+	.0634
6.6	-.937	.0430	.2684+	.0449
6.8	-.875	.0287	.2597+	.0285
7.0	-.746	.0422+	.6356$	.0378+
7.1	-.796	.7738+	.096$+	-.7177+
7.2	-.771	-.0341+	-.4634$	-.0309+
7.3	-.751	-.5924+	-.0858+	-.5279+
7.4	-.731	-.0727	-.1942+	-.0111
7.6	-.693	-.0253	-.4383+	-.0218
7.8	-.659	-.0386	-.7111+	-.0316
8.0	-.630	-.0511	-.0100	-.0406
8.2	-.602	-.0633	-.0131	-.0489
8.4	-.577	-.0752	-.0163	-.0566*
8.6	-.553	-.0866	-.0196	-.0636
8.8	-.534	-.0972	-.0227	-.0699
9.0	-.515	-.1064	-.0261	-.0763
9.2	-.496	-.1187	-.0294	-.0819
9.4	-.480	-.1288	-.0326	-.0872
9.6	-.465	-.1385	-.0358	-.0922
9.8	-.452	-.1480	-.0389	-.0969
10.0	-.438	-.1572	-.0420	-.1013
10.2	-.426	-.1659	-.0450	-.1054
10.4	-.415	-.1745	-.0479	-.1093
10.6	-.404	-.1828	-.0508	-.1130
10.8	-.394	-.1908	-.0536	-.1165
11.0	-.385	-.1986	-.0563	-.1198
11.2	-.377	-.2062	-.0589	-.1230
11.4	-.368	-.2135	-.0615	-.1260
11.6	-.360	-.2206	-.0640	-.1288
11.8	-.352	-.2275	-.0665	-.1315
12.0	-.346	-.2341	-.0688	-.1341
12.2	-.339	-.2407	-.0711	-.1366
12.4	-.333	-.2471	-.0734	-.1390
12.6	-.328	-.2527	-.0753	-.1411
12.8	-.321	-.2592	-.0777	-.1434
13.0	-.316	-.2650	-.0797	-.1455
13.2	-.311	-.2706	-.0817	-.1475

α₃ = 1.70

α₄	LAM 1	LAM 2	LAM 3	LAM 4
6.6	-1.064	.0580	.0000*	.0657
6.8	-1.057	.0525	.0000	.0588
7.0	-1.001	.0412	.1027+	.0441
7.2	-.935	.0275	.1513+	.0280
7.4	-.878	.0142	.1142+	.0138
7.5	-.852	.7546+	.0696+	.7179+
7.6	-.825	-.0250+	-.2601$	-.0232**
7.7	-.806	-.5469+	-.0619+	-.5000+
7.8	-.784	-.0119	-.1463+	-.0107
8.0	-.745	-.0245	-.3323+	-.0212
8.2	-.709	-.0367	-.5705+	-.0308
8.4	-.678	-.0487	-.8225+	-.0397
8.6	-.650	-.0603	-.0109	-.0478
8.8	-.622	-.0717	-.0138	-.0553*
9.0	-.598	-.0827	-.0167	-.0623
9.2	-.578	-.0933	-.0196	-.0688
9.4	-.557	-.1036	-.0226	-.0748
9.6	-.538	-.1136	-.0256	-.0804
9.8	-.521	-.1233	-.0286	-.0857
10.0	-.505	-.1329	-.0316	-.0907
10.2	-.489	-.1420	-.0346	-.0953
10.4	-.476	-.1509	-.0375	-.0997
10.6	-.463	-.1594	-.0403	-.1038
10.8	-.451	-.1677	-.0431	-.1077
11.0	-.440	-.1758	-.0458	-.1114
11.2	-.429	-.1837	-.0485	-.1149
11.4	-.419	-.1913	-.0511	-.1182
11.6	-.410	-.1988	-.0537	-.1214
11.8	-.401	-.2059	-.0562	-.1244
12.0	-.392	-.2128	-.0586	-.1272
12.2	-.384	-.2195	-.0610	-.1299
12.4	-.377	-.2261	-.0633	-.1325
12.6	-.369	-.2326	-.0656	-.1350
12.8	-.362	-.2388	-.0678	-.1374
13.0	-.356	-.2450	-.0700	-.1397
13.2	-.350	-.2508	-.0720	-.1419
13.4	-.344	-.2566	-.0741	-.1440
13.6	-.338	-.2622	-.0761	-.1460
13.8	-.333	-.2675	-.0780	-.1479*

$\alpha_3 = 1.80$

α_4	LAM 1	LAM 2	LAM 3	LAM 4
7.2	-1.045	.0417	.0000*	.0456
7.4	-1.039	.0367	.0000*	.0396
7.6	-1.007	.0284	.0378+	.0298
7.8	-.945	.0155	.0646+	.0155
7.9	-.918	.9177+	.0498+	.9006+
8.0	-.892	.2914+	.0193+	.2801+
8.1	-.868	-.3291+	-.0254+	-.3102+
8.2	-.846	-.9427+	-.0826+	-.8721+
8.4	-.804	-.0215	-.2289+	-.0192
8.6	-.767	-.0333	-.4103+	-.0288
8.8	-.733	-.0448	-.6190+	-.0376
9.0	-.702	-.0559	-.8489+	-.0456
9.2	-.675	-.0668	-.0109	-.0531
9.4	-.649	-.0774	-.0135	-.0601
9.6	-.625	-.0877	-.0162	-.0665
9.8	-.604	-.0978	-.0217	-.0726
10.0	-.583	-.1075	-.0244	-.0782
10.2	-.565	-.1169	-.0272	-.0835
10.4	-.548	-.1260	-.0289	-.0884
10.6	-.532	-.1349	-.0299	-.0931
10.8	-.517	-.1436	-.0327	-.0975
11.0	-.503	-.1520	-.0354	-.1016
11.2	-.490	-.1600	-.0380	-.1055
11.4	-.478	-.1679	-.0406	-.1092
11.6	-.467	-.1757	-.0432	-.1128
11.8	-.456	-.1831	-.0457	-.1161
12.0	-.445	-.1904	-.0482	-.1193
12.2	-.436	-.1974	-.0506	-.1223
12.4	-.427	-.2043	-.0530	-.1252
12.6	-.418	-.2109	-.0553	-.1279
12.8	-.410	-.2175	-.0576	-.1306
13.0	-.402	-.2238	-.0598	-.1331
13.2	-.395	-.2299	-.0619	-.1355
13.4	-.388	-.2359	-.0640	-.1378
13.6	-.381	-.2417	-.0661	-.1400
13.8	-.374	-.2473	-.0681	-.1421
14.0	-.368	-.2530	-.0701	-.1442
14.2	-.362	-.2583	-.0720	-.1461
14.4	-.357	-.2632	-.0737	-.1479

$\alpha_3 = 1.90$

α_4	LAM 1	LAM 2	LAM 3	LAM 4
8.0	-1.023	.0220	.0000*	.0230
8.2	-1.018	.0175	.0000	.0181
8.4	-.968	.6447+	.0150+	.6431+
8.5	-.946	.1239+	.0257+	.1215+
8.6	-.917	-.5444+	-.0657+	-.5220+
8.7	-.892	-.0113	-.1167+	-.0106
8.8	-.871	-.0171	-.2475+	-.0158
9.0	-.831	-.0284	-.4100+	-.0254
9.2	-.794	-.0395	-.5975+	-.0343
9.4	-.761	-.0503	.8046+	-.0424
9.6	-.731	-.0609	-.0103	-.0500
9.8	-.703	-.0712	-.0126	-.0570
10.0	-.679	-.0811	-.0150	-.0635
10.2	-.656	-.0907	-.0175	-.0692
10.4	-.634	-.1002	-.0200	-.0752
10.6	-.614	-.1093	-.0226	-.0805
10.8	-.595	-.1183	-.0251	-.0855
11.0	-.578	-.1269	-.0277	-.0902
11.2	-.562	-.1355	-.0302	-.0947
11.4	-.547	-.1437		-.0989
11.6	-.533	-.1515	-.0327	-.1028
11.8	-.520	-.1594	-.0352	-.1066
12.0	-.508	-.1665	-.0375	-.1100
12.2	-.495	-.1742	-.0401	-.1135
12.4	-.485	-.1811	-.0423	-.1166
12.6	-.474	-.1883	-.0448	-.1198
12.8	-.464	-.1950	-.0471	-.1227
13.0	-.455	-.2015	-.0493	-.1255
13.2	-.446	-.2080	-.0515	-.1282*
13.4	-.437	-.2142	-.0537	-.1307
13.6	-.429	-.2203	-.0558	-.1332
13.8	-.421	-.2262	-.0579	-.1355
14.0	-.414	-.2320	-.0599	-.1378
14.2	-.407	-.2376	-.0619	-.1399
14.4	-.400	-.2431	-.0638	-.1420
14.6	-.394	-.2485	-.0657	-.1440
14.8	-.388	-.2537	-.0676	-.1459
15.0	-.382	-.2589	-.0694	-.1478
15.2	-.376	-.2636	-.0711	-.1495

$\alpha_3 = 2.00$

α_4	LAM 1	LAM 2	LAM 3	LAM 4
8.6	-1.009	.8397+	.0000*	.8541+
8.8	-1.004	.4147+	.0000*	.4182+
8.9	-1.002	.2061+	.0001S*	.2070+
9.0	-.993	-.1081+	-.0407S$	-.1076+
9.1	-.974	-.5675+	-.7075$	-.5567+
9.2	-.950	-.0113	-.0272+	-.0109
9.4	-.905	-.0222	-.1012+	-.0207
9.6	-.865	-.0331	-.2125+	-.0298
9.8	-.828	-.0435	-.3537+	-.0381
10.0	-.796	-.0538	-.5187+	-.0458
10.2	-.766	-.0637	-.7027+	-.0529
10.4	-.738	-.0734	-.9016+	-.0595
10.6	-.713	-.0829	-.0111	-.0657
10.8	-.690	-.0920	-.0133	-.0714
11.0	-.670	-.1005	-.0154	-.0766
11.2	-.647	-.1097	-.0179	-.0819
11.4	-.629	-.1181	-.0202	-.0867
11.6	-.611	-.1264	-.0226	-.0912
11.8	-.595	-.1345	-.0249	-.0955
12.0	-.579	-.1423	-.0273	-.0995
12.2	-.565	-.1498	-.0296	-.1033
12.4	-.557	-.1555	-.0312	-.1062
12.6	-.539	-.1644	-.0342	-.1104
12.8	-.527	-.1715	-.0365	-.1137
13.0	-.515	-.1784	-.0388	-.1168
13.2	-.504	-.1851	-.0410	-.1198
13.4	-.495	-.1914	-.0431	-.1226
13.6	-.485	-.1979	-.0453	-.1254
13.8	-.475	-.2041	-.0474	-.1280
14.0	-.466	-.2101	-.0495	-.1305
14.2	-.458	-.2160	-.0515	-.1329
14.4	-.450	-.2216	-.0535	-.1351*
14.6	-.443	-.2271	-.0554	-.1373
14.8	-.436	-.2321	-.0571	-.1393
15.0	-.428	-.2380	-.0592	-.1415
15.2	-.422	-.2432	-.0610	-.1435*
15.4	-.415	-.2481	-.0628	-.1453*
15.6	-.409	-.2532	-.0646	-.1472
15.8	-.403	-.2580	-.0663	-.1489*

Table A.4 Fit of Johnson S_U Distribution

Values of $-\gamma$

$\sqrt{\beta_1}$ / β_2	0·05	0·10	0·15	0·20	0·25	0·30	0·35	0·40	0·45	0·50
3·2	0·3479	0·7373	1·228	1·940	3·189	6·389				
3·3	·2328	·4834	0·7747	1·143	1·656	2·477				
3·4	·1760	·3620	·5699	0·8166	1·133	1·569	2·236	3·484		
3·5	·1421	·2905	·4528	·6384	0·8620	1·148	1·546	2·146		
3·6	0·1196	0·2435	0·3776	0·5260	0·6997	0·9115	1·187	1·565	2·139	3·157
3·7	·1035	·2102	·3238	·4487	·5907	·7586	0·9681	1·238	1·614	2·188
3·8	·0914	·1853	·2845	·3921	·5125	·6515	·8197	1·028	1·302	1·687
3·9	·0820	·1661	·2542	·3490	·4536	·5723	·7127	0·8814	1·095	1·378
4·0	·0745	·1507	·2302	·3150	·4076	·5113	·6304	·7733	0·9470	1·169
4·1	0·0684	0·1381	0·2106	0·2875	0·3707	0·4629	0·5673	0·6902	0·8363	1·018
4·2	·0633	·1276	·1943	·2647	·3404	·4234	·5165	·6243	·7503	0·9031
4·3	·0589	·1188	·1806	·2456	·3151	·3907	·4747	·5708	·6814	·8132
4·4	·0552	·1112	·1689	·2294	·2937	·3632	·4397	·5265	·6251	·7407
4·5	·0519	·1046	·1588	·2153	·2752	·3396	·4100	·4891	·5780	·6811
4·6	0·0491	0·0989	0·1499	0·2031	0·2592	0·3192	0·3844	0·4564	0·5382	0·6311
4·7	·0466	·0938	·1421	·1923	·2451	·3014	·3622	·4288	·5040	·5886
4·8	·0444	·0893	·1352	·1828	·2327	·2857	·3426	·4048	·4744	·5520
4·9	·0424	·0852	·1290	·1743	·2216	·2717	·3254	·3836	·4484	·5202
5·0	·0406	·0816	·1234	·1666	·2117	·2592	·3099	·3648	·4254	·4922
5·1	0·0390	0·0783	0·1184	0·1597	0·2027	0·2480	0·2961	0·3479	0·4050	0·4674
5·2	·0374	·0752	·1138	·1534	·1946	·2378	·2836	·3328	·3866	·4453
5·3	·0361	·0725	·1096	·1477	·1872	·2285	·2723	·3191	·3696	·4255
5·4	·0348	·0700	·1057	·1424	·1804	·2201	·2620	·3066	·3547	·4076
5·5	·0337	·0676	·1022	·1376	·1742	·2123	·2525	·2952	·3411	·3913
5·6	0·0326	0·0655	0·0989	0·1331	0·1684	0·2052	0·2438	0·2848	0·3286	0·3765
5·7	·0316	·0635	·0958	·1290	·1631	·1986	·2358	·2752	·3172	·3629
5·8	·0307	·0616	·0930	·1251	·1582	·1925	·2284	·2663	·3066	·3504
5·9	·0298	·0599	·0904	·1215	·1536	·1868	·2215	·2581	·2967	·3385
6·0	·0290	·0583	·0879	·1182	·1493	·1815	·2151	·2504	·2879	·3278
6·1	0·0283	0·0568	0·0856	0·1151	0·1453	0·1766	0·2091	0·2433	0·2794	0·3180
6·2	·0276	·0553	·0835	·1121	·1415	·1719	·2035	·2366	·2716	·3088
6·3	·0269	·0540	·0814	·1094	·1380	·1676	·1983	·2304	·2643	·3002
6·4	·0263	·0527	·0795	·1067	·1347	·1635	·1933	·2245	·2574	·2921
6·5	·0257	·0515	·0777	·1043	·1315	·1596	·1887	·2190	·2509	·2846
6·6	0·0251	0·0504	0·0760	ρ·1020	0·1286	0·1560	0·1843	0·2138	0·2448	0·2775
6·7	·0246	·0493	·0743	·0998	·1258	·1525	·1802	·2089	·2391	·2709
6·8	·0241	·0483	·0728	·0977	·1231	·1492	·1762	·2043	·2337	·2646
6·9	·0236	·0473	·0713	·0957	·1206	·1461	·1725	·1999	·2285	·2586
7·0	·0232	·0464	·0699	·0938	·1182	·1432	·1690	·1957	·2237	·2530
7·1	0·0227	0·0455	0·0686	0·0920	0·1159	0·1404	0·1656	0·1918	0·2190	0·2476
7·2	·0223	·0447	·0673	·0903	·1137	·1377	·1624	·1880	·2147	·2426
7·3	·0219	·0439	·0661	·0887	·1116	·1352	·1594	·1844	·2105	·2377
7·4	0215	·0431	·0650	·0871	·1096	·1327	·1565	·1810	·2065	·2331
7·5	·0212	·0424	·0639	·0856	·1077	·1304	·1537	·1777	·2027	·2287
7·6	0·0208	0·0417	0·0628	0·0842	0·1059	0·1282	0·1510	0·1746	0·1991	0·2246
7·7	·0205	·0410	·0618	·0828	·1042	·1260	·1485	·1716	·1956	·2206
7·8	·0202	·0404	·0608	·0815	·1025	·1240	·1460	·1687	·1922	·2167
7·9	·0198	·0398	·0599	·0802	·1009	·1220	·1437	·1660	·1891	·2131
8·0	·0195	·0392	·0590	·0790	·0993	·1201	·1414	·1633	·1860	·2095
8·2	0·0190	0·0380	0·0572	0·0767	0·0964	0·1165	0·1371	0·1583	0·1802	0·2029
8·4	·0185	·0370	·0557	·0745	·0937	·1132	·1332	·1537	·1749	·1968
8·6	·0180	·0360	·0542	·0725	·0912	·1101	·1295	·1494	·1699	·1912
8·8	·0175	·0351	·0528	·0707	·0888	·1073	·1261	·1454	·1653	·1859
9·0	·0171	·0342	·0515	·0689	·0866	·1046	·1229	·1417	·1610	·1810
9·2	—	—	—	—	—	—	0·1199	0·1382	0·1570	0·1764
9·4	—	—	—	—	—	—	·1171	·1349	·1532	·1721

Source: N. L. Johnson, *Biometrika* (1965), pp. 547–558.

β_2 \ $\sqrt{\beta_1}$	0·55	0·60	0·65	0·70	0·75	0·80	0·85	0·90	0·95	1·00
3·8	2·284	3·383								
3·9	1·783	2·426								
4·0	1·469	1·906	2·621	4·104						
4·1	1·253	1·577	2·060	2·886						
4·2	1·096	1·349	1·705	2·254						
4·3	0·9748	1·182	1·460	1·860						
4·4	·8802	1·054	1·280	1·589						
4·5	·8033	0·9527	1·142	1·392						
4·6	0·7399	0·8705	1·033	1·240	1·522	1·931	2·602	4·029		
4·7	·6865	·8024	0·9434	1·121	1·353	1·676	2·167	3·038		
4·8	·6410	·7451	·8699	1·024	1·221	1·485	1·864	2·473		
4·9	·6017	·6962	·8083	0·9436	1·114	1·335	1·641	2·100		
5·0	·5675	·6539	·7550	·8761	1·025	1·215	1·469	1·832	2·407	3·538
5·1	0·5374	0·6170	0·7092	0·8154	0·9509	1·117	1·333	1·629	2·071	2·832
5·2	·5106	·5845	·6693	·7687	·8877	1·034	1·221	1·470	1·825	2·386
5·3	·4868	·5556	·6341	·7252	·8331	0·9642	1·128	1·342	1·635	2·072
5·4	·4653	·5298	·6028	·6869	·7856	·9039	1·050	1·236	1·484	1·839
5·5	·4459	·5066	·5749	·6530	·7419	·8515	0·9826	1·147	1·361	1·657
5·6	0·4283	0·4856	0·5498	0·6226	0·7067	0·8055	0·9243	1·071	1·259	1·510
5·7	·4122	·4665	·5270	·5953	·6735	·7647	·8732	1·006	1·172	1·390
5·8	·3975	·4491	·5063	·5706	·6437	·7284	·8282	0·9485	1·098	1·289
5·9	·3840	·4331	·4875	·5481	·6168	·6942	·7881	·8982	1·033	1·203
6·0	·3714	·4184	·4701	·5276	·5924	·6663	·7521	·8536	0·9765	1·129
6·1	0·3598	0·4049	0·4542	0·5088	0·5700	0·6396	0·7197	0·8138	0·9265	1·065
6·2	·3491	·3923	·4395	·4915	·5496	·6152	·6904	·7780	·8820	1·008
6·3	·3390	·3806	·4258	·4755	·5308	·5929	·6637	·7456	·8420	0·9581
6·4	·3297	·3697	·4131	·4607	·5134	·5724	·6392	·7161	·8060	·9132
6·5	·3209	·3595	·4013	·4470	·4973	·5535	·6150	·6892	·7733	·8729
6·6	0·3123	0·3500	0·3903	0·4341	0·4824	0·5359	0·5962	0·6646	0·7436	0·8364
6·7	·3046	·3410	·3799	·4221	·4684	·5197	·5770	·6419	·7164	·8033
6·8	·2973	·3326	·3702	·4109	·4554	·5045	·5592	·6209	·6914	·7730
6·9	·2904	·3247	·3611	·4004	·4433	·4904	·5427	·5951	·6683	·7453
7·0	·2839	·3172	·3524	·3905	·4318	·4772	·5273	·5835	·6470	·7198
7·1	0·2778	0·3101	0·3443	0·3812	0·4211	0·4648	0·5130	0·5666	0·6272	0·6962
7·2	·2719	·3034	·3366	·3723	·4110	·4531	·4995	·5509	·6087	·6744
7·3	·2664	·2967	·3293	·3640	·4014	·4421	·4868	·5362	·5915	·6541
7·4	·2611	·2907	·3224	·3561	·3924	·4318	·4749	·5224	·5754	·6352
7·5	·2561	·2849	·3159	·3486	·3838	·4220	·4636	·5094	·5593	·6175
7·6	0·2513	0·2795	0·3096	0·3415	0·3757	0·4127	0·4530	0·4972	0·5463	0·6010
7·7	·2467	·2742	·3037	·3347	·3680	·4039	·4429	·4857	·5328	·5855
7·8	·2423	·2692	·2980	·3283	·3607	·3956	·4334	·4747	·5203	·5709
7·9	·2381	·2645	·2926	·3221	·3537	·3876	·4244	·4644	·5084	·5571
8·0	·2341	·2599	·2871	·3163	·3470	·3801	·4158	·4546	·4971	·5442
8·1	0·2303	0·2556	0·2822	0·3106	0·3407	0·3729	0·4076	0·4452	0·4864	0·5309
8·2	·2266	·2514	·2774	·3053	·3346	·3660	·3998	·4364	·4763	·5203
8·3	·2230	·2473	·2729	·3001	·3288	·3594	·3923	·4279	·4667	·5092
8·4	·2196	·2435	·2685	·2952	·3232	·3531	·3852	·4199	·4575	·4987
8·5	·2163	·2397	·2643	·2904	·3179	·3471	·3784	·4122	·4488	·4888
8·6	0·2132	0·2362	0·2603	0·2859	0·3127	0·3413	0·3719	0·4048	0·4405	0·4793
8·7	·2101	·2327	·2564	·2812	·3078	·3358	·3657	·3978	·4325	·4702
8·8	·2072	·2294	·2526	·2770	·3031	·3305	·3597	·3910	·4248	·4616
8·9	·2044	·2262	·2490	·2730	·2985	·3253	·3539	·3845	·4175	·4533
9·0	·2016	·2231	·2455	·2691	·2941	·3204	·3484	·3783	·4105	·4454
9·2	0·1964	0·2172	0·2389	0·2617	0·2858	0·3111	0·3380	0·3666	0·3974	0·4305
9·4	·1915	·2117	·2328	·2548	·2778	·3025	·3283	·3558	·3852	·4169
9·6	—	—	—	·2483	·2706	·2944	·3193	·3457	·3739	·4042
9·8	—	—	—	·2422	·2639	·2868	·3109	·3363	·3634	·3925
10·0	—	—	—	·2365	·2575	·2798	·3030	·3275	·3537	·3816

Table A.4 Continued

β_2 \ $\sqrt{\beta_1}$	1·05	1·10	1·15	1·20	1·25	1·30	1·35	1·40	1·45	1·50
5·4	2·403	3·543								
5·5	2·099	2·874								
5·6	1·871	2·451								
5·7	1·692	2·149								
5·8	1·547	1·921	2·535	3·886						
5·9	1·428	1·741	2·224	3·124						
6·0	1·327	1·596	1·990	2·654						
6·1	1·241	1·475	1·805	2·325						
6·2	1·167	1·373	1·656	2·079	2·823	5·110				
6·3	1·102	1·285	1·531	1·885	2·459	3·712				
6·4	1·044	1·210	1·427	1·729	2·192	3·054				
6·5	0·9933	1·143	1·337	1·599	1·984	2·635				
6·6	0·9477	1·085	1·259	1·490	1·817	2·335				
6·7	·9066	1·032	1·190	1·396	1·679	2·106				
6·8	·8694	0·9857	1·130	1·315	1·563	1·924	2·525	3·929		
6·9	·8356	·9435	1·076	1·243	1·464	1·775	2·259	3·213		
7·0	·8046	·9053	1·028	1·180	1·379	1·650	2·055	2·768		
7·1	0·7762	0·8705	0·9840	1·124	1·303	1·544	1·890	2·455		
7·2	·7500	·8386	·9444	1·074	1·237	1·452	1·752	2·217		
7·3	·7258	·8093	·9083	1·028	1·178	1·372	1·636	2·029		
7·4	·7034	·7823	·8753	0·9871	1·125	1·301	1·537	1·875	2·430	3·669
7·5	·6825	·7573	·8450	·9495	1·077	1·238	1·450	1·746	2·202	3·088
7·6	0·6630	0·7342	0·8170	0·9151	1·034	1·182	1·374	1·636	2·024	2·705
7·7	·6448	·7126	·7910	·8834	·9945	1·132	1·307	1·540	1·877	2·426
7·8	·6278	·6924	·7670	·8542	·9584	1·086	1·246	1·457	1·752	2·211
7·9	·6117	·6736	·7445	·8272	·9251	1·044	1·192	1·384	1·646	2·037
8·0	·5967	·6559	·7236	·8020	·8945	1·006	1·143	1·319	1·554	1·893
8·1	0·5825	0·6393	0·7040	0·7786	0·8661	0·9706	1·099	1·260	1·473	1·772
8·2	·5690	·6237	·6857	·7568	·8397	·9382	1·058	1·207	1·401	1·667
8·3	·5563	·6089	·6684	·7364	·8152	·9083	1·020	1·159	1·337	1·576
8·4	·5443	·5950	·6521	·7172	·7923	·8805	0·9860	1·115	1·279	1·496
8·5	·5328	·5818	·6368	·6991	·7709	·8546	·9542	1·075	1·227	1·425
8·6	0·5220	0·5693	0·6223	0·6822	0·7507	0·8304	0·9246	1·038	1·180	1·361
8·7	·5099	·5574	·6085	·6661	·7318	·8078	·8972	1·004	1·136	1·304
8·8	·5018	·5461	·5955	·6510	·7140	·7866	·8716	0·9729	1·096	1·252
8·9	·4924	·5354	·5831	·6366	·6972	·7667	·8477	·9436	1·060	1·204
9·0	·4834	·5251	·5714	·6230	·6813	·7480	·8252	·9163	1·026	1·161
9·1	0·4748	0·5154	0·5601	0·6101	0·6663	0·7303	0·8042	0·8908	0·9943	1·121
9·2	·4666	·5060	·5494	·5978	·6520	·7136	·7843	·8669	·9651	1·084
9·3	·4587	·4971	·5393	·5861	·6385	·6977	·7656	·8445	·9377	1·050
9·4	·4511	·4885	·5295	·5749	·6256	·6827	·7480	·8234	·9122	1·019
9·5	·4439	·4803	·5202	·5642	·6133	·6685	·7312	·8036	·8882	0·9892
9·6	0·4369	0·4724	0·5112	0·5540	0·6016	0·6549	0·7154	0·7848	0·8657	0·9616
9·7	·4302	·4648	·5027	·5443	·5904	·6420	·7003	·7671	·8445	·9359
9·8	·4237	·4576	·4945	·5349	·5797	·6297	·6860	·7503	·8245	·9117
9·9	·4175	·4506	·4866	·5260	·5695	·6180	·6724	·7343	·8056	·8889
10·0	·4115	·4438	·4790	·5174	·5597	·6067	·6594	·7192	·7877	·8675
10·2	0·4001	0·4311	0·4646	0·5012	0·5413	0·5857	0·6352	0·6910	0·7546	0·8280
10·4	·3895	·4192	·4513	·4862	·5243	·5664	·6131	·6654	·7247	·7927
10·6	·3796	·4082	·4389	·4723	·5086	·5485	·5927	·6420	·6975	·7608
10·8	·3703	·3978	·4273	·4593	·4940	·5320	·5739	·6205	·6727	·7318
11·0	·3615	·3881	·4165	·4472	·4804	·5167	·5566	·6007	·6499	·7053
11·2	0·3533	0·3789	0·4063	0·4358	0·4678	0·5025	0·5404	0·5823	0·6289	0·6811
11·4	·3455	·3703	·3968	·4252	·4559	·4891	·5254	·5653	·6095	·6588
11·6	—	—	—	—	—	—	—	·5495	·5915	·6383
11·8	—	—	—	—	—	—	—	·5347	·5748	·6192
12·0	—	—	—	—	—	—	—	·5209	·5592	·6015

Table A.4 Continued

Values of $-\gamma$

β_2 \ $\sqrt{\beta_1}$	1·55	1·60	1·65	1·70	1·75	1·80	1·85	1·90	1·95	2·00
8·0	2·459	3·813								
8·1	2·240	3·192								
8·2	2·067	2·795								
8·3	1·924	2·510								
8·4	1·803	2·291								
8·5	1·698	2·115								
8·6	1·607	1·969	2·601	4·603						
8·7	1·527	1·846	2·365	3·582						
8·8	1·456	1·740	2·181	3·061						
8·9	1·392	1·647	2·030	2·719						
9·0	1·334	1·566	1·902	2·467						
9·1	1·282	1·493	1·793	2·269						
9·2	1·234	1·428	1·697	2·108						
9·3	1·190	1·370	1·613	1·973						
9·4	1·150	1·316	1·539	1·858	2·382	3·679				
9·5	1·112	1·268	1·472	1·758	2·208	3·147				
9·6	1·078	1·223	1·412	1·670	2·062	2·793				
9·7	1·046	1·182	1·357	1·592	1·938	2·535				
9·8	1·016	1·144	1·307	1·523	1·831	2·335				
9·9	0·9881	1·109	1·261	1·460	1·738	2·172				
10·0	·9619	1·076	1·219	1·403	1·656	2·036				
10·1	0·9374	1·046	1·180	1·351	1·582	1·920				
10·2	·9142	1·017	1·144	1·304	1·516	1·819	2·310	3·475		
10·3	·8924	0·9906	1·110	1·260	1·457	1·731	2·158	3·036		
10·4	·8718	·9655	1·079	1·220	1·402	1·652	2·028	2·726		
10·5	·8523	·9419	1·050	1·183	1·353	1·582	1·917	2·495		
10·6	0·8338	0·9196	1·022	1·148	1·307	1·518	1·820	2·313		
10·7	·8163	·8985	0·9963	1·115	1·265	1·461	1·734	2·163		
10·8	·7996	·8785	·9720	1·085	1·226	1·408	1·658	2·036		
10·9	·7837	·8596	·9490	1·057	1·190	1·360	1·590	1·927		
11·0	·7686	·8416	·9274	1·030	1·156	1·316	1·528	1·832	2·331	3·612
11·1	0·7541	0·8246	0·9069	1·005	1·124	1·274	1·472	1·748	2·185	3·131
11·2	·7403	·8083	·8874	0·9811	1·095	1·236	1·420	1·673	2·058	2·804
11·3	·7272	·7928	·8689	·9587	1·067	1·201	1·373	1·605	1·950	2·565
11·4	·7145	·7780	·8513	·9375	1·041	1·168	1·329	1·544	1·855	2·377
11·5	·7024	·7638	·8346	·9174	1·016	1·137	1·289	1·489	1·771	2·224
11·6	0·6907	0·7503	0·8186	0·8983	0·9928	1·108	1·251	1·438	1·696	2·096
11·7	·6796	·7373	·8034	·8801	·9708	1·080	1·216	1·391	1·629	1·985
11·8	·6688	·7248	·7888	·8628	·9499	1·054	1·183	1·347	1·568	1·889
11·9	·6585	·7129	·7749	·8463	·9300	1·030	1·152	1·307	1·512	1·804
12·0	·6486	·7014	·7615	·8306	·9112	1·007	1·124	1·270	1·461	1·728
12·1	0·6390	0·6904	0·7487	0·8155	0·8932	0·9851	1·096	1·235	1·414	1·660
12·2	·6297	·6798	·7364	·8011	·8761	·9644	1·071	1·202	1·370	1·598
12·3	·6208	·6696	·7246	·7873	·8597	·9447	1·046	1·171	1·330	1·542
12·4	·6122	·6598	·7132	·7740	·8441	·9260	1·024	1·143	1·293	1·490
12·5	·6039	·6503	·7023	·7613	·8291	·9081	1·002	1·115	1·258	1·443
12·6	0·5959	0·6411	0·6918	0·7491	0·8148	0·8910	0·9811	1·090	1·225	1·399
12·7	·5881	·6323	·6816	·7374	·8011	·8747	·9614	1·066	1·194	1·358
12·8	·5806	·6237	·6719	·7261	·7879	·8592	·9427	1·043	1·165	1·320
12·9	·5733	·6154	·6624	·7152	·7752	·8442	·9248	1·021	1·138	1·284
13·0	·5662	·6074	·6533	·7047	·7630	·8299	·9078	1·000	1·112	1·251
13·2	—	—	—	—	0·7400	0·8030	0·8758	0·9614	1·064	1·190
13·4	—	—	—	—	·7187	·7781	·8465	·9263	1·021	1·136
13·6	—	—	—	—	·6988	·7551	·8195	·8941	0·9822	1·088
13·8	—	—	—	—	·6802	·7336	·7945	·8646	·9466	1·045
14·0	—	—	—	—	·6628	·7136	·7712	·8373	·9141	1·005
14·2	—	—	—	—	0·6464	0·6949	0·7496	0·8121	0·8842	0·9690
14·4	—	—	—	—	·6311	·6774	·7295	·7886	·8566	·9359
14·6	—	—	—	—	·6166	·6609	·7106	·7668	·8310	·9055
14·8	—	—	—	—	·6029	·6454	·6929	·7464	·8072	·8774
15·0	—	—	—	—	·5900	·6308	·6763	·7273	·7851	·8514

Table A.4 Continued

Values of δ

β_2 \ $\sqrt{\beta_1}$	0.05	0.10	0.15	0.20	0.25	0.30	0.35	0.40	0.45	0.50
3·2	4·671	4·787	5·004	5·369	5·992	7·204				
3·3	3·866	3·927	4·036	4·208	4·469	4·875				
3·4	3·396	3·435	3·503	3·607	3·759	3·979	4·300	4·813		
3·5	3·081	3·108	3·156	3·227	3·328	3·467	3·663	3·943		
3·6	2·852	2·872	2·908	2·960	3·033	3·132	3·266	3·448	3·705	4·087
3·7	2·676	2·692	2·719	2·760	2·816	2·890	2·989	3·120	3·295	3·540
3·8	2·535	2·548	2·571	2·604	2·648	2·707	2·783	2·882	3·011	3·184
3·9	2·420	2·431	2·450	2·477	2·513	2·561	2·623	2·701	2·801	2·931
4·0	2·324	2·333	2·349	2·372	2·402	2·442	2·492	2·557	2·637	2·739
4·1	2·242	2·250	2·264	2·283	2·309	2·343	2·385	2·439	2·505	2·588
4·2	2·171	2·178	2·190	2·207	2·229	2·258	2·295	2·340	2·396	2·465
4·3	2·109	2·115	2·126	2·141	2·160	2·186	2·217	2·256	2·304	2·363
4·4	2·054	2·060	2·069	2·082	2·100	2·122	2·150	2·184	2·226	2·276
4·5	2·005	2·010	2·018	2·030	2·046	2·066	2·090	2·121	2·157	2·202
4·6	1·961	1·966	1·973	1·984	1·998	2·016	2·038	2·065	2·097	2·136
4·7	1·921	1·925	1·932	1·942	1·955	1·971	1·991	2·015	2·044	2·079
4·8	1·885	1·889	1·895	1·904	1·916	1·930	1·948	1·970	1·997	2·028
4·9	1·852	1·855	1·861	1·869	1·880	1·893	1·910	1·930	1·954	1·982
5·0	1·822	1·825	1·830	1·837	1·847	1·860	1·875	1·893	1·915	1·941
5·1	1·793	1·796	1·801	1·808	1·817	1·829	1·843	1·859	1·880	1·903
5·2	1·767	1·770	1·775	1·781	1·790	1·800	1·813	1·829	1·847	1·869
5·3	1·743	1·746	1·750	1·756	1·764	1·774	1·786	1·800	1·817	1·837
5·4	1·721	1·723	1·727	1·732	1·740	1·749	1·760	1·774	1·789	1·808
5·5	1·699	1·702	1·705	1·711	1·718	1·726	1·737	1·749	1·764	1·781
5·6	1·680	1·682	1·685	1·690	1·697	1·705	1·715	1·726	1·740	1·756
5·7	1·661	1·663	1·666	1·671	1·677	1·685	1·694	1·705	1·718	1·733
5·8	1·643	1·645	1·648	1·653	1·658	1·666	1·674	1·685	1·697	1·711
5·9	1·627	1·628	1·631	1·636	1·641	1·648	1·656	1·666	1·677	1·691
6·0	1·611	1·613	1·615	1·619	1·625	1·631	1·639	1·648	1·659	1·672
6·1	1·596	1·598	1·600	1·604	1·609	1·615	1·623	1·631	1·642	1·653
6·2	1·582	1·583	1·586	1·590	1·594	1·600	1·607	1·615	1·625	1·636
6·3	1·568	1·570	1·572	1·576	1·580	1·586	1·593	1·600	1·610	1·620
6·4	1·556	1·557	1·559	1·563	1·567	1·572	1·579	1·586	1·595	1·605
6·5	1·543	1·545	1·547	1·550	1·554	1·559	1·565	1·573	1·581	1·591
6·6	1·532	1·533	1·535	1·538	1·542	1·547	1·553	1·560	1·568	1·577
6·7	1·520	1·522	1·524	1·527	1·530	1·535	1·541	1·547	1·555	1·564
6·8	1·510	1·511	1·513	1·516	1·519	1·524	1·529	1·535	1·543	1·551
6·9	1·499	1·501	1·502	1·505	1·509	1·513	1·518	1·524	1·531	1·539
7·0	1·490	1·491	1·492	1·495	1·498	1·502	1·507	1·513	1·520	1·528
7·1	1·480	1·481	1·483	1·485	1·489	1·492	1·497	1·503	1·509	1·517
7·2	1·471	1·472	1·474	1·476	1·479	1·483	1·487	1·493	1·499	1·506
7·3	1·462	1·463	1·465	1·467	1·470	1·474	1·478	1·483	1·489	1·496
7·4	1·454	1·455	1·456	1·458	1·461	1·465	1·469	1·474	1·480	1·487
7·5	1·445	1·446	1·448	1·450	1·453	1·456	1·460	1·465	1·471	1·477
7·6	1·438	1·438	1·440	1·442	1·445	1·448	1·452	1·457	1·462	1·468
7·7	1·430	1·431	1·432	1·434	1·437	1·440	1·444	1·448	1·454	1·460
7·8	1·423	1·423	1·425	1·427	1·429	1·432	1·436	1·440	1·445	1·451
7·9	1·415	1·416	1·418	1·419	1·422	1·425	1·428	1·433	1·438	1·443
8·0	1·408	1·409	1·411	1·412	1·415	1·418	1·421	1·425	1·430	1·435
8·2	1·395	1·396	1·397	1·399	1·401	1·404	1·407	1·411	1·416	1·421
8·4	1·383	1·383	1·385	1·386	1·388	1·391	1·394	1·398	1·402	1·407
8·6	1·371	1·372	1·373	1·374	1·376	1·379	1·382	1·385	1·389	1·394
8·8	1·360	1·361	1·362	1·363	1·365	1·367	1·370	1·373	1·377	1·381
9·0	1·349	1·350	1·351	1·352	1·354	1·356	1·359	1·362	1·366	1·370
9·2	—	—	—	—	—	—	1·349	1·352	1·355	1·359
9·4	—	—	—	—	—	—	1·339	1·342	1·345	1·348

Table A.4 Continued

Values of δ

$\sqrt{\beta_1}$ / β_2	0·55	0·60	0·65	0·70	0·75	0·80	0·85	0·90	0·95	1·00
3·8	3·424	3·776								
3·9	3·105	3·346								
4·0	2·872	3·049	3·294	3·659						
4·1	2·694	2·830	3·013	3·269						
4·2	2·552	2·662	2·804	2·996						
4·3	2·436	2·526	2·641	2·791						
4·4	2·338	2·414	2·510	2·631						
4·5	2·255	2·320	2·401	2·502						
4·6	2·183	2·240	2·309	2·395	2·503	2·641	2·828	3·093		
4·7	2·120	2·170	2·231	2·304	2·395	2·511	2·662	2·868		
4·8	2·065	2·109	2·162	2·226	2·305	2·403	2·529	2·694		
4·9	2·015	2·055	2·100	2·159	2·227	2·312	2·418	2·555		
5·0	1·971	2·007	2·049	2·099	2·160	2·234	2·325	2·441	2·592	2·799
5·1	1·931	1·963	2·001	2·045	2·100	2·165	2·245	2·344	2·472	2·641
5·2	1·894	1·924	1·958	1·999	2·048	2·105	2·176	2·262	2·371	2·512
5·3	1·860	1·888	1·919	1·957	2·000	2·052	2·115	2·191	2·285	2·406
5·4	1·830	1·855	1·884	1·918	1·958	2·005	2·061	2·128	2·211	2·315
5·5	1·801	1·824	1·851	1·883	1·918	1·962	2·012	2·073	2·146	2·237
5·6	1·775	1·796	1·821	1·850	1·884	1·923	1·969	2·023	2·089	2·170
5·7	1·750	1·770	1·794	1·820	1·851	1·887	1·929	1·979	2·038	2·110
5·8	1·728	1·746	1·768	1·793	1·821	1·855	1·893	1·939	1·992	2·057
5·9	1·706	1·724	1·744	1·767	1·794	1·824	1·860	1·902	1·951	2·009
6·0	1·686	1·703	1·722	1·743	1·768	1·797	1·830	1·868	1·913	1·967
6·1	1·667	1·683	1·701	1·721	1·744	1·771	1·802	1·837	1·879	1·928
6·2	1·649	1·664	1·681	1·700	1·722	1·747	1·776	1·809	1·847	1·892
6·3	1·633	1·647	1·663	1·681	1·701	1·725	1·752	1·782	1·818	1·860
6·4	1·617	1·630	1·645	1·662	1·682	1·704	1·729	1·758	1·791	1·830
6·5	1·602	1·614	1·629	1·645	1·663	1·684	1·707	1·735	1·766	1·802
6·6	1·587	1·599	1·613	1·628	1·646	1·666	1·688	1·713	1·742	1·776
6·7	1·574	1·585	1·598	1·613	1·629	1·648	1·669	1·693	1·721	1·752
6·8	1·561	1·572	1·584	1·598	1·614	1·632	1·652	1·674	1·700	1·730
6·9	1·548	1·559	1·571	1·584	1·599	1·616	1·635	1·656	1·681	1·708
7·0	1·537	1·547	1·558	1·571	1·585	1·601	1·619	1·639	1·663	1·689
7·1	1·525	1·535	1·546	1·558	1·572	1·587	1·604	1·623	1·645	1·670
7·2	1·514	1·524	1·534	1·546	1·559	1·573	1·590	1·608	1·629	1·653
7·3	1·504	1·513	1·523	1·534	1·547	1·561	1·576	1·594	1·614	1·636
7·4	1·494	1·503	1·512	1·523	1·535	1·548	1·563	1·580	1·599	1·620
7·5	1·484	1·493	1·502	1·512	1·524	1·537	1·551	1·567	1·585	1·605
7·6	1·475	1·483	1·492	1·502	1·513	1·525	1·539	1·555	1·572	1·591
7·7	1·466	1·474	1·483	1·492	1·503	1·515	1·528	1·543	1·559	1·578
7·8	1·458	1·465	1·473	1·483	1·493	1·504	1·517	1·531	1·547	1·565
7·9	1·450	1·457	1·465	1·474	1·483	1·494	1·507	1·520	1·536	1·552
8·0	1·442	1·448	1·456	1·465	1·474	1·485	1·497	1·510	1·524	1·541
8·1	1·434	1·440	1·448	1·456	1·466	1·476	1·487	1·500	1·514	1·529
8·2	1·426	1·433	1·440	1·448	1·457	1·467	1·478	1·490	1·504	1·519
8·3	1·419	1·425	1·432	1·440	1·449	1·458	1·469	1·481	1·494	1·508
8·4	1·412	1·418	1·425	1·433	1·441	1·450	1·460	1·472	1·484	1·498
8·5	1·405	1·411	1·418	1·425	1·433	1·442	1·452	1·463	1·475	1·489
8·6	1·399	1·405	1·411	1·418	1·426	1·435	1·444	1·455	1·467	1·479
8·7	1·392	1·398	1·404	1·411	1·419	1·427	1·437	1·447	1·458	1·471
8·8	1·386	1·392	1·398	1·404	1·412	1·420	1·429	1·439	1·450	1·462
8·9	1·380	1·386	1·391	1·398	1·405	1·413	1·422	1·431	1·442	1·454
9·0	1·374	1·380	1·385	1·392	1·399	1·406	1·415	1·424	1·434	1·446
9·2	1·363	1·368	1·373	1·379	1·386	1·393	1·401	1·410	1·420	1·431
9·4	1·353	1·357	1·362	1·368	1·374	1·381	1·389	1·397	1·406	1·416
9·6	—	—	—	1·357	1·363	1·370	1·377	1·385	1·394	1·403
9·8	—	—	—	1·347	1·353	1·359	1·366	1·373	1·381	1·390
10·0	—	—	—	1·337	1·343	1·349	1·355	1·362	1·370	1·379

Table A.4 Continued

β_2 \ $\sqrt{\beta_1}$	1·05	1·10	1·15	1·20	1·25	1·30	1·35	1·40	1·45	1·50
5·4	2·450	2·632								
5·5	2·353	2·505								
5·6	2·270	2·400								
5·7	2·199	2·311								
5·8	2·136	2·234	2·362	2·530						
5·9	2·080	2·168	2·278	2·423						
6·0	2·031	2·109	2·206	2·331						
6·1	1·986	2·056	2·143	2·253						
6·2	1·945	2·009	2·087	2·184	2·309	2·476				
6·3	1·908	1·966	2·037	2·124	2·234	2·378				
6·4	1·875	1·928	1·992	2·070	2·168	2·294				
6·5	1·843	1·893	1·951	2·022	2·109	2·221				
6·6	1·815	1·860	1·914	1·978	2·057	2·156				
6·7	1·788	1·830	1·880	1·939	2·011	2·100				
6·8	1·763	1·803	1·849	1·903	1·969	2·049	2·151	2·281		
6·9	1·740	1·777	1·820	1·870	1·930	2·003	2·094	2·210		
7·0	1·719	1·753	1·793	1·840	1·895	1·962	2·044	2·148		
7·1	1·698	1·731	1·768	1·811	1·863	1·924	1·999	2·093		
7·2	1·679	1·710	1·745	1·785	1·833	1·890	1·958	2·043		
7·3	1·661	1·690	1·723	1·761	1·806	1·858	1·921	1·998		
7·4	1·644	1·671	1·703	1·738	1·780	1·829	1·887	1·958	2·046	2·157
7·5	1·628	1·654	1·683	1·717	1·756	1·802	1·856	1·921	2·001	2·102
7·6	1·613	1·637	1·665	1·697	1·734	1·776	1·827	1·887	1·960	2·051
7·7	1·598	1·622	1·648	1·678	1·713	1·753	1·800	1·856	1·923	2·006
7·8	1·584	1·607	1·632	1·660	1·693	1·731	1·775	1·827	1·889	1·965
7·9	1·571	1·593	1·616	1·644	1·675	1·710	1·751	1·800	1·858	1·928
8·0	1·559	1·579	1·602	1·628	1·657	1·691	1·730	1·775	1·829	1·894
8·1	1·547	1·566	1·588	1·613	1·640	1·672	1·709	1·752	1·802	1·862
8·2	1·535	1·554	1·575	1·598	1·625	1·655	1·690	1·730	1·777	1·833
8·3	1·524	1·542	1·562	1·585	1·610	1·639	1·671	1·709	1·754	1·806
8·4	1·514	1·531	1·550	1·571	1·596	1·623	1·654	1·690	1·732	1·781
8·5	1·504	1·520	1·538	1·559	1·582	1·608	1·638	1·672	1·711	1·758
8·6	1·494	1·510	1·527	1·547	1·569	1·594	1·623	1·655	1·692	1·736
8·7	1·484	1·500	1·517	1·536	1·557	1·581	1·608	1·639	1·674	1·715
8·8	1·475	1·490	1·507	1·525	1·545	1·568	1·594	1·623	1·657	1·695
8·9	1·467	1·481	1·497	1·514	1·534	1·556	1·580	1·608	1·640	1·677
9·0	1·458	1·472	1·487	1·504	1·523	1·544	1·568	1·594	1·625	1·660
9·1	1·450	1·463	1·478	1·495	1·513	1·533	1·556	1·581	1·610	1·643
9·2	1·442	1·455	1·469	1·485	1·503	1·522	1·544	1·568	1·596	1·628
9·3	1·435	1·447	1·461	1·476	1·493	1·512	1·533	1·556	1·583	1·613
9·4	1·427	1·440	1·453	1·468	1·484	1·502	1·522	1·545	1·570	1·599
9·5	1·420	1·432	1·445	1·459	1·475	1·492	1·512	1·534	1·558	1·586
9·6	1·413	1·425	1·437	1·451	1·466	1·483	1·502	1·523	1·546	1·573
9·7	1·407	1·418	1·430	1·443	1·458	1·474	1·492	1·513	1·535	1·560
9·8	1·400	1·411	1·423	1·436	1·450	1·466	1·483	1·503	1·524	1·549
9·9	1·394	1·404	1·416	1·428	1·442	1·458	1·474	1·493	1·514	1·538
10·0	1·388	1·398	1·409	1·421	1·435	1·450	1·466	1·484	1·504	1·527
10·1	1·382	1·392	1·403	1·414	1·428	1·442	1·458	1·475	1·495	1·516
10·2	1·376	1·386	1·396	1·408	1·420	1·434	1·450	1·467	1·485	1·506
10·3	1·371	1·380	1·390	1·401	1·414	1·427	1·442	1·458	1·477	1·497
10·4	1·365	1·374	1·384	1·395	1·407	1·420	1·435	1·450	1·468	1·488
10·5	1·360	1·369	1·378	1·389	1·401	1·413	1·427	1·443	1·460	1·479
10·6	1·355	1·363	1·373	1·383	1·394	1·407	1·420	1·435	1·452	1·470
10·7	1·349	1·358	1·367	1·377	1·388	1·400	1·414	1·428	1·444	1·462
10·8	1·345	1·353	1·362	1·372	1·382	1·394	1·407	1·421	1·437	1·454
10·9	1·340	1·348	1·357	1·366	1·377	1·388	1·401	1·414	1·429	1·446
11·0	1·335	1·343	1·352	1·361	1·371	1·382	1·394	1·408	1·422	1·439
11·2	1·326	1·334	1·342	1·351	1·360	1·371	1·383	1·395	1·409	1·424
11·4	1·318	1·325	1·332	1·341	1·350	1·360	1·371	1·383	1·396	1·411
11·6	—	—	—	—	—	—	—	1·372	1·384	1·398
11·8	—	—	—	—	—	—	—	1·361	1·373	1·386
12·0	—	—	—	—	—	—	—	1·351	1·363	1·375

Table A.4 Continued

Values of δ

β_2 \ $\sqrt{\beta_1}$	1·55	1·60	1·65	1·70	1·75	1·80	1·85	1·90	1·95	2·00
8·0	1·974	2·074								
8·1	1·936	2·028								
8·2	1·901	1·985								
8·3	1·869	1·947								
8·4	1·840	1·911								
8·5	1·812	1·879								
8·6	1·787	1·849	1·925	2·018						
8·7	1·763	1·821	1·891	1·978						
8·8	1·741	1·795	1·860	1·940						
8·9	1·720	1·771	1·831	1·906						
9·0	1·700	1·748	1·805	1·874						
9·1	1·682	1·727	1·780	1·844						
9·2	1·664	1·707	1·757	1·817						
9·3	1·648	1·688	1·735	1·792						
9·4	1·632	1·670	1·715	1·768	1·831	1·910				
9·5	1·617	1·653	1·696	1·745	1·805	1·879				
9·6	1·603	1·637	1·678	1·725	1·781	1·849				
9·7	1·589	1·622	1·660	1·705	1·758	1·822				
9·8	1·576	1·608	1·644	1·686	1·736	1·796				
9·9	1·564	1·594	1·629	1·669	1·716	1·772				
10·0	1·552	1·581	1·614	1·652	1·697	1·750				
10·1	1·541	1·569	1·600	1·636	1·679	1·729				
10·2	1·530	1·557	1·587	1·621	1·662	1·709	1·766	1·834		
10·3	1·520	1·545	1·574	1·607	1·646	1·691	1·744	1·809		
10·4	1·509	1·534	1·562	1·594	1·630	1·673	1·723	1·784		
10·5	1·500	1·523	1·550	1·581	1·616	1·656	1·704	1·761		
10·6	1·491	1·513	1·539	1·568	1·602	1·640	1·686	1·740		
10·7	1·482	1·504	1·528	1·556	1·588	1·625	1·669	1·720		
10·8	1·473	1·494	1·518	1·545	1·576	1·611	1·652	1·701		
10·9	1·465	1·485	1·508	1·534	1·564	1·597	1·637	1·683		
11·0	1·456	1·476	1·499	1·524	1·552	1·584	1·622	1·666	1·717	1·780
11·1	1·449	1·468	1·489	1·514	1·541	1·572	1·608	1·649	1·699	1·758
11·2	1·441	1·460	1·481	1·504	1·530	1·560	1·594	1·634	1·681	1·737
11·3	1·434	1·452	1·472	1·494	1·520	1·549	1·581	1·619	1·664	1·717
11·4	1·427	1·444	1·464	1·485	1·510	1·538	1·569	1·605	1·648	1·698
11·5	1·420	1·437	1·456	1·477	1·500	1·527	1·557	1·592	1·633	1·681
11·6	1·413	1·430	1·448	1·468	1·491	1·517	1·546	1·580	1·618	1·664
11·7	1·407	1·423	1·441	1·460	1·482	1·507	1·535	1·567	1·604	1·648
11·8	1·400	1·416	1·433	1·452	1·474	1·498	1·525	1·556	1·591	1·633
11·9	1·394	1·409	1·426	1·445	1·466	1·489	1·515	1·545	1·579	1·618
12·0	1·388	1·403	1·419	1·437	1·458	1·480	1·505	1·534	1·567	1·605
12·1	1·383	1·397	1·413	1·430	1·450	1·471	1·496	1·524	1·555	1·591
12·2	1·377	1·391	1·406	1·423	1·442	1·463	1·487	1·514	1·544	1·579
12·3	1·371	1·385	1·400	1·417	1·435	1·455	1·478	1·504	1·533	1·567
12·4	1·366	1·379	1·394	1·410	1·428	1·448	1·470	1·495	1·523	1·555
12·5	1·361	1·374	1·388	1·404	1·421	1·440	1·462	1·486	1·513	1·544
12·6	1·356	1·369	1·382	1·398	1·415	1·433	1·454	1·477	1·504	1·534
12·7	1·351	1·363	1·377	1·392	1·408	1·426	1·446	1·469	1·495	1·523
12·8	1·346	1·358	1·371	1·386	1·402	1·420	1·439	1·461	1·486	1·514
12·9	1·341	1·353	1·366	1·380	1·396	1·413	1·432	1·453	1·477	1·504
13·0	1·337	1·348	1·361	1·375	1·390	1·407	1·425	1·446	1·469	1·495
13·2	—	—	—	—	1·379	1·394	1·412	1·431	1·453	1·478
13·4	—	—	—	—	1·368	1·383	1·400	1·418	1·438	1·461
13·6	—	—	—	—	1·358	1·372	1·388	1·405	1·425	1·446
13·8	—	—	—	—	1·348	1·362	1·377	1·393	1·412	1·432
14·0	—	—	—	—	1·339	1·352	1·366	1·382	1·399	1·419
14·2	—	—	—	—	1·330	1·342	1·356	1·371	1·388	1·406
14·4	—	—	—	—	1·321	1·333	1·346	1·361	1·377	1·394
14·6	—	—	—	—	1·313	1·325	1·337	1·351	1·366	1·383
14·8	—	—	—	—	1·305	1·316	1·328	1·342	1·356	1·372
15·0	—	—	—	—	1·298	1·308	1·320	1·333	1·346	1·362

Table A.4 Continued

Table A.5 Percentage Points of Pearson Curves
Lower 1% points of the standardized deviate $(x_p - \mu)/\sigma$, (P = 0.01)
(Note that for positive skewness, i.e., $\mu_3 > 0$, the deviates in this table are negative.)

β_1 / β_2	0·00	0·01	0·03	0·05	0·10	0·15	0·20	0·30	0·40	0·50	0·60	0·70	0·80	0·90	1·00
1·8	1·70	—	—	—	—	—	—	—	—	—	—	—	—	—	—
2·0	1·87	1·77	1·69	1·62	—	—	—	—	—	—	—	—	—	—	—
2·2	2·01	1·91	1·83	1·76	1·64	1·55	1·45	—	—	—	—	—	—	—	—
2·4	2·12	2·03	1·95	1·89	1·77	1·68	1·59	1·43	—	—	—	—	—	—	—
2·6	2·21	2·12	2·05	1·99	1·88	1·79	1·70	1·55	1·41	—	—	—	—	—	—
2·8	2·27	2·19	2·13	2·08	1·98	1·89	1·81	1·66	1·52	1·39	—	—	—	—	—
3·0	2·33	2·25	2·19	2·14	2·05	1·97	1·90	1·76	1·62	1·50	1·38	—	—	—	—
3·2	2·37	2·29	2·24	2·19	2·11	2·03	1·96	1·84	1·71	1·59	1·48	1·37	1·26	—	—
3·4	2·40	2·33	2·28	2·24	2·16	2·09	2·02	1·90	1·79	1·68	1·57	1·46	1·36	1·26	—
3·6	2·43	2·36	2·31	2·27	2·20	2·13	2·07	1·96	1·86	1·76	1·65	1·55	1·45	1·35	1·26
3·8	2·45	2·39	2·34	2·30	2·23	2·17	2·11	2·01	1·91	1·82	1·72	1·62	1·53	1·43	1·34
4·0	2·47	2·41	2·36	2·33	2·26	2·20	2·15	2·05	1·96	1·87	1·78	1·69	1·60	1·51	1·42
4·2	2·49	2·43	2·38	2·35	2·28	2·23	2·18	2·09	2·00	1·92	1·83	1·75	1·66	1·58	1·49
4·4	2·50	2·44	2·40	2·37	2·31	2·25	2·21	2·12	2·04	1·96	1·88	1·80	1·72	1·64	1·56
4·6	2·51	2·46	2·42	2·38	2·32	2·27	2·23	2·15	2·07	2·00	1·92	1·84	1·77	1·70	1·62
4·8	2·52	2·47	2·43	2·40	2·34	2·29	2·25	2·17	2·10	2·03	1·96	1·88	1·81	1·74	1·67
5·0	2·53	2·48	2·44	2·41	2·36	2·31	2·27	2·19	2·12	2·06	1·99	1·92	1·85	1·79	1·72

Upper 1% points of the standardized deviate $(x_p - \mu)/\sigma$, (P = 0.99)

β_1 / β_2	0·00	0·01	0·03	0·05	0·10	0·20	0·20	0·30	0·40	0·50	0·60	0·70	0.80	0·90	1·00
1·8	1·70	—	—	—	—	—	—	—	—	—	—	—	—	—	—
2·0	1·87	1·95	2·00	2·03	—	—	—	—	—	—	—	—	—	—	—
2·2	2·01	2·10	2·15	2·18	2·22	2·24	2·25	—	—	—	—	—	—	—	—
2·4	2·12	2·20	2·25	2·28	2·33	2·36	2·38	2·40	—	—	—	—	—	—	—
2·6	2·21	2·28	2·33	2·36	2·42	2·45	2·48	2·51	2·52	—	—	—	—	—	—
2·8	2·27	2·34	2·39	2·43	2·48	2·52	2·55	2·59	2·61	2·63	—	—	—	—	—
3·0	2·33	2·40	2·44	2·48	2·53	2·56	2·59	2·64	2·68	2·70	2·71	—	—	—	—
3·2	2·37	2·44	2·48	2·51	2·56	2·60	2·63	2·68	2·72	2·75	2·77	2·79	2·80	—	—
3·4	2·40	2·47	2·51	2·54	2·59	2·63	2·66	2·71	2·75	2·79	2·82	2·84	2·86	2·87	—
3·6	2·43	2·49	2·53	2·56	2·61	2·65	2·68	2·74	2·78	2·81	2·85	2·87	2·90	2·91	2·93
3·8	2·45	2·51	2·55	2·58	2·63	2·67	2·70	2·75	2·80	2·83	2·87	2·90	2·92	2·95	2·97
4·0	2·47	2·53	2·57	2·60	2·65	2·68	2·71	2·77	2·81	2·85	2·88	2·91	2·94	2·97	2·99
4·2	2·49	2·54	2·58	2·61	2·66	2·69	2·73	2·78	2·82	2·86	2·89	2·92	2·95	2·98	3·01
4·4	2·50	2·56	2·59	2·62	2·67	2·70	2·73	2·78	2·83	2·86	2·90	2·93	2·96	2·99	3·02
4·6	2·51	2·57	2·60	2·63	2·68	2·71	2·74	2·79	2·83	2·87	2·90	2·94	2·97	3·00	3·03
4·8	2·52	2·58	2·61	2·64	2·68	2·72	2·75	2·80	2·84	2·87	2·91	2·94	2·97	3·00	3·03
5·0	2·53	2·58	2·62	2·64	2·69	2·72	2·75	2·80	2·84	2·88	2·91	2·95	2·97	3·00	3·03

Source: E. S. Pearson and H. O. Hartley, *Biometrika Tables for Statisticians*, Vol. 1, 3rd ed., Cambridge University Press, London, 1966, Table 42.

Lower 2.5% points of the standardized deviate $(x_p - \mu)/\sigma$, (P = 0.025)
(Note that for positive skewness, i.e., $\mu_3 > 0$, the deviates in
this table are negative.)

β_2 \ β_1	0·00	0·01	0·03	0·05	0·10	0·15	0·20	0·30	0·40	0·50	0·60	0·70	0·80	0·90	1·00
1·8	1·65	—	—	—	—	—	—	—	—	—	—	—	—	—	—
2·0	1·76	1·68	1·62	1·56	—	—	—	—	—	—	—	—	—	—	—
2·2	1·83	1·76	1·71	1·66	1·57	1·49	1·41	—	—	—	—	—	—	—	—
2·4	1·88	1·82	1·77	1·73	1·65	1·58	1·51	1·39	—	—	—	—	—	—	—
2·6	1·92	1·86	1·82	1·78	1·71	1·64	1·58	1·47	1·37	—	—	—	—	—	—
2·8	1·94	1·89	1·85	1·82	1·76	1·70	1·65	1·55	1·45	1·35	—	—	—	—	—
3·0	1·96	1·91	1·87	1·84	1·79	1·74	1·69	1·60	1·52	1·42	1·33	—	—	—	—
3·2	1·97	1·93	1·89	1·86	1·81	1·77	1·72	1·65	1·57	1·49	1·40	1·32	1·24	—	—
3·4	1·98	1·94	1·90	1·88	1·83	1·79	1·75	1·68	1·61	1·54	1·46	1·39	1·31	1·23	—
3·6	1·99	1·95	1·91	1·89	1·85	1·81	1·77	1·71	1·65	1·58	1·51	1·44	1·38	1·30	1·23
3·8	1·99	1·95	1·92	1·90	1·86	1·82	1·79	1·73	1·67	1·62	1·56	1·49	1·43	1·36	1·29
4·0	1·99	1·96	1·93	1·91	1·87	1·84	1·81	1·75	1·70	1·64	1·59	1·53	1·47	1·41	1·35
4·2	2·00	1·96	1·93	1·91	1·88	1·84	1·82	1·76	1·72	1·67	1·62	1·56	1·51	1·45	1·40
4·4	2·00	1·96	1·94	1·92	1·88	1·85	1·83	1·78	1·73	1·69	1·64	1·59	1·54	1·49	1·44
4·6	2·00	1·96	1·94	1·92	1·89	1·86	1·83	1·79	1·75	1·70	1·66	1·62	1·57	1·52	1·47
4·8	2·00	1·97	1·94	1·93	1·89	1·87	1·84	1·80	1·76	1·72	1·68	1·64	1·59	1·55	1·50
5·0	2·00	1·97	1·94	1·93	1·90	1·87	1·85	1·81	1·77	1·73	1·69	1·65	1·61	1·57	1·53

Upper 2.5% points of the standardized deviate $(x_p - \mu)/\sigma$, (P = 0.975)

β_2 \ β_1	0·00	0·01	0·03	0·05	0·10	0·15	0·20	0·30	0·40	0·50	0·60	0·70	0·80	0·90	1·00
1·8	1·65	—	—	—	—	—	—	—	—	—	—	—	—	—	—
2·0	1·76	1·82	1·86	1·89	—	—	—	—	—	—	—	—	—	—	—
2·2	1·83	1·89	1·93	1·96	2·00	2·04	2·06	—	—	—	—	—	—	—	—
2·4	1·88	1·94	1·98	2·01	2·05	2·08	2·11	2·15	—	—	—	—	—	—	—
2·6	1·92	1·97	2·01	2·03	2·08	2·11	2·14	2·18	2·22	—	—	—	—	—	—
2·8	1·94	1·99	2·03	2·05	2·09	2·13	2·15	2·20	2·24	2·27	—	—	—	—	—
3·0	1·96	2·01	2·04	2·06	2·10	2·13	2·16	2·21	2·25	2·28	2·32	—	—	—	—
3·2	1·97	2·02	2·05	2·07	2·11	2·14	2·16	2·21	2·25	2·29	2·32	2·35	2·38	—	—
3·4	1·98	2·02	2·05	2·07	2·11	2·14	2·16	2·21	2·25	2·28	2·32	2·35	2·38	2·41	—
3·6	1·99	2·02	2·05	2·07	2·11	2·14	2·16	2·20	2·24	2·28	2·31	2·34	2·37	2·41	2·44
3·8	1·99	2·03	2·05	2·07	2·11	2·13	2·16	2·20	2·24	2·27	2·30	2·33	2·36	2·40	2·43
4·0	1·99	2·03	2·05	2·07	2·11	2·13	2·15	2·19	2·23	2·26	2·29	2·32	2·35	2·38	2·42
4·2	2·00	2·03	2·05	2·07	2·10	2·13	2·15	2·19	2·22	2·25	2·28	2·31	2·34	2·37	2·40
4·4	2·00	2·03	2·05	2·07	2·10	2·13	2·15	2·18	2·22	2·25	2·28	2·31	2·33	2·36	2·39
4·6	2·00	2·03	2·05	2·07	2·10	2·12	2·14	2·18	2·21	2·24	2·27	2·30	2·32	2·35	2·38
4·8	2·00	2·03	2·05	2·07	2·10	2·12	2·14	2·17	2·21	2·23	2·26	2·29	2·31	2·34	2·37
5·0	2·00	2·03	2·05	2·07	2·09	2·12	2·14	2·17	2·20	2·23	2·25	2·28	2·30	2·33	2·35

Table A.5 Continued

Table A.6 Coefficients$\{\alpha_{n-i+1}\}$ for the W Test for Normality, for n = 2(1)50

i \ n	2	3	4	5	6	7	8	9	10	
1	0·7071	0·7071	0·6872	0·6646	0·6431	0·6233	0·6052	0·5888	0·5739	
2	—	·0000	·1677	·2413	·2806	·3031	·3164	·3244	·3291	
3	—	—	—	·0000	·0875	·1401	·1743	·1976	·2141	
4	—	—	—	—	—	·0000	·0561	·0947	·1224	
5	—	—	—	—	—	—	—	·0000	·0399	

i \ n	11	12	13	14	15	16	17	18	19	20
1	0·5601	0·5475	0·5359	0·5251	0·5150	0·5056	0·4968	0·4886	0·4808	0·4734
2	·3315	·3325	·3325	·3318	·3306	·3290	·3273	·3253	·3232	·3211
3	·2260	·2347	·2412	·2460	·2495	·2521	·2540	·2553	·2561	·2565
4	·1429	·1586	·1707	·1802	·1878	·1939	·1988	·2027	·2059	·2085
5	·0695	·0922	·1099	·1240	·1353	·1447	·1524	·1587	·1641	·1686
6	0·0000	0·0303	0·0539	0·0727	0·0880	0·1005	0·1109	0·1197	0·1271	0·1334
7	—	—	·0000	·0240	·0433	·0593	·0725	·0837	·0932	·1013
8	—	—	—	—	·0000	·0196	·0359	·0496	·0612	·0711
9	—	—	—	—	—	—	·0000	·0163	·0303	·0422
10	—	—	—	—	—	—	—	—	·0000	·0140

i \ n	21	22	23	24	25	26	27	28	29	30
1	0·4643	0·4590	0·4542	0·4493	0·4450	0·4407	0·4366	0·4328	0·4291	0·4254
2	·3185	·3156	·3126	·3098	·3069	·3043	·3018	·2992	·2968	·2944
3	·2578	·2571	·2563	·2554	·2543	·2533	·2522	·2510	·2499	·2487
4	·2119	·2131	·2139	·2145	·2148	·2151	·2152	·2151	·2150	·2148
5	·1736	·1764	·1787	·1807	·1822	·1836	·1848	·1857	·1864	·1870
6	0·1399	0·1443	0·1480	0·1512	0·1539	0·1563	0·1584	0·1601	0·1616	0·1630
7	·1092	·1150	·1201	·1245	·1283	·1316	·1346	·1372	·1395	·1415
8	·0804	·0878	·0941	·0997	·1046	·1089	·1128	·1162	·1192	·1219
9	·0530	·0618	·0696	·0764	·0823	·0876	·0923	·0965	·1002	·1036
10	·0263	·0368	·0459	·0539	·0610	·0672	·0728	·0778	·0822	·0862
11	0·0000	0·0122	0·0228	0·0321	0·0403	0·0476	0·0540	0·0598	0·0650	0·0697
12	—	—	·0000	·0107	·0200	·0284	·0358	·0424	·0483	·0537
13	—	—	—	—	·0000	·0094	·0178	·0253	·0320	·0381
14	—	—	—	—	—	—	·0000	·0084	·0159	·0227
15	—	—	—	—	—	—	—	—	·0000	·0076

Source: S. Shapiro and M. Wilk, Analysis of variance test for normality, *Biometrika 52*, 1965.

i \ n	31	32	33	34	35	36	37	38	39	40
1	0·4220	0·4188	0·4156	0·4127	0·4096	0·4068	0·4040	0·4015	0·3989	0·3964
2	·2921	·2898	·2876	·2854	·2834	·2813	·2794	·2774	·2755	·2737
3	·2475	·2463	·2451	·2439	·2427	·2415	·2403	·2391	·2380	·2368
4	·2145	·2141	·2137	·2132	·2127	·2121	·2116	·2110	·2104	·2098
5	·1874	·1878	·1880	·1882	·1883	·1883	·1883	·1881	·1880	·1878
6	0·1641	0·1651	0·1660	0·1667	0·1673	0·1678	0·1683	0·1686	0·1689	0·1691
7	·1433	·1449	·1463	·1475	·1487	·1496	·1505	·1513	·1520	·1526
8	·1243	·1265	·1284	·1301	·1317	·1331	·1344	·1356	·1366	·1376
9	·1066	·1093	·1118	·1140	·1160	·1179	·1196	·1211	·1225	·1237
10	·0899	·0931	·0961	·0988	·1013	·1036	·1056	·1075	·1092	·1108
11	0·0739	0·0777	0·0812	0·0844	0·0873	0·0900	0·0924	0·0947	0·0967	0·0986
12	·0585	·0629	·0669	·0706	·0739	·0770	·0798	·0824	·0848	·0870
13	·0435	·0485	·0530	·0572	·0610	·0645	·0677	·0706	·0733	·0759
14	·0289	·0344	·0395	·0441	·0484	·0523	·0559	·0592	·0622	·0651
15	·0144	·0206	·0262	·0314	·0361	·0404	·0444	·0481	·0515	·0546
16	0·0000	0·0068	0·0131	0·0187	0·0239	0·0287	0·0331	0·0372	0·0409	0·0444
17	—	—	·0000	·0062	·0119	·0172	·0220	·0264	·0305	·0343
18	—	—	—	—	·0000	·0057	·0110	·0158	·0203	·0244
19	—	—	—	—	—	—	·0000	·0053	·0101	·0146
20	—	—	—	—	—	—	—	—	·0000	·0049

i \ n	41	42	43	44	45	46	47	48	49	50
1	0·3940	0·3917	0·3894	0·3872	0·3850	0·3830	0·3808	0·3789	0·3770	0·3751
2	·2719	·2701	·2684	·2667	·2651	·2635	·2620	·2604	·2589	·2574
3	·2357	·2345	·2334	·2323	·2313	·2302	·2291	·2281	·2271	·2260
4	·2091	·2085	·2078	·2072	·2065	·2058	·2052	·2045	·2038	·2032
5	·1876	·1874	·1871	·1868	·1865	·1862	·1859	·1855	·1851	·1847
6	0·1693	0·1694	0·1695	0·1695	0·1695	0·1695	0·1695	0·1693	0·1692	0·1691
7	·1531	·1535	·1539	·1542	·1545	·1548	·1550	·1551	·1553	·1554
8	·1384	·1392	·1398	·1405	·1410	·1415	·1420	·1423	·1427	·1430
9	·1249	·1259	·1269	·1278	·1286	·1293	·1300	·1306	·1312	·1317
10	·1123	·1136	·1149	·1160	·1170	·1180	·1189	·1197	·1205	·1212
11	0·1004	0·1020	0·1035	0·1049	0·1062	0·1073	0·1085	0·1095	0·1105	0·1113
12	·0891	·0909	·0927	·0943	·0959	·0972	·0986	·0998	·1010	·1020
13	·0782	·0804	·0824	·0842	·0860	·0876	·0892	·0906	·0919	·0932
14	·0677	·0701	·0724	·0745	·0765	·0783	·0801	·0817	·0832	·0846
15	·0575	·0602	·0628	·0651	·0673	·0694	·0713	·0731	·0748	·0764
16	0·0476	0·0506	0·0534	0·0560	0·0584	0·0607	0·0628	0·0648	0·0667	0·0685
17	·0379	·0411	·0442	·0471	·0497	·0522	·0546	·0568	·0588	·0608
18	·0283	·0318	·0352	·0383	·0412	·0439	·0465	·0489	·0511	·0532
19	·0188	·0227	·0263	·0296	·0328	·0357	·0385	·0411	·0436	·0459
20	·0094	·0136	·0175	·0211	·0245	·0277	·0307	·0335	·0361	·0386
21	0·0000	0·0045	0·0087	0·0126	0·0163	0·0197	0·0229	0·0259	0·0288	0·0314
22		—	·0000	·0042	·0081	·0118	·0153	·0185	·0215	·0244
23	—	—	—	—	·0000	·0039	·0076	·0111	·0143	·0174
24	—	—	—	—	—	—	·0000	·0037	·0071	·0104
25	—	—	—	—	—	—	—	—	·0000	·0035

Table A.6 Continued

Table A.7 Percentage Points of the W test for n = 3(1)50

Level

n	0·01	0·02	0·05	0·10	0·50	0·90	0·95	0·98	0·99
3	0·753	0·756	0·767	0·789	0·959	0·998	0·999	1·000	1·000
4	·687	·707	·748	·792	·935	·987	·992	ι·996	·997
5	·686	·715	·762	·806	·927	·979	·986	·991	·993
6	0·713	0·743	0·788	0·826	0·927	0·974	0·981	0·986	0·989
7	·730	·760	·803	·838	·928	·972	·979	·985	·988
8	·749	·778	·818	·851	·932	·972	·978	·984	·987
9	·764	·791	·829	·859	·935	·972	·978	·984	·986
10	·781	·806	·842	·869	·938	·972	·978	·983	·986
11	0·792	0·817	0·850	0·876	0·940	0·973	0·979	0·984	0·986
12	·805	·828	·859	·883	·943	·973	·979	·984	·986
13	·814	·837	·866	·889	·945	·974	·979	·984	·986
14	·825	·846	·874	·895	·947	·975	·980	·984	·986
15	·835	·855	·881	·901	·950	·975	·980	·984	·987
16	0·844	0·863	0·887	0·906	0·952	0·976	0·981	0·985	0·987
17	·851	·869	·892	·910	·954	·977	·981	·985	·987
18	·858	·874	·897	·914	·956	·978	·982	·986	·988
19	·863	·879	·901	·917	·957	·978	·982	·986	·988
20	·868	·884	·905	·920	·959	·979	·983	·986	·988
21	0·873	0·888	0·908	0·923	0·960	0·980	0·983	0·987	0·989
22	·878	·892	·911	·926	·961	·980	·984	·987	·989
23	·881	·895	·914	·928	·962	·981	·984	·987	·989
24	·884	·898	·916	·930	·963	·981	·984	·987	·989
25	·888	·901	·918	·931	·964	·981	·985	·988	·989
26	0·891	0·904	0·920	0·933	0·065	0·982	0·985	0·988	0·989
27	·894	·906	·923	·935	·965	·982	·985	·988	·990
28	·896	·908	·924	·936	·966	·982	·985	·988	·990
29	·898	·910	·926	·937	·966	·982	·985	·988	·990
30	·900	·912	·927	·939	·967	·983	·985	·988	·900
31	0·902	0·914	0·929	0·940	0·967	0·983	0·986	0·988	0·990
32	·904	·915	·930	·941	·968	·983	·986	·988	·990
33	·906	·917	·931	·942	·968	·983	·986	·989	·990
34	·908	·919	·933	·943	·969	·983	·986	·989	·990
35	·910	·920	·934	·944	·969	·984	·986	·989	·990
36	0·912	0·922	0·935	0·945	0·970	0·984	0·986	0·989	0·990
37	·914	·924	·936	·946	·970	·984	·987	·989	·990
38	·916	·925	·938	·947	·971	·984	·987	·989	·990
39	·917	·927	·939	·948	·971	·984	·987	·989	·991
40	·919	·928	·940	·949	·972	·985	·987	·989	·991
41	0·920	0·929	0·941	0·950	0·972	0·985	0·987	0·989	0·991
42	·922	·930	·942	·951	·972	·985	·987	·989	·991
43	·923	·932	·943	·951	·973	·985	·987	·990	·991
44	·924	·933	·944	·952	·973	·985	·987	·990	·991
45	·926	·934	·945	·953	·973	·985	·988	·990	·991
46	0·927	0·935	0·945	0·953	0·974	0·985	0·988	0·990	0·991
47	·928	·936	·946	·954	·974	·985	·988	·990	·991
48	·929	·937	·947	·954	·974	·985	·988	·990	·991
49	·929	·937	·947	·955	·974	·985	·988	·990	·991
50	·930	·938	·947	·955	·974	·985	·988	·990	·991

Source: S. Shapiro and M. Wilk, Analysis of variance test for normality, *Biometrika 52*, 1965.

Table A.8 Percentage Points of W-Exponential

n	.005	.01	.025	.05	.10	.50	.90	.95	.975	.99	.995
3	.2519	.2538	.2596	.2697	.2915	.5714	.9709	.9926	.9981	.9997	.99993
4	.1241	.1302	.1434	.1604	.1891	.3768	.7514	.8581	.9236	.9680	.9837
5	.0845	.0905	.1048	.1187	.1442	.2875	.5547	.6682	.7590	.8600	.9192
6	.0610	.0665	.0802	.0956	.1173	.2276	.4292	.5089	.5842	.6775	.7501
7	.0514	.0591	.0700	.0810	.0986	.1874	.3474	.4162	.4852	.5706	.6426
8	.0454	.0512	.0614	.0710	.0852	.1625	.2934	.3497	.4033	.4848	.5428
9	.0404	.0442	.0537	.0633	.0751	.1415	.2553	.3005	.3454	.4015	.4433
10	.0369	.0404	.0487	.0568	.0678	.1225	.2178	.2525	.2879	.3391	.3701
11	.0339	.0380	.0447	.0528	.0616	.1112	.1934	.2265	.2619	.3039	.3314
12	.0311	.0358	.0410	.0494	.0567	.1009	.1723	.2019	.2364	.2716	.2978
13	.0287	.0337	.0382	.0460	.0528	.0925	.1563	.1829	.2113	.2422	.2642
14	.0265	.0317	.0362	.0428	.0496	.0847	.1417	.1647	.1862	.2131	.2315
15	.0247	.0298	.0344	.0398	.0466	.0778	.1285	.1485	.1669	.1926	.2123
16	.0233	.0280	.0326	.0374	.0438	.0728	.1187	.1355	.1542	.1770	.1931
17	.0222	.0264	.0310	.0352	.0412	.0684	.1099	.1257	.1423	.1614	.1794
18	.0212	.0250	.0294	.0332	.0388	.0640	.1015	.1164	.1311	.1483	.1668
19	.0203	.0238	.0278	.0314	.0368	.0600	.0935	.1071	.1199	.1374	.1452
20	.0196	.0227	.0264	.0302	.0352	.0570	.0884	.1002	.1121	.1286	.1369
21	.0190	.0217	.0250	.0290	.0337	.0540	.0839	.0948	.1054	.1198	.1288
22	.0185	.0208	.0238	.0278	.0323	.0516	.0794	.0894	.0988	.1118	.1213
23	.0181	.0201	.0230	.0266	.0310	.0492	.0749	.0836	.0933	.1043	.1142
24	.0177	.0194	.0224	.0256	.0298	.0468	.0704	.0788	.0882	.0984	.1071
25	.0173	.0188	.0218	.0248	.0286	.0447	.0668	.0749	.0836	.0927	.1000
26	.0169	.0182	.0213	.0240	.0274	.0426	.0636	.0712	.0791	.0885	.0948
27	.0165	.0177	.0208	.0232	.0264	.0407	.0606	.0678	.0747	.0843	.0896
28	.0161	.0172	.0203	.0225	.0256	.0391	.0576	.0649	.0706	.0801	.0859
29	.0157	.0168	.0198	.0219	.0249	.0377	.0555	.0621	.0671	.0759	.0822
30	.0153	.0164	.0193	.0213	.0242	.0364	.0536	.0593	.0643	.0719	.0786
31	.0149	.0160	.0188	.0207	.0235	.0352	.0518	.0569	.0615	.0686	.0753
32	.0145	.0156	.0183	.0201	.0229	.0340	.0491	.0547	.0591	.0661	.0722
33	.0141	.0152	.0178	.0195	.0223	.0329	.0475	.0527	.0573	.0636	.0691
34	.0137	.0148	.0173	.0190	.0217	.0319	.0459	.0507	.0555	.0611	.0660
35	.0133	.0144	.0168	.0185	.0211	.0309	.0444	.0488	.0537	.0588	.0639
36	.0129	.0141	.0164	.0180	.0205	.0300	.0429	.0470	.0519	.0567	.0608

Source: S. Shapiro and M. Wilk, An analysis of variance test for the exponential distribution, *Technometrics 14*, 1972.

n	.005	.01	.025	.05	.10	.50	.90	.95	.975	.99	.995
37	.0125	.0138	.0160	.0176	.0200	.0291	.0414	.0454	.0501	.0546	.0578
38	.0122	.0135	.0156	.0172	.0195	.0283	.0400	.0440	.0483	.0525	.0553
39	.0120	.0133	.0152	.0168	.0190	.0275	.0386	.0426	.0465	.0512	.0531
40	.0118	.0131	.0148	.0164	.0186	.0267	.0375	.0414	.0447	.0499	.0510
41	.0116	.0129	.0144	.0161	.0182	.0260	.0364	.0402	.0430	.0476	.0493
42	.0114	.0127	.0140	.0158	.0178	.0253	.0355	.0389	.0417	.0464	.0482
43	.0112	.0125	.0137	.0155	.0174	.0248	.0346	.0379	.0405	.0452	.0471
44	.0110	.0123	.0134	.0152	.0170	.0243	.0338	.0369	.0394	.0440	.0460
45	.0108	.0121	.0131	.0149	.0166	.0238	.0329	.0359	.0385	.0423	.0449
46	.0106	.0119	.0129	.0146	.0162	.0233	.0320	.0349	.0376	.0416	.0438
47	.0104	.0117	.0127	.0143	.0158	.0228	.0311	.0340	.0367	.0394	.0427
48	.0103	.0115	.0125	.0141	.0155	.0223	.0303	.0332	.0358	.0382	.0416
49	.0102	.0113	.0123	.0139	.0152	.0218	.0295	.0324	.0349	.0371	.0405
50	.0101	.0111	.0122	.0137	.0149	.0213	.0288	.0317	.0340	.0360	.0394
51	.0100	.0109	.0120	.0135	.0147	.0209	.0282	.0310	.0331	.0349	.0383
52	.0099	.0107	.0119	.0133	.0145	.0205	.0276	.0303	.0323	.0341	.0373
53	.0097	.0106	.0118	.0131	.0143	.0201	.0270	.0296	.0315	.0332	.0363
54	.0095	.0104	.0116	.0129	.0141	.0197	.0264	.0289	.0307	.0329	.0353
55	.0094	.0103	.0115	.0127	.0139	.0193	.0258	.0282	.0299	.0321	.0343
56	.0093	.0102	.0113	.0125	.0137	.0189	.0252	.0275	.0292	.0313	.0333
57	.0092	.0101	.0112	.0123	.0135	.0185	.0247	.0268	.0285	.0306	.0324
58	.0091	.0100	.0110	.0121	.0133	.0182	.0242	.0262	.0279	.0301	.0318
59	.0090	.0098	.0109	.0119	.0131	.0179	.0238	.0257	.0274	.0296	.0312
60	.0089	.0095	.0108	.0117	.0129	.0176	.0234	.0252	.0270	.0291	.0306
61	.0088	.0093	.0107	.0115	.0127	.0173	.0230	.0247	.0266	.0286	.0301
62	.0087	.0092	.0105	.0113	.0125	.0170	.0226	.0242	.0262	.0281	.0296
63	.0086	.0091	.0104	.0112	.0123	.0167	.0222	.0238	.0257	.0276	.0291
64	.0085	.0090	.0102	.0111	.0121	.0164	.0218	.0234	.0252	.0271	.0286
65	.0084	.0089	.0101	.0109	.0119	.0161	.0215	.0230	.0247	.0266	.0281
66	.0082	.0088	.0099	.0108	.0117	.0159	.0211	.0225	.0242	.0261	.0276
67	.0081	.0087	.0098	.0107	.0115	.0157	.0207	.0221	.0237	.0256	.0271
68	.0080	.0086	.0096	.0105	.0114	.0155	.0204	.0217	.0232	.0251	.0266
69	.0079	.0085	.0095	.0104	.0113	.0152	.0198	.0213	.0227	.0246	.0261
70	.0078	.0084	.0094	.0103	.0111	.0150	.0194	.0209	.0222	.0241	.0256
71	.0077	.0083	.0093	.0102	.0109	.0147	.0191	.0205	.0218	.0237	.0251
72	.0076	.0082	.0092	.0101	.0108	.0145	.0188	.0201	.0214	.0232	.0246

n	.005	.01	.025	.05	.10	.50	.90	.95	.975	.99	.995
73	.0075	.0081	.0091	.0100	.0107	.0143	.0185	.0198	.0211	.0228	.0241
74	.0074	.0080	.0090	.0098	.0106	.0141	.0182	.0195	.0208	.0224	.0236
75	.0073	.0079	.0089	.0097	.0105	.0139	.0179	.0192	.0205	.0220	.0231
76	.0073	.0078	.0088	.0096	.0104	.0137	.0176	.0189	.0202	.0217	.0227
77	.0072	.0077	.0087	.0095	.0103	.0135	.0173	.0186	.0199	.0214	.0223
78	.0071	.0077	.0086	.0093	.0101	.0134	.0170	.0183	.0196	.0211	.0219
79	.0070	.0076	.0085	.0092	.0100	.0132	.0168	.0180	.0193	.0208	.0215
80	.0070	.0075	.0084	.0091	.0099	.0131	.0166	.0177	.0190	.0205	.0211
81	.0069	.0074	.0083	.0090	.0098	.0129	.0164	.0175	.0187	.0202	.0207
82	.0068	.0074	.0082	.0088	.0097	.0128	.0162	.0173	.0184	.0199	.0203
83	.0067	.0073	.0081	.0087	.0096	.0126	.0160	.0170	.0181	.0196	.0199
84	.0067	.0073	.0080	.0086	.0095	.0125	.0158	.0168	.0178	.0193	.0196
85	.0066	.0072	.0079	.0085	.0094	.0123	.0156	.0166	.0174	.0190	.0193
86	.0066	.0071	.0078	.0085	.0093	.0122	.0154	.0164	.0172	.0187	.0190
87	.0065	.0071	.0077	.0084	.0092	.0120	.0152	.0162	.0170	.0184	.0187
88	.0065	.0070	.0077	.0084	.0091	.0119	.0150	.0160	.0168	.0181	.0185
89	.0064	.0070	.0076	.0083	.0090	.0117	.0148	.0158	.0166	.0179	.0183
90	.0064	.0069	.0075	.0082	.0089	.0116	.0147	.0156	.0164	.0176	.0181
91	.0063	.0068	.0075	.0082	.0088	.0114	.0145	.0154	.0162	.0173	.0179
92	.0063	.0068	.0074	.0081	.0087	.0113	.0143	.0153	.0160	.0171	.0177
93	.0062	.0067	.0073	.0081	.0086	.0112	.0141	.0151	.0158	.0168	.0175
94	.0062	.0067	.0073	.0080	.0085	.0110	.0139	.0149	.0156	.0165	.0173
95	.0061	.0066	.0072	.0079	.0084	.0109	.0138	.0147	.0154	.0163	.0171
96	.0061	.0065	.0072	.0078	.0083	.0108	.0136	.0145	.0153	.0161	.0169
97	.0060	.0065	.0071	.0077	.0082	.0107	.0134	.0143	.0152	.0159	.0167
98	.0060	.0064	.0070	.0076	.0081	.0105	.0133	.0142	.0151	.0157	.0165
99	.0059	.0064	.0070	.0075	.0080	.0104	.0132	.0140	.0150	.0155	.0163
100	.0059	.0063	.0069	.0074	.0079	.0103	.0131	.0139	.0149	.0153	.0161

Table A.8 Continued

Table A.9 Percentiles for Kolmogorov-Smirnov Test

Sample Size (N)	Percentile		
	0.90	0.95	0.99
1	0.950	0.975	0.995
2	0.776	0.842	0.929
3	0.642	0.708	0.828
4	0.564	0.624	0.733
5	0.510	0.565	0.669
6	0.470	0.521	0.618
7	0.438	0.486	0.577
8	0.411	0.457	0.543
9	0.388	0.432	0.514
10	0.368	0.410	0.490
11	0.352	0.391	0.468
12	0.338	0.375	0.450
13	0.325	0.361	0.433
14	0.314	0.349	0.418
15	0.304	0.338	0.404
16	0.295	0.328	0.392
17	0.286	0.318	0.381
18	0.278	0.309	0.371
19	0.272	0.301	0.363
20	0.264	0.294	0.356
25	0.24	0.27	0.32
30	0.22	0.24	0.29
35	0.21	0.23	0.27

Source: F. J. Massey, *J. Amer. Stat. Assoc. 46*, 1951, p. 70.

Table A.10 Percentage Points for W_m^2 and A_m^2 Test for Test of
Normality and Exponentiality

Statistic		Percentage point					
	0.85	0.90	0.95	0.975	0.99	0.995	0.9975
W_m^2 Normal	0.091	0.104	0.126	0.148	0.178	0.201	0.224
W_m^2 Exponential	0.148	0.175	0.222	0.271	0.338	0.390	0.442
A_m^2 Normal	0.561	0.631	0.752	0.873	1.035	1.159	1.283
A_m^2 Exponential	0.916	1.062	1.321	1.591	1.959	2.244	2.534

Source: M. A. Stephens, private correspondence.

EQUATION FOR $E(y^4)$
IN THE PROPAGATION-OF-MOMENTS TECHNIQUE

Note: For purposes of simplification of this lengthy expression, μ_{ik} represents $E(X_i - \mu_i)^k$ and hence $\mathrm{var}(X_i) = \mu_{i2}$.

$$
E(Y^4) = \sum_{i=1}^{n} [b_i^4 \mu_{i4} + b_{ii}^4 \mu_{i8} + 4b_i^3 b_{ii} \mu_{i5} + 4b_i b_{ii}^3 \mu_{i7}
$$

$$
+ 6b_i^2 b_{ii}^2 \mu_{i6}] + \sum_i \sum_{\substack{j \\ i<j}} \{6b_i^2 b_j^2 \mu_{i2} \mu_{j2} + 6b_{ii}^2 b_{jj}^2 \mu_{i4} \mu_{j4}
$$

$$
+ b_{ij}^4 \mu_{i4} \mu_{j4} + 12 b_{ij} [b_i^2 b_j \mu_{i3} \mu_{j2} + b_i b_j^2 \mu_{i2} \mu_{j3}]
$$

$$
+ 12 b_{ij} b_{ii} b_{jj} [b_{ii} \mu_{i5} \mu_{j3} + b_{jj} \mu_{i3} \mu_{j5}]
$$

$$
+ 12 b_{ii} b_{jj} [b_i^2 \mu_{i4} \mu_{j2} + b_j^2 \mu_{i2} \mu_{j4} + 2 b_i b_j \mu_{i3} \mu_{j3}]
$$

$$
+ 6b_{ij}^2 [b_i^2 \mu_{i4} \mu_{j2} + b_j^2 \mu_{i2} \mu_{j4} + 2 b_i b_j \mu_{i3} \mu_{j3}]
$$

$$
+ 6b_{ij}^2 [b_{ii}^2 \mu_{i6} \mu_{j2} + b_{jj}^2 \mu_{i2} \mu_{j6} + 2 b_{ii} b_{jj} \mu_{i4} \mu_{j4}]
$$

$$
+ 12 b_{ij} [b_j b_{ii} (b_j \mu_{i3} \mu_{j3} + 2 b_i \mu_{i4} \mu_{j2}) + b_i b_{jj} (b_i \mu_{i3} \mu_{j3}
$$

$$
+ 2 b_j \mu_{i2} \mu_{j4})] + 12 b_{ij} [b_i b_{jj} (b_{jj} \mu_{i2} \mu_{j5} + 2 b_{ii} \mu_{i4} \mu_{j3})
$$

$$
+ b_j b_{ii} (b_{ii} \mu_{i5} \mu_{j2} + 2 b_{jj} \mu_{i3} \mu_{j4})] + 12 b_{ij}^2 [b_{ii} (b_i \mu_{i5} \mu_{j2}
$$

$$+ b_j \mu_{i4} \mu_{j3}) + b_{jj}(b_i \mu_{i3}\mu_{j4} + b_j \mu_{i2}\mu_{j5})]\}$$

$$+ \sum_{\substack{i=1 \\ j\neq i}}^{n} \sum_{j=1}^{n} \{4b_{ii}^3 b_{jj}\mu_{i6}\mu_{j2} + 4b_{ii}b_j^3\mu_{i2}\mu_{j3}$$

$$+ 12b_i b_{ii} b_j^2 \mu_{i3}\mu_{j2} + 12b_i b_{ii}^2 b_{jj}\mu_{i5}\mu_{j2}$$

$$+ 12b_i b_{ii} b_{jj}^2 \mu_{i3}\mu_{j4} + 4b_i b_{ij}^3\mu_{i4}\mu_{j3} + 4b_{ii} b_{ij}^3\mu_{i5}\mu_{j3}$$

$$+ 6b_{ii}^2 b_j^2 \mu_{i4}\mu_{j2}\} + \sum_{i=1}^{n-2} \sum_{j=i+1}^{n-1} \sum_{k=j+1}^{n} \{[12b_{ii}^2 b_{jj}b_{kk}$$

$$+ 6b_{ij}^2 b_{ik}^2 + 12b_{ii}(b_{kk}b_{ij}^2 + b_{jj}b_{ik}^2) + 6b_{ii}^2 b_{jk}^2]\mu_{i4}\mu_{j2}\mu_{k2}$$

$$+ [12b_{ii}b_{jj}^2 b_{kk} + 6b_{ij}^2 b_{jk}^2 + 12b_{jj}(b_{kk}b_{ij}^2 + b_{ii}b_{jk}^2)$$

$$+ 6b_{jj}^2 b_{ik}^2]\mu_{i2}\mu_{j4}\mu_{k2} + [12b_{ii}b_{jj}b_{kk}^2 + 6b_{ik}^2 b_{jk}^2$$

$$+ 12b_{kk}(b_{ii}b_{jk}^2 + b_{jj}b_{ik}^2) + 6b_{kk}^2 b_{ij}^2]\mu_{i2}\mu_{j2}\mu_{k4}$$

$$+ [12b_{ij}^2 b_{ik}b_{jk} + 24b_{ii}b_{jj}b_{kk}b_{ij} + 4b_{kk}b_{ij}^3$$

$$+ 24b_{ii}b_{kk}b_{ik}b_{jk}]\mu_{i3}\mu_{j3}\mu_{k2} + [12b_{ij}^2 b_{ik}b_{jk}$$

$$+ 24b_{ii}b_{jj}b_{kk}b_{ik} + 4b_{jj}b_{ik}^3 + 24b_{ii}b_{kk}b_{ij}b_{jk}]\mu_{i3}\mu_{j2}\mu_{k3}$$

$$+ [12b_{ij}b_{ik}b_{jk}^2 + 24b_{ii}b_{jj}b_{kk}b_{jk} + 4b_{ii}b_{jk}^3$$

$$+ 24b_{jj}b_{ij}b_{ij}b_{ik}]\mu_{i2}\mu_{j3}\mu_{k3} + [12b_{ij}b_{ik}b_{jk}$$

$$+ 24b_{ii}b_{jj}b_{kk}b_{jk} + 4b_{ii}b_{jk}^3 + 24b_{jj}b_{kk}b_{ij}b_{ik}]\mu_{i2}\mu_{j3}\mu_{k3}$$

$$+ 24[b_{ii}b_{jj}b_{kk} + b_{ij}b_{ik}b_{jk}][b_i \mu_{i3}\mu_{j2}\mu_{k2} + b_j \mu_{i2}\mu_{j3}\mu_{k2}$$

$$+ b_k \mu_{i2} \mu_{j2} \mu_{k3}] + 12[b_i b_{jk}^2 \mu_{i2} (b_{ij} \mu_{j3} \mu_{k2} + b_{ik} \mu_{j2} \mu_{k3})$$

$$+ b_j b_{ik}^2 \mu_{j2} (b_{ij} \mu_{i3} \mu_{k2} + b_{jk} \mu_{i2} \mu_{k3}) + b_k b_{ij}^2 \mu_{k2} (b_{ik} \mu_{i3} \mu_{j2}$$

$$+ b_{jk} \mu_{i2} \mu_{j3})] + 12[b_{ii} b_{jk}^2 \mu_{i3} (b_{ij} \mu_{j3} \mu_{k2} + b_{ik} \mu_{j2} \mu_{k3})$$

$$+ b_{jj} b_{ik}^2 \mu_{j3} (b_{ij} \mu_{i3} \mu_{k2} + b_{jk} \mu_{i2} \mu_{k3}) + b_{kk} b_{ij}^2 \mu_{k3} (b_{ik} \mu_{i3} \mu_{j2}$$

$$+ b_{jk} \mu_{i2} \mu_{j3})] + 24 b_{ij} b_{ik} b_{jk} [b_{ii} \mu_{i4} \mu_{j2} \mu_{k2}$$

$$+ b_{jj} \mu_{i2} \mu_{j4} \mu_{k2} + b_{kk} \mu_{i2} \mu_{j2} \mu_{k4}] + \mu_{i2} \mu_{j2} \mu_{k2} [12(b_{ii} b_{jj} b_k^2$$

$$+ b_{ii} b_{kk} b_j^2 + b_{jj} b_{kk} b_i^2) + 6(b_i^2 b_{jk}^2 + b_j^2 b_{ik}^2 + b_k^2 b_{ij}^2)$$

$$+ 24(b_{ij} b_{ik} b_j b_k + b_{ij} b_{jk} b_i b_k + b_{ik} b_{jk} b_i b_j)$$

$$+ 24(b_i b_j b_{kk} b_{ij} + b_i b_k b_{jj} b_{ik} + b_j b_k b_{ii} b_{jk})]$$

$$+ \mu_{i3} \mu_{j2} \mu_{k2} [24 b_j b_{ij} b_{ii} b_{kk} + 24 b_k b_{ik} b_{ii} b_{jj} + 12 b_i b_{jk}^2 b_{ii}$$

$$+ 24 b_j b_{ik} b_{jk} b_{ii} + 24 b_k b_{ij} b_{jk} b_{ii} + 12 b_i b_{ik}^2 b_{jj}$$

$$+ 12 b_i b_{ij}^2 b_{kk}] + \mu_{i2} \mu_{j3} \mu_{k2} [24 b_i b_{ij} b_{jj} b_{kk} + 24 b_k b_{jk} b_{ii} b_{jj}$$

$$+ 12 b_j b_{ik}^2 b_{jj} + 24 b_i b_{ik} b_{jk} b_{jj} + 24 b_k b_{ij} b_{ik} b_{jj}$$

$$+ 12 b_j b_{jk}^2 b_{ii} + 12 b_j b_{ij}^2 b_{kk}] + \mu_{i2} \mu_{j2} \mu_{k3} [24 b_i b_{ik} b_{jj} b_{kk}$$

$$+ 24 b_j b_{jk} b_{ii} b_{kk} + 12 b_k b_{ij}^2 b_{kk} + 24 b_i b_{ij} b_{jk} b_{kk}$$

$$+ 24 b_j b_{ij} b_{ik} b_{kk} + 12 b_k b_{jk}^2 b_{ii} + 12 b_k b_{ik}^2 b_{jj}]\} \qquad \text{(B.1)}$$

$$+ \sum_{i=1}^{n-3} \sum_{j=i+1}^{n-2} \sum_{k=j+1}^{n-1} \sum_{m=k+1}^{n} \mu_{i2} \mu_{j2} \mu_{k2} \mu_{m2}$$

$$\times \ [24(b_{ii}b_{jj}b_{kk}b_{mm} + b_{ij}b_{ik}b_{jm}b_{km} + b_{ij}b_{im}b_{jk}b_{km}$$

$$+ \ b_{ik}b_{im}b_{jk}b_{jm} + b_{ii}b_{jk}b_{jm}b_{km} + b_{jj}b_{ik}b_{im}b_{km}$$

$$+ \ b_{kk}b_{ij}b_{im}b_{jm} + b_{mm}b_{ij}b_{ik}b_{jk}) + 12(b_{ii}b_{jj}b_{km}^{2}$$

$$+ \ b_{ii}b_{kk}b_{jm}^{2} + b_{ii}b_{mm}b_{jk}^{2} + b_{jj}b_{kk}b_{im}^{2} + b_{jj}b_{mm}b_{ik}^{2}$$

$$+ \ b_{kk}b_{mm}b_{ij}^{2} + 6(b_{ij}^{2}b_{km}^{2} + b_{ik}^{2}b_{jm}^{2} + b_{im}^{2}b_{jk}^{2})]$$

Abramowitz, M., 78, 103
Aitken, A., 228, 265
Amos, D., 221,223
Anderson, T., 253, 254, 265
Andrews, D., 240, 265
Antle, C., 76, 103
Aroian, L., 61, 103

Bain, L., 76, 103
Bartlett, M., 95, 103
Bennett, C., 11, 60
Birnbaun, Z., 68, 69, 104
Bishop, Y., 143,159
Blom, G., 230, 265
Boag, J., 100,103
Bray, T., 289,326
Bury, K., 103

Choi, S., 84, 103
Clark, V., 61, 70, 71, 73, 76,
    77, 78, 80, 81, 83-86, 98,
    99, 100, 103
Cochran, W., 259, 266
Cox, N., 299, 303, 304, 326
Craig, A., 11, 55, 56, 59, 60,
    108, 160
Cramér, H., 55, 60, 92, 103,
    253, 266

Dahiya, R., 124, 125, 159, 259,
    266
Daniel, S., 221, 223
Darling, D., 253, 254, 265
David, H., 53, 60, 228, 266

Dixon, W., 110, 159
Dubey, S., 95, 103
Dudewiez, E., 178, 180, 223
Duguay, M., 320, 326

Epstein, B., 66, 86, 103

Feller, W., 11, 60, 63, 103, 106
    159
Filliben, J., 247, 266
Fisher, R., 159
Fox, B., 147, 148, 159
Franklin, N., 11, 60
Fraser, D., 240, 266
Freund, J., 11, 60, 196, 223, 229
    266, 269
Francia, R., 247, 266

Gross, A., 61, 70, 71, 73, 76-78,
    80, 81, 83-86, 96, 98-100, 103,
    110, 124, 125, 148, 159
Goodman, L., 142, 159
Gumbel, E., 86, 103
Gurland, J., 259, 260

Hahn, G., 11, 60, 61, 85, 104,
    106, 159, 269, 300, 326
Haight, F., 106, 159
Hald, A., 327, 328
Harter, H., 148, 159
Hartley, H., 78, 104, 107, 108,
    110, 148, 160, 347
Hastings, C., 178, 223

361